Beneficiation of Phosphates

TECHNOLOGY AND SUSTAINABILITY

Edited by

Patrick Zhang
Jan Miller
Hassan El-Shall
Regis Stana

Published by the

Society for Mining, Metallurgy, and Exploration, Inc.

Society for Mining, Metallurgy, and Exploration, Inc.
8307 Shaffer Parkway
Littleton, Colorado, USA 80127
(303) 973-9550 / (800) 763-3132
www.smenet.org

ISBN-13: 978-0-87335-254-3
ISBN-10: 0-87335-254-8

Library of Congress Cataloging-in-Publication Data
Beneficiation of phosphates : technology and sustainability / edited by Patrick Zhang ... [et al.].
p. cm.
Includes bibliographical references and index.
ISBN-13: 978-0-87335-254-3
ISBN-10: 0-87335-254-8
1. Flotation. 2. Ore-dressing. 3. Phosphates. I. Zhang, Patrick.

TN523.B428 2006
622'.364--dc22

2006050576

Contents

Foreword

The cooperation between SME and the Florida Institute of Phosphate Research (FIPR) has played a significant role in elevating the phosphate beneficiation conference to a world-class event.

Since the December 2001 conference, the market conditions have improved significantly for the phosphate industry. The Cargill-IMC merger with Mosaic has made the industry more profitable in Florida. However, the political and environmental pressures have also been mounting, as new mining permits become ever more difficult to obtain. It is, therefore, not surprising that this conference has an environmental theme. While Professor Ralston's keynote speech addresses resource utilization efficiency from the fundamental aspects of fine particle separation, Dr. Hendrix's speech tries to raise the awareness of sustainable mining.

It is becoming evident that flocculation-based technology holds promise in dealing with the issue of waste clay/settling ponds. The interest has grown in developing flotation processes that use smaller amount of reagents, particularly fuel oil. In this endeavor, the collaboration among academia, reagent suppliers, and the industry is vital. Our conference is fostering this type of collaboration.

We are grateful to the members of the organizing and advisory committees, and the authors and reviewers for their efforts in making this conference a success and the proceedings a quality product. We are perhaps unique by having the conference papers peer reviewed and published as a hardcover, chaptered book.

Karen Mikkola of Mosaic and Cao Xianjun of CLDRI deserve special recognition for using their influence in getting many participants to the conference. Credit is due to Kevin Korpics of Engineering Conference International and his staff for handling the logistics. Rain Dong of Lakeland did a marvelous job of editing and formatting all the papers. Jane Olivier of SME and her staff are acknowledged for their assistance in publishing this book. This series of conference would not have been possible without the financial support of FIPR and the help of many FIPR staff. The unwavering support of Paul Clifford, the executive director of FIPR, is highly appreciated. The PowerPoint presentations went smoothly to the credit of Mike Bogan of FIPR.

All participants recommended continuing the conference regularly, with a majority suggesting every three years. We will continue to work hard to fulfill that goal.

Patrick Zhang
Jan Miller
Hassan El-Shall
Regis Stana

CONFERENCE ORGANIZATION

J. Zhang
Chairman
FIPR
Bartow, Florida

H. El-Shall
Co-Chairman
University of Florida
Gainesville, Florida

J. Miller
Co-Chairman
University of Utah
Salt Lake City, Utah

R. Stana
Co-Chairman
R Square S
Lakeland, Florida

ORGANIZING COMMITTEE

B. Moudgill
University of Florida
Gainesville, Florida

G. Luttrell
Virginia Tech
Blacksburg, Virginia

D. Gossett
CF Industries, Inc.
Bartow, Florida

P. Somasundaran
Columbia University
New York, New York

K. Mikkola
Mosaic
Mulberry, Florida

J. Shoniker
PCS Phosphate
White Springs, Florida

R. Wiegel
Lakeland, Florida

G. Gruber
Jacobs Engineering Group
Lakeland, Florida

Y.X. Han
Northeastern University
Shenyang, China

ADVISORY COMMITTEE

J. Hendrix
University of Nebraska
Lincoln, Nebraska

A. Peres
UFMG
Relo Horizonte, Brazil

M. Mankosa
Eriez Magnetics
Erie, Pennsylvania

D. Tao
University of Kentucky
Lexington, Kentucky

R.H. Yoon
Virginia Tech
Blacksburg, Virginia

J. Ralston
University of South Australia
Mawson Lakes, Australia

P. Clifford
FIPR
Bartow, Florida

L.S. Leal Filho
University Sao Paulo
Sao Paulo, Brazil

M. Fuerstenau
University of Nevada, Reno
Reno, Nevada

SECTION 1

Theoretical Understanding

CHAPTER 1

Multiple Bubble Particle Capture: Critical Role of Hydrodynamics and Interfacial Science in Flotation Rate Constant Prediction

J. Ralston,* D. Fornasiero,* D. Snoswell,* and J. Duan*

ABSTRACT

The flotation rate constants for mineral particles may be predicted, as a function of particle size, in a flotation cell under known hydrodynamic conditions using a flotation model that incorporates contributions from the efficiencies of collision, attachment and stability between particles and bubbles. The frequency of collision is also taken into account, along with the turbulent dissipation energy.

For micron and submicron size particles, selective separation by flotation is difficult due to their very low collision efficiency. Selective aggregation, where the interfacial charge and potential may be manipulated, is an alternative or addition to flotation. pH and salt concentration regimes may then be identified where highly selective separations may be achieved for dispersions up to at least 2 wt%.

INTRODUCTION

Flotation rate constant control and the selective separation of particles are critically important in mineral processing. The basic processes that govern the rate of recovery, or collection efficiency (E_{coll}), of particles in a flotation cell are well understood and include the consecutive subprocesses of particle-bubble collision, attachment and stability, represented by efficiencies E_c, E_a, E_S, respectively (Derjaguin and Dukhin, 1961).

$$E_{coll} = E_c . E_a . E_s \tag{1}$$

The rate of particle recovery in the froth (or froth efficiency) can also be incorporated in equation 1; however, in the present study, it is assumed that it is unity as the froth is very shallow with depth of only a few centimetres (Mular and Musara, 1991). Collision is dominated by bulk hydrodynamics inside the flotation cell (e.g., bubble velocity and turbulence), while attachment is dominated by the interfacial behaviour between the particle and bubble (e.g., particle hydrophobicity influences thin film drainage). As for stability, its efficiency depends on both hydrodynamics and interfacial events. Several models can be used for the calculation of collision, attachment and stability efficiencies; however, their applicability is restricted to the approximations (e.g., fluid flow conditions, mobility of the bubble surface) inherent to each model (e.g., Dai et al., 2000). The validity of equation 1 has been demonstrated in studies

*Ian Wark Research Institute, The ARC Special Research Centre for Particle and Material Interfaces, University of South Australia, Mawson Lakes, Adelaide S.A., Australia.

involving, mostly, single mineral flotation experiments in simple flotation columns. With these models, the collection efficiency, or flotation rate constant, of well-defined particles has been calculated as a function of particle size and contact angle, as well as bubble size and velocity (e.g., Dai et al., 1998; Crawford and Ralston, 1988). With mixed minerals or ores, no easy and reliable technique exists for the measurement of the contact angle of a particular mineral in an ore. Recent studies have shown that it is now possible to obtain this information using time of flight secondary ion mass spectroscopy, ToF-SIMS, correlated with independent measurements of contact angle (Piantadosi et al., 2000; Pyke et al., 2001; Piantadosi and Smart, 2002).

The first objective of this present work is to show how the flotation rate constants of particles as a function of particle size and contact angle, bubble size and velocity, and level of turbulence inside flotation cell may be calculated. These calculations are then compared with the flotation rate constants of chalcopyrite particles in an ore measured in a Rushton flotation cell of well-known hydrodynamic characteristics. The contact angle of the chalcopyrite particles in the ore is determined by ToF-SIMS, using an appropriate calibration procedure.

The second objective of this work is to consider the selective aggregation of micron and sub-micron size particles, whose flotation efficiency is rather poor. This poor flotation response is due to a low collision frequency. Selective aggregation may be considered as an alternative, or a precursor step, to flotation. If the ore body is composed of very finely disseminated minerals, all of the liberated minerals in the feed will be present as very small micron or submicron size particles.

For example, the discovery of high grade rutile ores may make it necessary to extract colloidal sized rutile, finely dispersed in sandstone deposits such as those found in Sweden, Peru and elsewhere (Morad, 1986). Selective aggregation techniques may prove to be an appropriate separation technique for ores with colloidal grain sizes.

Table 1 Rushton Turbine Flotation Cell

Cell Dimensions	Impeller Specifications
Volume = 2.25×10^{-3} m^3 Diameter = 0.144 m Height:Diameter = 1:1 Baffle:Diameter = 1:10	6 blade Rushton Turbine Diameter = 0.048 m Width:Height:Diameter = 5:4:20

Commercially viable separation techniques must often be performed at high solid concentrations. It is therefore important to investigate the effect of solid concentration on the separation process. In this study, separations at solid concentrations from 0.03 wt% to 2 wt% are investigated, spanning a range of conditions from those found in the laboratory experiments, up to industrially relevant conditions.

EXPERIMENTAL SECTION

Flotation Experiment

The characteristics of the Rushton turbine flotation cell used in this study are given in Table 1 (Pyke et al., 2002). The flotation cell is equipped with a stirring impeller whose speed can be controlled and measured accurately using an optical tachometer. Pressurized air was used to generate gas bubbles through a stainless steel frit with pore size of 13 mm at the bottom of the flotation cell.

For the flotation experiments, a Mount Isa Mines (MIM) copper ore was used. The ore contains 4 wt% chalcopyrite, 6 wt% iron sulfide minerals (predominantly as pyrite) and 82 wt% non-sulfide gangue minerals, quartz and dolomite. The ore (1,000 grams) was ground in a stainless steel Galigher mill with mild steel rods for 20 minutes at pH 12.5 to produce a feed of particle size, d80, of around 84 mm. Conditioning and flotation were conducted in the Rushton flotation cell at pH 11.5. After addition of 20 g/t of frother (DOWFROTH250), a pre-flotation was conducted for 8 minutes in the absence of collector to remove the carbonaceous pyrite present in the ore. Then, after addition of

50 g/t of Aero 3477 dialkyl dithiophosphate (DTP) as collector and 5 g/t of DOWFROTH250, concentrates were collected after 1, 2, 5, and 10 minutes of flotation (cumulative). Particles in each concentrate and tail sample were then separated into 13 different sub-samples of sizes ranging between 2 and 150 mm using sieves, a Warman cyclosizer and a pre-cyclone. Chemical analysis of these samples was performed by Analabs Pty. Ltd.

A similar flotation experiment was conducted with clean quartz particles of size range similar to that in the ore to determine the level of particle entrainment. The entrainment contribution was subtracted from the particle recovery in each size fraction to obtain the true flotation recovery of the particles in the ore.

Experimental Flotation Rate Constant

Assuming that the flotation of particles of diameter, i, follows a first-order rate reaction, the recovery of these particles, $R_{t,i}$, at time, t, to a first approximation is given by the following expression:

$$R_{t,i} = R_{max,i}\,(1-\exp(-k_i\,t)) \qquad (2)$$

where k_i is the flotation rate constant of particles of diameter, i, and $R_{max,i}$ is their flotation recovery at an infinite time.

A nonlinear least square regression was used to calculate k_i and $R_{max,i}$ from the best fit of the curve of experimental flotation recoveries versus time for each particle size range using equation 2. These flotation rate constants are referred to in the text as experimental flotation rate constants.

Laser Doppler Velocimetry

Measurements of fluid velocity and velocity fluctuations were made with a laser doppler velocimetry (LDV) at different positions inside the Rushton flotation cell at the same gas flow rate and stirring speed as in the flotation experiments but in the absence of ore and gas bubbles (Pyke et al., 2003).

The average fluid velocity, V_{fl}, was calculated from the LDV measurements of velocity vectors in the X, Y, and Z directions, V_x, V_y, and V_z, respectively, using the expression (Wu and Patterson, 1989):

$$V_{fl} = \sqrt{V_x^2 + V_y^2 + V_z^2} \qquad (3)$$

The turbulent dissipation energy, ε, can be calculated from the expression (Wu and Patterson, 1989):

$$\varepsilon = A\,q^{3/2}\,L^{-1} \qquad (4)$$

where A = 0.85 (Wu and Patterson, 1989; Jenne and Reuss, 1999); L is a measure of the turbulent length macroscale, generally assumed constant and equal to half the turbine depth (0.005 m in this case); and q is the turbulent kinetic energy calculated from the root mean square of the experimental fluctuations of the turbulent fluid velocities in the three directions, $\overline{U}_x$, $\overline{U}_y$, and $\overline{U}_z$, using the expression

$$q = \frac{1}{2}\left(\overline{U}_x^2 + \overline{U}_y^2 + \overline{U}_z^2\right) \qquad (5)$$

Time-of-Flight Secondary Ion Mass Spectrometry (ToF-SIMS)

ToF-SIMS spectra were obtained using a PHI Model 2100 TRIF II system equipped with a gallium liquid-metal ion gun (LMIG) in pulsed mode. The surface of insulating samples was charged compensated using an electron flood gun. The wet mineral samples prepared under the same conditions as in the flotation experiments were mounted onto a piece of indium foil and introduced into the spectrometer preparation chamber. Prior to analysis the samples were outgassed under vacuum for around 12 hours. The primary beam was produced using a gallium gun with a current of 600 pA. An acceleration voltage of 25 kV was used to produce high spatial resolution with a spot size of 0.2 mm. In static mode, the analysis is confined to the top two monolayers. Imaging of the sample involved mapping for positive and negative ions of interest (e.g., DTP, SO_{3-}, etc.). Up to 96 chalcopyrite particles from the concentrate sample collected after 2 min of flotation were analyzed by ToF-SIMS to obtain statistical information on the amount of hydrophobic species (such as DTP) and hydrophilic species (such as hydroxide, oxide, sulphate and sulphite) on their surface (Piantadosi et al., 2000; Piantadosi and Smart, 2002).

Bubble Size Determinations

A University of Cape Town (UCT) bubble size analyzer was used to determine the bubble size in the flotation cell. The UCT bubble size analyzer captures the bubbles and draws them through a glass capillary (diameter 0.4 mm). Diodes measure the change in refractive index between the water and gas. Using these readings and the volume of gas collected in a burette, the bubble size distribution can be obtained. Up to 4,000 bubbles were typically measured using the analyzer over a period of 90 sec, permitting a statistically significant number of bubbles to be counted. Measurements were performed in duplicate and the analyzer data were then compared with photographs.

Agreement between the two techniques was very satisfactory. The mean bubble diameter was 0.97 mm, with a standard deviation of 0.4.

Aggregation Experiments

High purity Gel-Tec Incorporated silica particles with diameters between 0.74 and 1.03 microns were used. Titanium dioxide particles with diameters between 0.27 and 0.33 microns were synthesized. Standard synthetic and colloid science experimental procedures were used and have been described elsewhere (Snoswell, 2003a).

RESULTS AND DISCUSSION

Chalcopyrite Flotation

The flotation rate constant may be expressed by

$$k = 2.39\,\frac{G_{fr}}{d_b \cdot V_{cell}}\left[\frac{0.33\,\varepsilon^{\frac{4}{9}}\,d_b^{\frac{7}{9}}}{\upsilon^{\frac{1}{3}}}\left(\frac{\Delta\rho_b}{\rho_{fl}}\right)^{\frac{2}{3}}\cdot\frac{1}{v_b}\right]E_c \cdot E_a \cdot E_s \qquad (6)$$

mechanical term primary turbulence term elementary processes

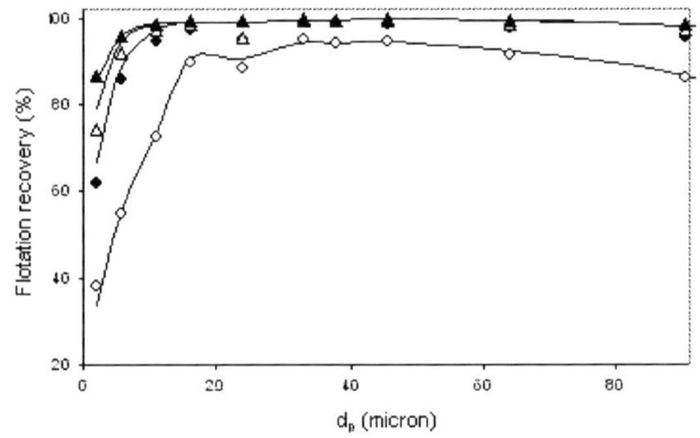

Figure 1 Experimental (symbols) and calculated (lines) flotation recoveries of chalcopyrite as a function of particle diameter and flotation times of 1 (bottom), 2, 5 and 10 (top) min. A first order rate equation was used in the calculation of recoveries ([DTP]=50 g/t; stirring speed=720 rpm; Gfr=5.5×10^{-3} m^3/s; db=0.97 mm). [Reproduced with permission from *International Journal of Mineral Processing*, 72, 227–237 (2003).]

A full derivation and extended term definition is given elsewhere (Pyke et al.; Duan et al., 2003). (The experimental flotation recoveries for copper as a function of particle diameter and flotation time (after removing the entrainment contribution) are compared in Figure 1 with the flotation recoveries calculated using equation 2. The satisfactory agreement between the experimental and calculated curves validates the use of a first order rate reaction to describe the flotation process. The flotation rate constants, ki, obtained from this fitting exercise are shown in Figure 2. The trend in flotation rate constant with particle diameter is similar to that observed previously for copper minerals (Lynch et al., 1981) with maximum values obtained for particle diameters near 40 microns. The lower flotation rate constant values of finer and coarser particles have been attributed to the low collision efficiency and low stability efficiency, respectively, of these particles with gas bubbles.

Fluid Velocity and Energy Dissipation in the Rushton Flotation Cell

Figure 3 shows the value of the fluid resultant velocity, V_{fl}, and turbulent dissipation energy, ε, in the Rushton flotation cell measured by laser Doppler velocimetry at a fixed distance of

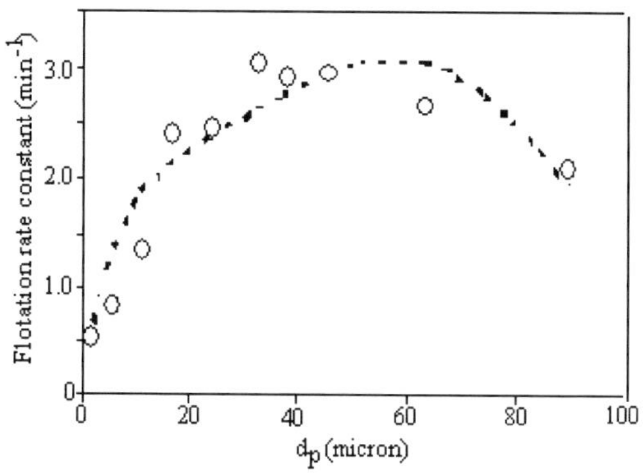

Figure 2 Experimental (symbols) and calculated (broken line) flotation rate constants of chalcopyrite as a function of particle diameter. Equations 9 to 22 were used in the calculations of the flotation rate constants ([DTP]= 50 g/t; stirring speed=720 rpm; Gfr=5.5×10^{-3} m^3/s; A=0.05; B=0.6; vb=0.2 m s^{-1}; e=14 m^2/s^3; p=4.1 g cm^{-3}; db=0.97 mm). [Reproduced with permission from *International Journal of Mineral Processing*, 72, 227–237 (2003).]

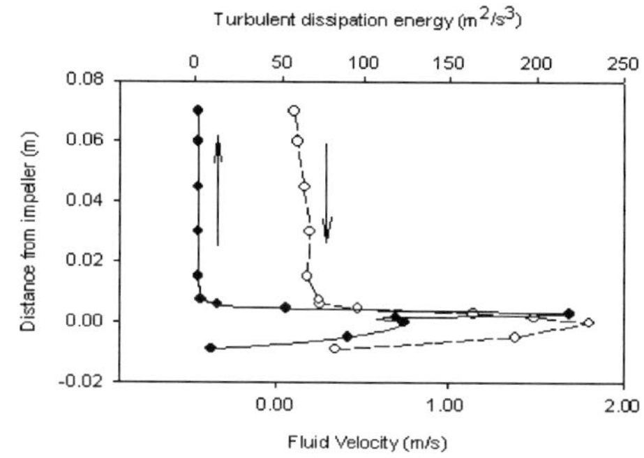

Figure 3 Fluid velocity, V_{fl} (empty circles) and turbulent dissipation energy, ε (filled circles) versus vertical distance from the impeller in the Rushton flotation cell and at a fixed distance of 0.026 m away from the impeller axis (stirring speed=720 rpm). [Reproduced with permission from *International Journal of Mineral Processing*, 72, 227–237 (2003).]

0.026 m away from the impeller shaft and at different heights above and below the impeller. The results indicate that the value of fluid velocity and energy dissipation is extremely high near the impeller, but small and constant from a height of 0.01 m above the impeller. These measured values are in agreement with values reported in the literature in similar Rushton flotation cells (Wu and Patterson, 1989; Rutherford, 1996) and Denver cells (Jordan and Scheiner, 1990).

ToF-SIMS Results

The ToF-SIMS results demonstrated that the DTP collector is selectively adsorbed on the chalcopyrite particles in the MIM ore. However, large differences in the amount and distribution of DTP or oxidation products were observed on the surface of the chalcopyrite particles. This problem was overcome by sampling a large number of these particles to produce a statistical result. Also, the results consistently indicated that the differences in surface chemistry between three size fractions of chalcopyrite (<10 μm; 20–30 μm; >50 μm) are not statistically significant.

A recent study has shown that ToF-SIMS hydrophobic/hydrophilic indices such as the ratio between the ToF-SIMS signals of DTP and SO_{3-} on the surface of a mineral are a good representation of the surface hydrophobicity of that mineral (SO_{3-} is the fragment associated with sulphate, representing the hydrophilic species, while DTP represents the hydrophobic collector). Indeed, the flotation recovery, advancing contact angle and ToF-SIMS DTP/SO_{3-} index of chalcopyrite particles in single mineral experiments were shown to increase with DTP concentration (Pyke et al., 2001).

Using this relationship between advancing contact angle and ToF-SIMS DTP/SO_{3-} index established for chalcopyrite particles in single mineral experiments, an advancing contact angle of 71° ±2° for the chalcopyrite particles in the MIM ore was predicted from a measured ToF-SIMS DTP/SO^{3-} index of 9.7×10^{-3} for those particles.

Calculation of Flotation Rate Constants

Equations 1 to 6 were used to calculate the flotation rate constant as a function of particle diameter (Duan, Fornasiero and Ralston, 2003). A value of 4.1 g cm^{-3} was used for the density of chalcopyrite (rp).

The calculated flotation rate constant versus particle diameter curves were then compared with the experimental curve in Figure 2 and a non-linear least-squares program was used to optimize the values of the bubble velocity, vb, and turbulent dissipation energy, ε, until the best fit of the experimental curve was obtained. From these calculations, values of 0.2 m s^{-1} and 14 m^2 s^{-3} were obtained for the bubble velocity and turbulent dissipation energy, respectively. Both these calculated values correlate well with data obtained in a region 0.005 to 0.01 m above or below the impeller inside the flotation cell (Figure 3). These relatively low values suggest that the mean vessel turbulent dissipation energy, rather than the local dissipation energy. determines the mean stability of the bubble-particle aggregates (Schulze, 1977).

Overall, the values of the calculated flotation rate constants in Figure 2 are in a satisfactory agreement with the experimental data. More importantly, the characteristic shape of the flotation rate constant versus particle size curve is reproduced in these calculations.

Selective Aggregation of Silica and Titania

Selective aggregation involves the aggregation of one particle species in a mixed colloidal dispersion. In a binary colloidal system of titania and silica, it is possible that three distinct aggregation processes can occur. Homo-aggregation refers to the aggregation of particles of the same type, and in this case can describe titania-titania aggregation, or silica-silica aggregation. Hetero-aggregation refers to the aggregation of dissimilar particles and in this case describes titania-silica aggregation. In this present investigation, the aim is to achieve the homo-aggregation of titania particles while preventing the homo-aggregtion of silica and the hetero-aggregation of titania and silica. When sufficiently larger titania aggregates or flocs are formed, they should then settle out under gravity and become physically separated from the silica that remains colloidally dispersed.

To prevent hetero-aggregation, one must ensure that the different particles have both a significant surface potential *and* the same sign conferring adequate electrostatic repulsion and a stable colloidal dispersion. The surface potential of both silica and titania varies with pH, and, for both titania and silica to be negatively charged, a pH above the iso-electric point of titania (pH 6.1) in this present study must be chosen. In addition, the solubility of silica increases significantly above pH 9 (Iler, 1979); therefore, a pH between 6.1 and 9 should be selected. For a given pH within this range, the electrolyte concentration required to achieve selective aggregation is then determined by reference to the relevant colloid stability curves, obtained through the classic log (stability ratio) versus log (electrolyte orientation) approach (Gregory, 1989). The success of this method relies on a difference in colloidal stability between the particle types at the chosen solution conditions.

Stability curves provide a relative measure of aggregation rate at various electrolyte concentrations and can be determined both experimentally and theoretically. Based on our previous studies (Snoswell, 2003a), experimental, colloid stability curves describing silica and titania have been determined pH 8.4. They have been described with a DLVO model using experimental zeta potential data, as described in detail elsewhere (Snoswell et al., 2004). These curves permit the prediction of homo-aggregation rates in a mixed suspension and provide a tool for choosing specific electrolyte concentrations required to achieve selective aggregation of titania at pH 8.4. By referring to the relative positions of the titania and silica stability curves given, choosing an electrolyte concentration within the region *between* the two critical coagulation concentrations (ccc) should result in selective aggregation of the particle with the smaller ccc. However, to avoid silica-titania hetero-aggregation, a third colloid stability curve describing this situation is also required. A

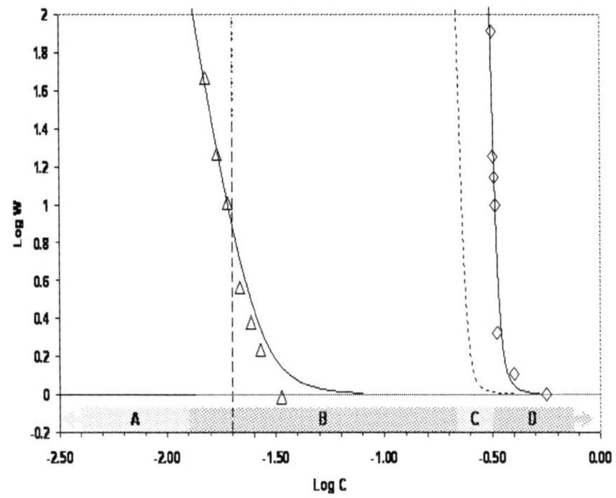

Figure 4 Experimental stability curves for the homo-aggregation of titania (triangles) and silica spheres (diamonds) at pH 8.4

stability curve for the hetero-aggregation of titania and silica can be calculated using fitted model parameters obtained for the homo-aggregation of titania and silica (Figure 4). In Figure 4, solid lines indicate calculated theoretical stability curves for titania and silica spheres. The dotted line indicates the theoretical stability curve for titania/silica hetero-aggregation. The vertical dashed line indicates the electrolyte concentration chosen for selective aggregation. Letters A, B, C and D indicate aggregation regimes described in the text.

Assuming a constant charge boundary condition, low surface potentials and small separations, the electrostatic interaction energy between silica and titania particles can be calculated using equation 7 (Elimelech et al., 1995).

$$V_R = \frac{\pi a_1 a_2 \varepsilon}{(a_1 + a_2)}\left(\frac{kT}{z\alpha}\right)^2 (\Phi_1^2 + \Phi_2^2)\left[\frac{2\Phi_1\Phi_2}{\Phi_1^2 + \Phi_2^2}\ln\left(\frac{1+e^{-\kappa h}}{1-e^{-\kappa h}}\right) - \ln\left(1 - e^{-2\kappa h}\right)\right] \quad (7)$$

where subscripts 1 and 2 denote silica and titania, respectively, and

a = particle radius
ε = dielectric constant of intervening medium
z = valency of electrolyte ions (assumes a 1:1 electrolyte)
k = Boltzmann constant
T = Absolute temperature
α = Charge of one electron
κ = Debye length
h = separation between surfaces

$\Phi = \frac{z\alpha\psi_d}{kT}$ where ψ_d the zeta is potential, determined through electrophoretic mobility determinations (Hunter, 1979).

The van der Waals interaction for two different particles of radii a_1 and a_2 is calculated using equation 8:

$$V_A = -\frac{A a_1 a_2}{6h(a_1 + a_2)} \quad (8)$$

where A is the Hamaker constant.

In this case, the Hamaker constant is calculated for a silica/water/titania interaction.

Assuming the subscripts 1, 2, and 3 denote silica, titania and water, respectively then A_{132} can be approximated by equation 9 (Elimelech et al., 1995).

$$A_{132} = \left(\sqrt{A_{12}} - \sqrt{A_{33}}\right)\left(\sqrt{A_{22}} - \sqrt{A_{33}}\right) \quad (9)$$

Since estimates of A_{33} relating to water can be found from the literature (4.35×10^{-20}J), (Bargeman and Vader, 1972) and A_{131} and A_{232} were determined previously from experiment, A_{11} and A_{22} can be deduced using equations 10 and 11 (Elimelech et al., 1995).

$$A_{131} = \left(\sqrt{A_{11}} - \sqrt{A_{33}}\right)^2 \quad (10)$$

Using this method, a value of 4.0×10^{-21}J is obtained for A_{132}. The corresponding calculated stability curve for the hetero-aggregation of silica and titania is presented.

$$A_{232} = \sqrt{(A_{22} A_{33})}^2 \quad (11)$$

There are four distinct regions of stability for a mixed titania/silica suspension that depend on electrolyte concentration. In region A, all aggregation rates are very slow and the entire suspension is stable. At the electrolyte concentrations indicated in region B, the stability curves predict that only titania-titania homo-aggregation will occur. At the higher concentrations in region C, titania-silica hetero-aggregation becomes significant, and, finally, in region D, silica-silica homo-aggregation also occurs and the entire suspension becomes unstable. For the purposes of separating titania from the mixed suspension, an electrolyte concentration within region B must therefore be selected.

Within region B, the electrolyte concentration can be chosen to achieve the aggregation of titania in either the "fast" or "slow" aggregation regimes. It is known that aggregates formed under slow or "reaction limited" conditions have higher fractal dimensions than those formed under diffusion limited conditions (Gregory, 1989). Therefore, since higher fractal dimensions indicate a closer packing of particles, aggregation within the slow regime should be chosen to both enhance the aggregate density and settling rate, and minimize the physical entrapment of silica within aggregate voids. With these factors in mind, an electrolyte concentration of 20 mM was chosen, as indicated. Selective and highly efficient separations were achieved, as shown in Figure 5.

CONCLUSIONS

The flotation rate constants of chalcopyrite particles in a complex sulphide ore as a function of particle diameter can be predicted using a flotation model that includes the contributions from the efficiencies of collision, attachment and stability between particles and bubbles as well as their frequency of collision. These calculated flotation rate constants were in satisfactory agreement with the flotation rate constants for chalcopyrite particles in an ore determined under turbulent flow conditions in a Rushton flotation cell. In particular, the relatively low flotation rate constants of fine and coarse chalcopyrite particles were reproduced and have been attributed to the low efficiency of collision and stability of these particles with bubbles, respectively. For these calculations, only the values of mean turbulent dissipation energy, bubble velocity and particle contact angle are required. The water advancing contact angle of the chalcopyrite particles in the ore was independently determined using a technique based upon time of flight secondary ion mass spectroscopy. As for the turbulent dissipation energy, its optimum value used in the calculations corresponds to those measured in a region of relatively low turbulence inside the flotation cell close to the impeller. The optimum bubble velocity was found to be rather close to the bubble rising velocity, giving a good indication of the relative bubble velocity with respect to the "fluid" in the flotation cell.

The separation of colloidal titania and silica is a potentially important industrial process. By consulting stability curves for these particles, solution conditions have been chosen to achieve this separation. The application of this knowledge is demonstrated to qualitatively predict the outcome of a set of separation experiments.

Chemical and visual evidence indicate that a very efficient separation process for colloidal titania and silica can be achieved. This separation has been demonstrated over a range of solid concentrations, including concentrations over 1 wt%, which are relevant to industrial processing. Entrapment of silica particles is noted to decrease significantly as the titania aggregates experience a transition from "free" to hindered settling, resulting in a significant increase in separation efficiency.

ACKNOWLEDGMENTS

The financial support for this work from the University of South Australia, the Australian Mineral Industries Research Association and the Australian Research Council, through the Special Research Centres Scheme, is gratefully acknowledged.

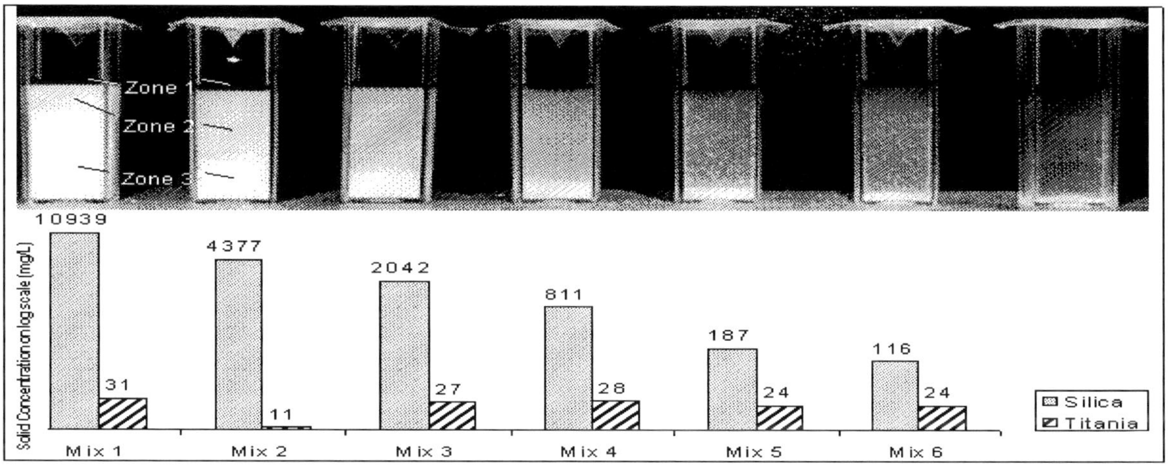

Figure 5 Titania/silica colloidal mixtures 75 minutes after the addition of KCl electrolyte (upper photo). Histogram of zone 2 solid concentrations (lower diagram). Mixtures arranged from highest solid concentration (Mix 1, left) to lowest solid concentration (Mix 7, right). Note: Mix 7 analysis not available.

REFERENCES

Abrahamson, J., 1975. Collision rates of small particles in a vigorously turbulent fluid. *Chem. Eng. Sci.*, 30, 1371–1379.

Bargeman, D., and Vader, F.V. 1972. Van der Waals forces between immersed particles. *J. Electroanal. Chem. Interfacial Electrochem.*, 37:45.

Barringer, E.A., and Bowen, H.K. 1982. Formation, packing and sintering of monodisperse TiO_2 powders. *Communications of the American Ceramic Society*, December 1982: 199:201.

Barringer, E.A., and Bowen, H.K. 1985. High-Purity, Monodisperse TiO_2 Powders by Hydrolysis of Titanium Tetraethoxide. 1. Synthesis and Physical Properties. *Langmuir*, 1: 414:420.

Crawford, R., and Ralston, J. 1988. The influence of particle and contact angle in mineral flotation. *Int. J. Miner. Process.*, 23, 1–24.

Dai, Z., Dukhin, S., Fornasiero, D., and Ralston, J. 1998. The Inertial Hydrodynamic Interaction of Particles and Rising Bubbles with Mobile Surfaces. *J. Colloid Interface Sci.* 197, 275–292.

Dai, Z., Fornasiero, D. and Ralston, J. 1999. Particle-Bubble Attachment in Mineral Flotation. *J. Colloid Interface Sci.* 217, 70–76.

Dai, Z., Fornasiero, D., and Ralston, J. 2000. Particle bubble collision models–A review. *Adv. Colloid Interface Sci.* 85, 231–256.

Derjaguin, B.V., and Dukhin, S.S. 1961. Theory of flotation of small and medium size particles. *Trans. Inst. Min. Metall.*, 70, 221–246.

Dobby, G.S., and Finch, J.A. 1987. Particle size dependence in flotation derived from a fundamental model of the capture process. *Int. J. Miner. Process.* 21, 241–260.

Duan, J., Fornasiero, D., and Ralston, J. 2003. Calculation of the flotation rate constant of chalcopyrite particles in an ore. *Int. J. Miner. Process.* 72, 227–237.

Dukhin, S.S., Kretschmar, G., and Miller, R. 1995. Dynamics of Adsorption at Liquid Interfaces: Theory, Experiment, Application [D. Mobius and R. Miller, Editors]. *Studies in Interface Science*. Elsevier.

Duineveld, P.C. 1995, The rise velocity and shape of bubbles in pure water at high Reynolds number. *J. Fluid Mech.*, 292, 325–332.

Eddy, M. 1991. *Wastewater Engineering*. McGraw-Hill Inc., New York.

Elimelech, M., Gregory, J., Jia, X., and Williams, R. 1995. *Particle Deposition and Aggregation*. Butterworth-Heinemann, Oxford.

Gorain, B.K., Franzidis, J-P, and Manlapig, E.V. 1997. Studies on impeller type, impeller speed and air flow rate in an industrial scale flotation cell. Part 4. Effect of bubble surface area flux on flotation performance. *Minerals Engineering*, 10, 367–379.

Gorain, B.K., Franzidis, J-P, and Manlapig, E.V. 1996. Studies on impeller type, impeller speed and air flow rate in an industrial scale flotation cell. Part 3. Effect on superficial gas velocity. *Minerals Engineering*, 9, 639–654.

Gregory, J. 1989. Fundamentals of Flocculation. *Critical Reviews in Environmental Control.* 19(3): 185:230.

Huh, C., and Scriven, L.E.J. 1971. Hydrodynamic model of steady movement of a solid/liquid/fluid contact line. *J. Colloid Interface Sci.*, 35(1), 85–101.

Hui, S. 2001. Hydrodynamics of coarse-particle flotation. PhD Thesis, University of Newcastle.

Iler, R.K. 1979. *The Chemistry of Silica: Solubility, Polymerisation, Colloid and Surface Properties, and Biochemistry*. Wiley and Sons, New York.

Jameson, G.J., Nam, S., and Young, M.M. 1977. Physical factors affecting recovery rates in flotation. *Miner. Sci. Engng*. 9(3), 103–118.

Jenne, M., and Reuss, M. 1999. A Critical Assessment on the Use of Turbulence Models for the Simulation of Turbulent Liquid Flow Induced by a Rushton-Turbine in Baffled Stirred Tank Reactors. *Chem. Eng. Sci.*, 54, 3921–3941.

Jordan, C.E., and Spears, D.R. 1990. Evaluation of a Turbulent Flow Model for Fine Bubble and Fine Particle Flotation. *Minerals and Metallurgical Processing*, 65–73.

Jordan, C.E., and Scheiner, B.J. 1990. Velocity Profile Measurements within a Laboratory Flotation Cell. *Fluid/Particle Separation Journal*, 3, 71–75.

Jowett, A. 1980. Chapter 37 in Fine particle processing. (Somasundaran, P., Editor) In: *Proc. Int. Symp. Fine Particle Processing*. AIME, AIMM, New York.

Li, D., Fitzpatrick, J.A. and Slattery, J.C. 1990. Rate of Collection of Particles by Flotation. *Ind. Engng. Chem. Res.*, 29(6), 955–967.

Lynch, A.J., Johnson, N.J., Manlapig, E.V., and Thorne, C.G. 1981. *Mineral and Coal Flotation Circuits*. Elsevier, Amsterdam.

Millennium. 2004. TiO_2 Fundamentals, http://www.millenniumchem.com/Products+and+Services/Products+by+Type/Titanium+Dioxide+Paint+and+Coatings/r_TiO2+Fundamentals/Titanium+Dioxide+Manufacturing+Processes_EN.htm. Millennium Chemicals Inc.

Morad, S. 1986. SEM study of authigenic rutile, anatase and brookite in Proterozoic sandstones from Sweden. *Sedimentary Geology*, 46(1–2): 77–89.

Mular, A.L., and Musara, W.T. 1991. Batch Column Flotation: Rate Data Measurements. *Proceedings of an International Conference on Column Flotation*, Sudbury, Ontario, Canada, 1, 63–74.

Piantadosi, C., Jasieniak, M., Skinner, W.M., and Smart, R.C. 2000. Statistical comparison of surface species in flotation concentrates and tails from TOF-SIMS evidence. *Mineral Engineering*, 13(13), 1377–1394.

Piantadosi, C., and Smart, R.St.C. 2002. Statistical Comparison of Hydrophobic and Hydrophilic Species on Galena and Pyrite Particles in Flotation Concentrates and Tails from ToF-SIMS Evidence. *International Journal of Mineral Processing*, 64: 43–54.

Pugh, R.H. 1992. Selective coagulation of colloidal mineral particles. In: J.S. Laskowski and J. Ralston (Editors). *Colloid Chemistry in Mineral Processing*. Elsevier, Amsterdam, pp. 243.

Pyke, B.L., Fornasiero, D., Grano, S.R., and Ralston, J. 2001. Prediction of the flotation rate of chalcopyrite particles in a copper ore. Edited by J.A. Finch, S.R. Rao, and L. Huang. *Interactions in Mineral Processing, COM 2001*.

Pyke, B.L., Fornasiero, D., and Ralston, J. 2003. Bubble particle heterocoagulation under turbulent conditions. *Journal of Colloid and Interface Science*, 265, 141–151.

Ralston, J. 1992. Chapter 6 in *Colloid Chemistry in Minerals Processing* [Laskowski and Ralston, Editors], Elsevier, Amsterdam.

Ralston, J., and Dukhin, S.S. 1999. The interaction between particles and bubbles. *Colloids Surf.*, A, 151: 3–14.

Roskill. 2003. Titanium Minerals and Pigments, www.roskill.com/reports/titaniummap. Roskill Information Services.

Rutherford, K., Lee, K.C., Mahmoudi, S.M.S., and Yianneskis, M. 1996. Hydrodynamic characteristics of dual Rushton impeller stirred vessels. *AIChE Journal*, 42(2), 332–346.

Sam, A., Gomez, C.O., and Finch, J.A. 1996. Axial velocity profiles of single bubbles in water/frother solutions. *Int. J. Miner. Process.*, 47, 177–196.

Schubert, H., and Bischofberger, C. 1978. On the hydrodynamics of flotation machines. *Int. J. Miner. Process.*, 5, 131–142.

Schulze, H.J. 1993. Chapter 7. Flotation as a heterocoagulation process: Possibilities of calculating the probability of flotation, in surfactant science series (coagulation and flocculation). 47:321–364. Marcel Dekker, New York.

Snoswell, D.R.E. 2003a. Colloid Stability and the Influence of Dissolved Gas. *J. Phys. Chem. B*, 107: 2986–2994.

Snoswell, D.R.E. 2003b. *The influence of Surface Heterogeneity and Solution Composition on the Colloid Stability of SiO_2 and TiO_2 Dispersions.* University of South Australia, Adelaide.

Sutherland, K.L. 1948. Kinetics of the flotation process. *J. Phys. Coll.*, 52, 394–425.

Thomas, J.C. 1991. Photon Correlation Spectroscopy: Technique and Instrumentation, SPIE Photon Correlation Spectroscopy: Multicomponent Systems.

Wu, H., and Patterson, G.K. 1989. Laser-doppler measurements of trubulent-flow parameters in a stirred mixer. *Chem. Eng. Sci.*, 44(10), 2207–2221.

Ye, Y., and Miller, J.D. 1988. Bubble/particle contact time in the analysis of coal flotation. *Coal Prep.* 5, 147–166.

CHAPTER 2

Selective Adsorption of Starches Onto Ca-Bearing Surfaces

Armando C. de Araujo,* Rogerio C. Guimarães,† Antonio E.C. Peres,* George W. Poling,‡ and Paulo R.M. Viana*

ABSTRACT

Starches are extensively employed in several minerals processing applications acting, in most cases, as a selective depressant in anionic and cationic flotation circuits. Adsorption of starch macromolecules onto different surfaces takes place under diverse mechanisms. The presence of surfactants also plays an important role upon adsorption of the natural polymers. In the present work, adsorption isotherms of starch fractions (amylose and amylopectin) as well as the whole polymer (potato and tapioca starches) were obtained for two substrates: calcium phosphate and silica. The shape of the adsorption isotherms and the adsorption maxima for both substrates differ significantly in each case. The adsorption mechanisms in the case of high affinity adsorption isotherms obtained for the Ca-bearing surfaces indicate the possibility of chelation of starch rings onto Ca sites at the surfaces. In the case of silica, a typical low affinity adsorption isotherm is displayed, indicating that only secondary chemical bonds are formed upon adsorption. This striking different behavior of starch adsorbed on these two surface types promoted settling at very fast rates for surfaces where starch is firmly bonded to the substrate surface. Selective depression on flotation of minerals representing these model surfaces was assessed by microflotation experiments in both cationic and anionic systems.

INTRODUCTION

Starches and starch derivatives are utilized in the processing of different ores, such as: in the reverse flotation of iron ore,[1–7] in the direct flotation of carbonatitic phosphate ores,[8–10] in the depression of coals[11–12] and as flocculants in the processing of bauxite and iron ores.[13] They are also used as selective flocculants in the selective desliming prior to flotation in Tilden iron ore concentrator, in Michigan, USA.[14] Starches also have been extensively tested in many investigations, especially in the field of selective flocculation of phosphate ores[15–20] and iron ores.[21–36]

Although considerable efforts have already been placed in the area involving fundamental studies in starch/mineral systems, the reasons behind the selectivity obtained for the adsorption of these reagents on certain minerals, their depression action as well as their role as selective flocculants are still not completely understood.

*Federal University of Minas Gerais, Brazil.

†Bunge Fertilizantes, Brazil.

‡The University of British Columbia, Canada.

Chemically classified as a polysaccharide, starch is a widespread natural polymer formed by the condensation of α-d(+)glucose molecules. It occurs in nature as grains whose size and shape are functions of the parent plant. There is an intensive literature available on this subject.[37–46]

In general, starches contain between 20 to 25% of linear fraction called amylose and between 75 to 80% of a branched fraction known as amylopectin. The ratio between amylopectin and amylose can however vary considerably in some cases.[37–46] Cooke et al.[1] observed that, although all different starches tested adsorbed on hematite, there was a considerable difference in the amount adsorbed as a function of the type of starch used. This difference could be attributed either to the rations amylose/amylopectin in the starch samples used or to differences in the level and quality of impurities. Iwasaki and Lai[4] observed a greater adsorption of amylopectin on hematite as compared to amylose, in contradiction to the results of Cooke et al.[1] More recently, Khosla and co-workers[43–44] showed that in terms of moles of polymer adsorbed by unit area of hematite surface, the adsorption density of amylose was greater than that of amylopectin. However, if the adsorption was calculated as weight per unit area (mg/m^2), the adsorption density of amylopectin is considerably higher, as expected because of the large differences in molecular mass between the two starch fractions. Amylose is, on average, 20 times less bulky than amylopectin. The effect of starch fraction on the depression of hematite was also found to conform to the most important role of amylopectin, as reported by Pinto et al.[54]

Most of the effort in the study of the interaction mechanisms between starch and mineral surfaces has been concentrated in the hematite/starch system.[1–7, 21–36, 46, 54] Other systems that have also received attention include starch/calcite,[43–45, 47] starch/fluorite/barite/calcite,[47] starch/quartz,[5, 22, 48, 51–53] and starch/galena/scheelite/talc/apatite.[49] Somasundaran[45] proposed for the starch/calcite system the formation of strong chemical bonds involving Ca sites on the surface of calcite and the adsorbed starch macromolecules. Khosla and co-workers[43–44] also concluded that the interaction between starch and calcite is of a chemical nature. Calorimetric studies for the calcite/starch system and hematite/starch system demonstrated that adsorption reactions taking place are exothermic. Liu and Laskowski[51, 52] and Liu et al.,[53] studying quartz modified by the adsorption of metallic cations also have proposed a chemical mechanism for the interaction between modified quartz and starch.

In the present work, results are presented for the adsorption of starches on apatite and quartz surfaces. The main objective of this research was to characterize the mode of interaction between apatite and starch as well as to shed some light on the selectivity observed in the practice of selective desliming of sedimentary phosphate ores with dominant siliceous gangue when using starches as (selective) flocculants.

EXPERIMENTAL

Minerals

Samples of pure fluorapatite from Monteiro, Brazil, and quartz from Canada, were employed. These samples were characterized by x-ray diffraction and infrared spectroscopy. For details see reference.[48]

Reagents

Tapioca starch (commercial grade from Thailand), potato starch (reagent grade from Fisher Scientific), potato amylopectin and potato amylose (both from BDH, also reagent grade) were utilized in the present work. All starch samples were analyzed by infrared spectroscopy. Sodium oleate and dodecylamine hydrochloride were obtained with Fisher Scientific and Eastman Kodak Co., respectively. Other chemicals used (all ACS certified) are listed in reference.[48]

Equipment and Techniques

Mineral samples were ground to a very fine powder by a sequence of dry grinding stages, all without the use of any metallic grinding media or surfaces. Multipoint BET N2 adsorption technique (Quantasorb apparatus from Quantachrome) gave the following specific surface areas: quartz = 0.53 m^2/g and fluorapatite = 1.04m^2/g.

Adsorption tests were carried out at 294K. Mineral samples weighing 1.5g were placed in closed glass vials. Fifty milliliters of starch solution at the desired concentration were then added to the vials, already at the pre-established pH value. For each adsorption isotherm, all vials required were shaken together for 60 minutes. Solids were then separated by filtration. The filtrate was analyzed for its starch residual concentration by a modified and standardized KI/I_2 method.[48]

Electrical conductivity measurements of solutions containing starch and Ca ionic species were performed in order to clarify some aspects regarding the interaction between these solutes. Infrared spectra were obtained in a Perkin Elmer double beam, optical null spectrophotometer, model 283-B.

Microflotation was performed in duplicate tests using a modified Hallimond tube. Quartz and fluorapatite samples were sized between 0.212mm and 0.150mm. Medical air was used at a flow rate of 40cm^3 per minute.

RESULTS AND DISCUSSION

Infrared spectra of all starch samples indicate the presence of all characteristic vibrations for starch macromolecules.[48] Further analysis of the spectra reveals the presence of other vibrations that can be correlated to trace impurities. For instance, the spectrum of potato amylose presents vibrations in a region characteristic of angular deformation of NH groups, possibly present as associated proteins. The same vibrations are absent in the spectrum of amylopectin but are present in the other starches. Another impurity detected in all starch samples indicated the presence of carboxylic groups (COOH). Phosphate anion vibrations were also observed, especially in the tapioca starch sample.

Figure 1a depicts the adsorption isotherms, at pH 10, for tapioca and potato starches, amylose and amylopectin on fluorapatite. The simple observation of this figure reveals that amylopectin is more strongly adsorbed than amylose. The experimental isotherm for amylopectin shows a plateau region reached at very low equilibrium concentrations. It also shows that for low initial concentrations the polymer is completely adsorbed (equilibrium concentration is so low that it cannot be measured by the employed technique). Adsorption isotherms of the shape seen in Figure 1 have been associated with the adsorption of polymers with high affinity for the surface.[51] Figure 1b also shows the adsorption isotherms obtained for the same starches on quartz surfaces.

The following aspects of the adsorption isotherms shown in Figure 1 deserve to be highlighted:

- For both tapioca and potato starches, the shape of the isotherms are more similar to the shape obtained for amylopectin than that obtained for amylose.
- When compared, the shapes of the isotherms on quartz and fluorapatite are extremely diverse, probably reflecting that different adsorption mechanisms operate in each system.
- On average, tapioca and potato starches contain 17% and 25% of amylose, respectively.[42] In the adsorption isotherms presented in Figure 1, the greater proportion of amylopectin in tapioca starch can explain, at least partially, its larger adsorption in comparison to potato starch.
- Regarding the adsorption on quartz, the isotherms show a different shape and the extent of adsorption is much smaller. To explain this difference in adsorption based only on electrostatic repulsion, as originally proposed by Balajee and Iwasaki,[22] may not be satisfactory. Both minerals at pH 10 present negatively charged surface. Zeta potential using Smoluchowski equation indicates -30mV for fluorapatite and -45mV for quartz.[48] The magnitude of the zeta potential of the two minerals would not justify such diverse behavior by itself.

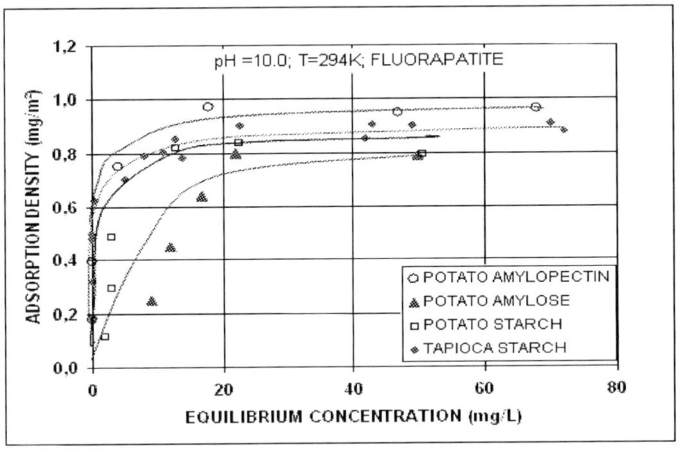

Figure 1(a) Starch adsorption isotherms on fluorapatite

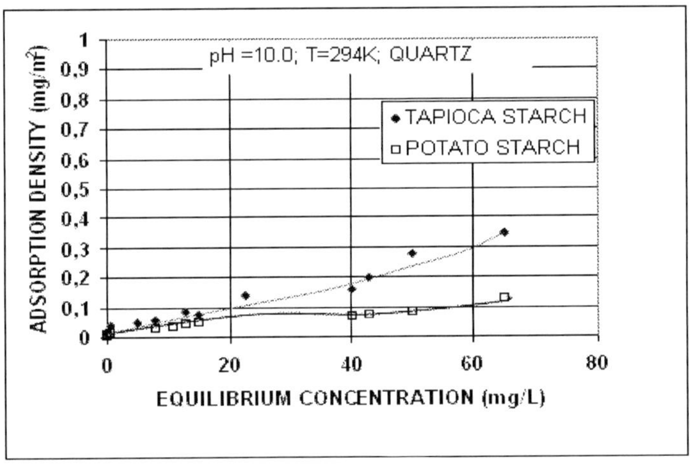

Figure 1(b) Starch adsorption isotherms on quartz (y-axis scale maintained in order to make this figure directly comparable to Figure 1(a))

The small relevance of the electrostatic forces operating as a major component in the overall adsorption mechanism is also shown in Figure 2. In this figure, the effect of pH on the adsorption of tapioca starch on fluorapatite is presented. The adsorption density does not decrease as pH is increased from 6 to 11 (and, consequently, the negative value of the zeta potential is also increased once the measured iep (pH) for this sample occurs at pH 4.0). Conversely, the adsorption density increases at the pH range aforementioned and only starts to decrease above pH 11. This result demonstrates that, if important at all, electrostatic forces are not the main reasons for the observed difference in behavior for quartz and fluorapatite regarding the adsorption of starches.

Another driving force generally considered as relevant for the adsorption of starches is due to the ability of these polymers to bind with surfaces via hydrogen bonds,[(5, 22, 47)] for example).

Hydrogen bonds most likely participate effectively in the overall adsorption mechanism. They may be used to explain the greater adsorption of

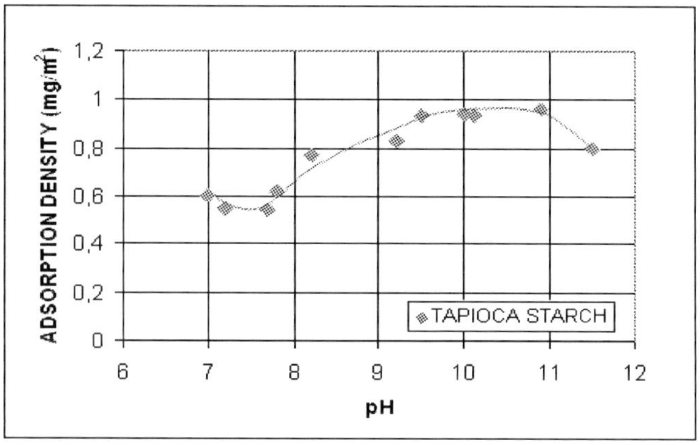

Figure 2 pH effect on the adsorption of tapioca starch on fluorapatite

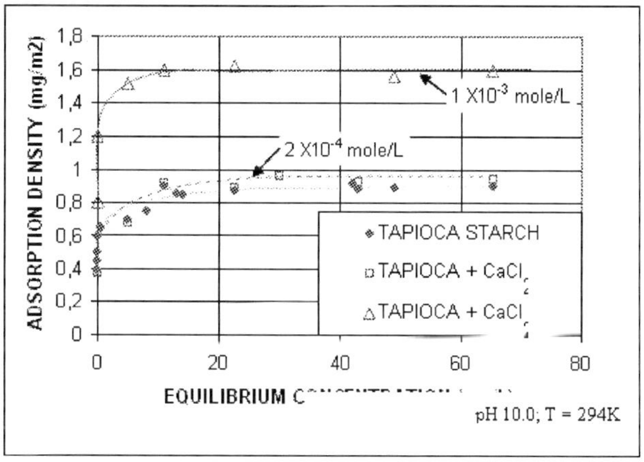

Figure 3 Effect of Ca ionic species on the adsorption of tapioca starch on fluorapatite

amylopectin on fluorapatite as compared to the smaller adsorption of amylose.

A third type of force that can be involved in the adsorption mechanism is of a chemical nature. Somasundaran[45] proposed that a chemical interaction was operative in the calcite/corn starch system. One of the ways to explain this chemical interaction is to suppose that ionic impurities present in all starch samples react with metallic sites at the surface of the minerals. Figure 3 shows the importance of the presence of Ca ionic species on the adsorption of tapioca starch on fluorapatite at pH 10. One can observe that the amount adsorbed in the presence of the highest level of Ca tested is almost twice the amount adsorbed in its absence. One should also consider that the adsorption on quartz in the presence of Ca ionic species is also enhanced. Therefore, one possible conclusion at this point is that the high affinity displayed by all starches for the surface of fluorapatite is, most probably, a result of some sort of chemical interaction involving Ca sites at the surface of this mineral.

To investigate in more detail such interaction, a series of tests involving the measurement of

Table 1 Interaction between starch and Ca species measured by the electrical conductivity of the solutions (pH 10.0 ± 0.2, $CaCl_2$ = 1 × 10^{-3} mole/L)

		Conductivity (μmho/cm)		
Starch sample	Conc. [mg/L]	Measured	Calculated	% Variation
Potato Amylose	50	800	950	-15.8
	75	890	1040	-14.4
Potato Amylopectin	50	790	940	-16.0
	75	875	1060	-17.5
Potato Starch	60	1175	1425	-17.5
	85	1350	1625	-16.9
Tapioca Starch	50	875	1160	-24.6
	75	975	1300	-25.0

Note: Calculated values refer to the condition of no interaction between the two solutes

the changes in solution electrical conductivity when starch solutions are contacted with Ca chloride solution was performed. The results are summarized in Table 1. The technique employed in this case was modified from one originally proposed by Khosla et al.[43]

Table 1 clearly shows that there is an interaction between starch and Ca ions in solution. This interaction is approximately equivalent for amylopectin and potato starch, greater for tapioca starch and smaller for amylose. Sulkan et al. (50) obtained that the amount of Ca bound by corn and potato starches was approximately 18%, using a completely different approach. These results corroborate the hypothesis for a chemical mechanism operating in the system involving a calcium phosphate mineral surface.

Chemical mechanisms were also proposed by Liu and Laskowski[51, 52] when studying the adsorption of starches on quartz surfaces modified by the previous adsorption of metal ions.

The importance of this selective behavior in adsorption displayed by starches regarding fluorapatite surface, and also regarding Ca-bearing surfaces (calcite, for instance), is paramount. Starches have been used as selective flocculants for siliceous phosphate ores in laboratory scale with considerable success.[19] They are used industrially as selective depressants in the direct flotation of igneous phosphate ores with associated carbonate gangue (calcite and dolomite). The results obtained for the adsorption behavior of these natural polymers on fluorapatite and quartz surfaces demonstrate that the selectivity reached in real systems is due to a strong interaction taking place between starch macromolecules and specific sites on the mineral surfaces.

Figure 4 depicts the flotation depression action of starches on anionic flotation (with sodium oleate) of fluorapatite at pH 10.5. Low starch dosages are sufficient to depress apatite floatability with tapioca starch, potato starch and amylopectin depressing fluorapatite flotation more strongly than amylose. This result again confirms that amylopectin rich starches interact in a different fashion with the Ca-bearing surface of apatite than the linear, low molecular mass, starch fraction of amylose.

Figure 5 shows a similar depressant action of starches on the cationic flotation (with

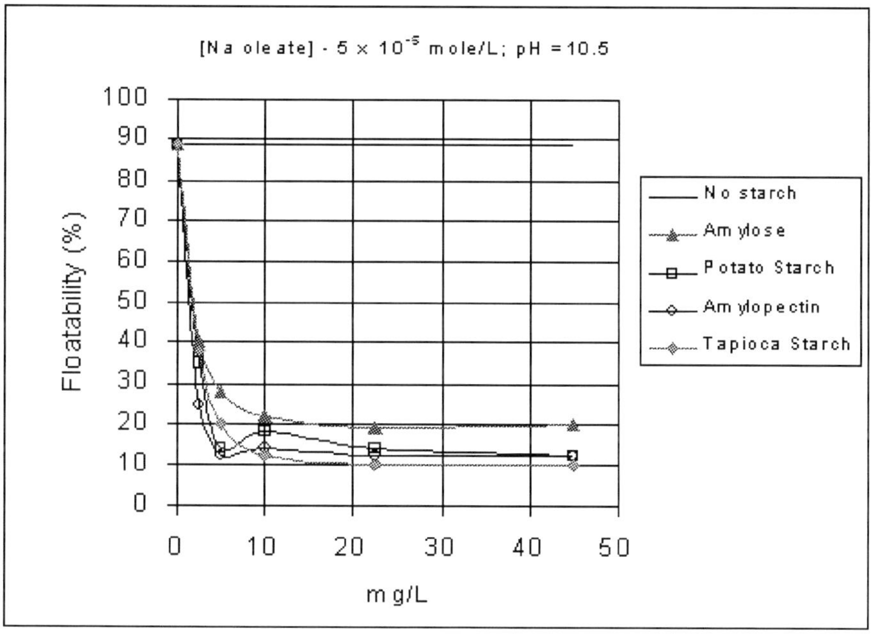

Figure 4 Depression by starches of sodium oleate flotation of fluorapatite

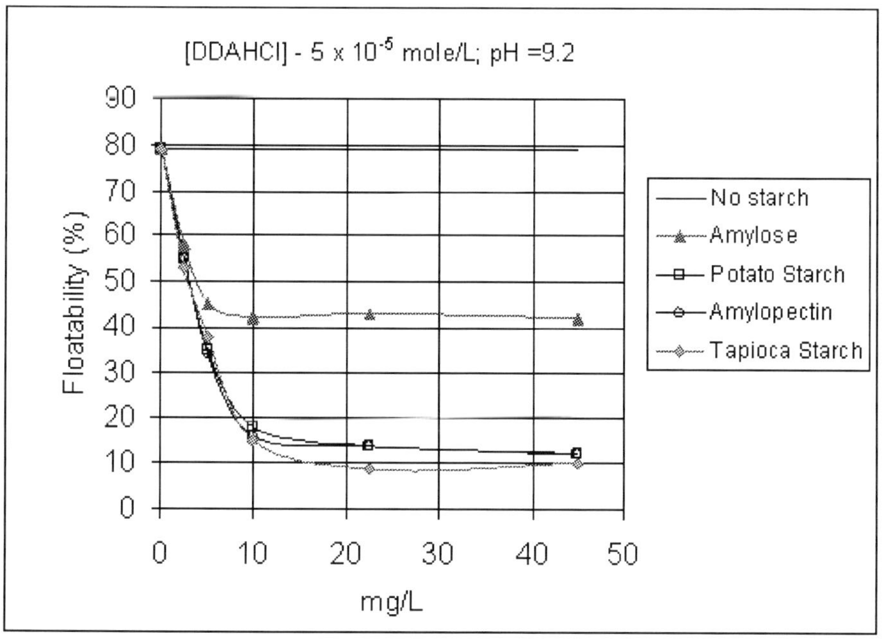

Figure 5 Depression by starches of dodecylamine flotation of fluorapatite

dodecylamine hydrochloride) of fluorapatite at pH 9.2. The shape of the starch concentration/floatability curves is very similar to that obtained for sodium oleate. Low starch dosages are able to decrease significantly apatite floatability. However, the difference in behavior between the linear amylose fraction is enhanced in the cationic flotation case. Here amylose was able to depress apatite floatability only to a 40% level in comparison to the 20% level obtained for sodium oleate microflotation (see Figure 4). This difference suggests that the interaction between the collector species and starches are also important in the depression mechanism. One should take into consideration that the branched macromolecules would provide extra room for trapping collector species in comparison to the linear and low molecular mass amylose.

Tapioca starch flocculation tests performed with fluorapatite sample with all particles passing a 0.038mm opening sieve displayed high settling rates (mudline monitoring). For a dosage of 5mg/L, the settling rate reached values above 10cm/minute. The maximum settling rate of 25cm/minute was achieved, however, with amy lopectin at a dosage of 20mg/L. On the other hand, a 20cm/minute settling rate maximum with amylose was reached only for a 200mg/L concentration. These findings again confirm the higher affinity of the branched starch fraction towards the Ca-bearing surface in comparison to the linear amylose fraction.

CONCLUSION

The adsorption mechanism of starches on fluorapatite involves forces of chemical nature. Electrostatic forces (either attraction or repulsion) are not important, and if present in the overall adsorption mechanism, they likely play only a minor role.

The presence of Ca ionic species in solution increases considerably the adsorption of starches on apatite. Amylose is adsorbed to a smaller extension than amylopectin on apatite. The smaller adsorption of amylose and its lower molecular mass hinders it flocculant action on apatite.

Starch adsorption on quartz is much smaller than that observed on apatite. The preferential adsorption on apatite is responsible for the selectivity obtained when starch is used as selective flocculant for the selective desliming of phosphate ores with associated siliceous minerals gangue.

The results indicate the need to control the concentration of Ca ionic species in solution in order to obtain selective adsorption in apatite/quartz systems.

The highest affinity displayed by tapioca starch may be related to the high content of impurities in this sample in comparison to other starch sample and starch fractions employed in the current investigation. Carboxylic and phosphate groups may represent the most relevant impurities associated to the highest affinity of adsorption observed for tapioca starch.

Microflotation tests showed the ability of starches to depress cationic and anionic flotation of apatite. In both cases, amylose was a weaker depressant in comparison to whole starches and to the branched amylopectin fraction.

ACKNOWLEDGMENTS

The authors are indebted to CNPq (Brazil) through its grant for the Millenium Project-water a mineral vision.

REFERENCES

1. Cooke, S.R.B. et al., *Trans. AIME*, 193, 607 (1952)
2. Chang, C.S. et al., *Trans. AIME*, 196, 1282 (1953)
3. Chang, C.S. et al., *Trans. AIME*, 199, 922 (1954)
4. Iwasaki, I. and Lai, R.W., *Trans. AIME*, 232, 362 (1965)
5. Iwasaki, I. and Lipp, R.J., *Water Pollution Control Research Series*, Minneapolis, 108 pp. (1971)
6. Peres, A.E.C. and Correa, M.I., *Minerals Engineering*, V. 9, N. 12, 1227 (1996)

7. Viana, P.R.M. and Souza, H., II Congreso Latino-Americano de Flotacion, Anales, 19–23, V. 2, Concepcion (1985)

8. Hout, R., *International Journal of Mineral Processing*, 9, 353 (1982)

9. Silva, J.M. and Coelho, E.M, Conferencia Latino Americana de Roca Fosfórica, Anales, Cochabamba, 159–168 (1984)

10. Assis, S.M. et al., XI Encontro Nacional de Tratamento de Minérios e Hidrometalurgia Anais, V. I, 87, Natal (1985)

11. Im, C.M. and Aplan, F.F., 1st Australian Coal Preparation Conference, Newcastle, 183 (1981)

12. Miller, J.D., et al., *Colloids and Surfaces*, 8, 137 (1983)

13. Gardner, G.R. and Kenneth, R.B., *Trans AIME*, 134, 146 (1939)

14. Paananen, A.D. and Turcotte, W.A., *Mining Engng.*, August, 1224 (1939)

15. Coelho, E.M., *Mineração e Metalurgia*, 418, 28 (1980)

16. Oliveira, J.F. and Gomes, J.I.A., Projeto Floculação Seletiva de Fosfato de Patos de Minas, Project Report, No. 15, CETEM, Rio de Janeiro (1981)

17. Oliveira, J.F. and Gomes, J.I.A., *1st Southern Hemisphere Meeting on Mineral Technology, Proceedings*, V. 1, 139 (1982)

18. Coelho, E.M., Full Professor Dissertation, 232 pp. (1984)

19. Araujo, A.C. et al., *CIM Bulletin*, November, 42 (1986)

20. Schemeling, U.F.M. et al., Ferrominerales, ILAFA, 6 (1979)

21. Frommer, D.M., Proceedings of the VIII International Mineral Processing Congress, Leningrad, paper D-9 (1968)

22. Balajee, S.R. and Iwasaki, I., *Trans. AIME*, 244, 401 (1969)

23. Iwasaki, I., *Trans. AIME*, 232, 383 (1965)

24. Colombo, A.F. and Frommer, D.M., in: *Flotation, A.M. Memorial Volume*, V. 2, AIME, 1285 (1976)

25. Colombo, A.F., Seminar on the beneficiation of lean phosphate with carbonate gangue, XI International Mineral Processing Congress, Cagliari (1975)

26. Colombo, A.F., *Beneficiation of Mineral Fines*, AIME/National Science Foundation, New York, 237 (1979)

27. Colombo, A.F., in: *Fine Particle Processing*, V. II, AIME, New York, 1034 (1980)

28. Chen, X.Y. and Leja, J., unpublished results, University of British Columbia, Vancouver (1980)

29. Aburashid, A.R. and Smith, M.R., XIV International Mineral Processing Congress, paper IX-I, Toronto (1982)

30. Gururaj, B. et al., *International Journal of Mineral Processing*, 11, 285 (1985)

31. Zuleta, M. et al., in: *Reagents in the minerals industry, IMM*, Roma, 277 (1984)

32. Zuleta, M. et al., *Proceedings of the XV International Mineral Processing Congress*, V. II, 346, Cannes (1985)

33. Hanuamantha Rao, K. and Narashimanm K.S., *International Journal of Mineral Processing*, 14, 67 (185)

34. Haseman, J.S., U.S. Patent No. 2660303 (1953)

35. Davenport, J.E. et al., *Ind. Engineering Chemistry*, 8, 527 (1969)

36. Baudet, G. et al., U.S. Patent No. 4235709 (1980)

37. Radley, J.A., *Starch and its derivatives*, Vol. I, 3rd ed., Chapman & Hall, London (1953)

38. Whistler, R.L. and Paschall, E.F. (eds.), *Starch: Chemistry and Technology*, Academic Press, New York (1965)

39. Whistler, R.L. (ed.), *Methods in Carbohydrate Chemistry*, Vol. IV, Academic Press, New York (1964)

40. Greenwood, C.T., *The Carbohydrates Chemistry and Biochemistry*, 2nd ed., Vol. IIB, Academic Press, New York (1980)

41. BeMiller, J.N., *Industrial Gums—Polysaccharides and Their Derivatives*, Academic Press, New York (1973)

42. Whistler, R.L. et al. (eds.), *Starch: Chemistry and Technology*, 2nd ed., Academic Press, New York (1984)

43. Khosla, N.K. and Biswas, A.K., *Colloids and Surfaces*, 8, 321 (1984)

44. Khosla, N.K. and Biswas, A.K., *Colloids and Surfaces*, 9, 219 (1984)

45. Somasundaran, P., *Journal of Colloid and Interface Science*, 31, 557 (1969)

46. Hanna, H.S., *Recent Advances in Science and Technology of Materials*, Vol. I, Plenun Press, New York, 365 (1973)

47. Steemberg, E. and Harris, P.J., *South African Journal of Chemistry*, 37, 85 (1984)

48. Araujo, A.C., Ph.D. Thesis, University of British Columbia, 357 pp. Vancouver (1988)

49. Lipatov, Y.A. and Sergeeva, L.M., *Adsorption of Polymers*, John Wiley & Sons, Jerusalem (1974)

50. Sulkan, G. et al., *Starch*, 31, 125 (1979)

51. Liu, Q. and Laskowski, J.S., *International Journal of Mineral Processing*, 26, 297 (1989)

52. Liu, Q. and Laskowski, J.S., *International Journal of Mineral Processing*, 27, 147 (1989)

53. Liu, Q. et al., *International Journal of Mineral Processing*, 60, 229 (2000)

54. Pinto, C.L.L. et al., *Minerals Engineering*, V. 5, N. 3–5, 469 (1992)

CHAPTER 3

The Role of Hydrodynamics in Coarse Apatite Flotation

L.S. Leal Filho,* W.J. Rodrigues,* O.A. de Lima,* and L.A.F. de Barros†

ABSTRACT

This work aims to contribute towards a better understanding of the importance of hydrodynamics in flotation response of coarse (particle diameter=d_p=0.25mm) apatite. Microflotation tests were carried out with apatite (contact angle~85–88°) in the presence of sodium oleate (100mg/dm^3) at pH=10.5 versus Froude Number (Fr), Reynolds Number (Re) and mean energy dissipation (ε). The results have shown a maximum recovery (R>90%) when 0.3<Fr<0.8; Re~4×10^3 and e~0.1W/kg. For more quiescent hydrodynamic conditions (Fr<0.3; Re<4×10^3 and ε<0.1W/kg), recovery of apatite decreased due to a lack in particle suspension; whereas, for the most severe hydrodynamic conditions (Fr>2.0; Re>10^4 and ε>1.0W/kg) the recovery was almost nil due to particle-bubble detachment. Flotation kinetics of apatite was studied in laboratory scale with phosphate ore from Tapira-MG under strong hydrodynamic conditions that characterize Denver laboratory cell (Fr=2.6; Re=1.2×10^5). The results indicated that coarse particles of apatite (d_p>0.15mm) float at a lower rate than finer particles (0.15mm>d_p>0.037mm) and, further, the flotation kinetics of coarse particles does not fit a first order model. Those results suggest that coarse particles demand either more time or more suitable hydrodynamic conditions to float efficiently. To illustrate the role of hydrodynamics in coarse apatite flotation, a successful modification was accomplished in industrial scale.

INTRODUCTION

The four biggest Brazilian phosphate flotation mills reject about 0.2 million of metric tons per year of coarse apatite (d_p≥0.15mm). As a matter of consequence, valuable material is wasted and a greater amount of solids is disposed in the environment. The lower recovery of coarse apatite may be due to either deficiency in particle liberation or no suitable hydrodynamic conditions under which flotation is usually carried out. This chapter attempts to contribute towards a better understanding of the importance of hydrodynamics in coarse particle flotation.

BACKGROUND

Laboratory and in-plant investigations attempting to correlate particle size with flotation

*University of São Paulo, Polytechnic School, Department of Mining and Petroleum Engineering, São Paulo, Brazil.

†Fosfertil SA, Tapira-MG, Brazil.

response have established that very coarse particles do not float efficiently. Theoretical treatments of the problem pose an important practical dilemma: the forces that bring particle and bubble together during collision may jeopardize the stability of the aggregate, promoting particle detachment. Unlike the bubble-particle attachment process, which is related to the surface chemistry of the particle, particle-bubble collision and aggregate stability constitute processes, which are largely affected by the physical environment in the flotation cell, where hydrodynamics plays a major role (Leal Filho et al., 2003; Jewett, 1980; Rodrigues et al., 2001; Schulze, 1977; Ralston et al., 1999).

In mechanical cells, to collide with bubbles and float, particles must be kept in suspension which could be considered as a result of transport phenomena caused mainly by the turbulence (macroturbulence) created by the impeller rotation (N) and also by the motion of the ascending air bubbles inside the vessel. This fact indicates that flotation is realized under highly turbulent flow conditions (Reynolds numbers of industrial-scale impellers lay in the range 10^6–10^7) and a better understanding of the phenomenon requires the utilization of the statistical turbulence theory, where an extensive simplification of complicated laws is typical for the application in mineral processing (Schubert, 1999).

To characterize the state of particle suspension, Zwietering (Zwietering, 1958) proposed the so-called one-second criterion: individual particles do not remain settled at the bottom of the stirred vessel for more than 1 second. Zwietering's pristine approach has been modified by other researchers in order to take into account the presence of air in the flotation cell (Schulze, 1984). A critical value for impeller speed ($N_{Critical}$) above which the one-second criterion is valid may be experimentally determined. This value is also dependent upon the airflow rate at which flotation is carried out.

Unlike particle suspension, which is largely controlled by macroturbulence, the dispersion of air in the pulp, the turbulent particle-bubble collisions and detachment are controlled by microturbulence. Intensity and structure of microturbulence depend only upon energy dissipation rate and kinematic viscosity of a fluid system. For the mean dissipation rate (ε) in a gassed stirred tank containing a mass M of a fluid and having a power input P, use equation 1. However, close to the impeller, maximum local values of dissipation rate (ε_{max}) may be 10–30 times greater than the mean value (ε) (Schubert, 1999; Schulze, 1984).

$$\varepsilon = P/M \tag{1}$$

Using Komolgorov's theory of isotropic microturbulence, local fluctuations of energy dissipation that actually occur inside the flotation cell could be disregarded in order to simplify the approach. Accordingly, the mean value of other variables that are relevant to hydrodynamics of coarse particle flotation may be determined by using the same rationale: turbulent velocity, vortex acceleration, shear stresses, radii of eddies, etc. (Schubert, 1999; Schulze, 1984). Nevertheless, one cannot fail to notice that "new" available techniques suitable for in-situ measurements (piezoelectric probes, acoustic Doppler velocimetry) enhanced by computational techniques (like CFD) have been gaining attention of scientists and engineers concerned with this subject. The trend for research in hydrodynamics of flotation is towards those techniques (Leal Filho et al., 2002).

MATERIALS AND METHODS

Handpicked crystals of igneous apatite (specific gravity=3,080kg/m^3) from Monteiro-PB were dry ground minus 48 mesh Tyler (0.295mm) and screened. The mass fraction –0.295mm +0.208mm was purified by dense liquids, followed by magnetic separation in Frantz isodynamic separator. The purified sample was thoroughly washed in ketone, ethanol and water before flotation experiments. Glass spheres (specific gravity=2,480kg/m^3) were screened and the mass fraction –0.295mm+0.208mm was used for further flotation experiments. For both apatite and glass spheres, the particle diameter (d_p) was assumed to be d_p=0.248mm. This value is the geometric mean of the screen apertures (0.295mm and 0.208mm) that were used to classify the solids.

Sodium oleate (analytical grade) was used as collector for apatite and ether amine acetate (Flotigan-EDA supplied by Clariant do Brasil) was used as collector for glass spheres. Sodium hydroxide analytical grade was used for pH regulation.

The contact angle (θ) of apatite particles and glass spheres, after conditioning with collector solution, was determined by the Washburn method (Subramanyan et al., 1999). Apatite was conditioned with sodium oleate solution ($100mg/dm^3$) at pH=10.5 for 1 minute; whereas, glass spheres were conditioned with ether amine acetate ($75mg/m^3$) at pH=10 for 1 minute as well. The magnitude of the measured contact angles fell in the range $85° < \theta < 88°$ for both solid species.

Pulp viscosity (μ_p) was calculated by equation 2, proposed by Roscoe (Roscoe, 1952), where μ_w is the dynamic viscosity of water and Φ is the fraction of pulp volume occupied by the solids.

$$\mu_p = \mu_w (1-\Phi)^{-2.5} \quad (2)$$

Flotation experiments were carried out in a microflotation cell (diameter=32.2mm; height=94.2mm) whose shape and geometry resembles the traditional modified Hallimond tube. Agitation was accomplished by a mechanical stirring system coupled to a rotational speed control and torque meter. The floated material was manually skimmed off the cell with a small ladle. Gas (nitrogen) was blown into the microflotation cell through a sintered glass plate (mean pore radius=0.26mm). Bubble diameter (d_b=0.37mm) was estimated by the method described by Manqiu (Manqiu et al., 1988) based on gas hold-up and gas flow rate fed into the system. According to Yianatos (Yanatos et al., 1988), the estimated value for d_b is within ±15% of the actual value.

Impellers (cylindroid-type, paddle and Rushton turbine) showing length=15mm and width=6.52mm were used in flotation experiments. Impeller rotational speed (N) varied within a very wide range: $0.07s^{-1}$–49.33^{s-1}. All flotation experiments were carried out with 1 gram of solids suspended in solutions of collectors: ether amine acetate ($75mg/dm^3$ at pH=10) for glass spheres and sodium oleate ($100mg/dm^3$ at pH=10.5) for apatite. Surface tension of both collector solutions (γ) was measured by Wilhelmy plate method. For sodium oleate solution ($100mg/dm^3$), γ=32mN/m at 20°C and for ether amine acetate solution ($75mg/dm^3$), γ=36 mN/m at 20°C. After 1 minute of collector conditioning time, gas (N_2) was introduced in the system at a flow rate of $8.33 \times 10^{-4} dm^3/s$ (gas hold-up ~3%) during 60s of flotation time. After each microflotation test, the floated and non-floated products were filtered, dried and weighed. The flotation response (recovery) was calculated on a percentage basis by dividing the mass of the floated product by the mass of the feed.

To describe the hydrodynamic conditions inside the flotation cell, the diameter of the impeller (D), impeller rotational speed (N) and physical properties of the pulp were combined to give dimensionless numbers: Reynolds (Re) and Froude (Fr). Information about them is presented in Table 1.

The power Number (Po) is a hydrodynamic parameter related to the macro variables: N, D, ρ (pulp density) and P (power). Since the power measured at the impeller's shaft was lower than the sensitivity of the available torque meter, Po was assessed by empirical equations proposed by Nagata (Nagata, 1975) that are valid solely for paddle-type impellers operating at Re>10. To solve the problem for the actual geometry (cylindroid) of the impeller used in our flotation experiments, it was necessary to make further corrections that are fully described by Rodrigues (Rodrigues et al., 2001). The rationale used to

$$f \approx [C_D \text{ (cylinder of d/W=2.3)}] / [C_D \text{ (for rectangular plate of d/W=2.3)}] \approx 0.60 \quad (5)$$

$$Po = Po_{cylinder} \approx f\,Po_{Paddle} \quad (6)$$

Table 1 Hydrodynamic dimensionless parameters referred to macro (equipment) system of variables

Parameters	Equations	Remarks
Reynolds Number-Re (Ratio between inertial and viscous forces)	$Re = (ND^2 \rho)/\mu_p$ (3)	ρ = pulp density μ_p = pulp dynamic viscosity g = gravity acceleration N = Impeller speed D = Impeller diameter
Froude Number-Fr (Ratio between inertial and gravity forces)	$Fr = N^2D/g$ (4)	

make those corrections is based on the existing analogy of impeller's Power Number (Po) to Drag Coefficient (C_D) of immersed bodies (cylinders versus rectangular plates) that have a similar geometry and the same ratio length/width (L/W) for rectangular plates or length/diameter (L/d) for cylinders (Uhl et al., 1966). Since our cylindroid impeller bears a ratio L/W of 2.3, it was necessary to perform a numerical interpolation from the CD values of cylinders and rectangular plates found in the literature (Sissom et al., 1972; Knudsen et al., 1958) and calculate a correction factor f exemplified by equation 5 for d/W=2.3. The magnitude of Po related to our cylinder-type impeller was, thus, calculated by equation 6. Where $Po_{cylinder}$ and Po_{Paddle} are the Power Number referred to cylinder-type or paddle-type impellers, respectively; since f is the correction factor and assumes the value f≈0.60 for d/W=2.3. Power number was used to determine the power consumption (P) at each microflotation test, by means of equation 7. Mean energy dissipation (ε) was determined by equation 1, where P is the net power consumption of the impeller's shaft, Po is power number, N is the rotational speed of impeller, D is the impeller diameter and r is pulp density.

$$P = Po\, N^3 D^5 \rho \qquad (7)$$

Kinetic flotation tests were carried out in Denver laboratory flotation cell (D=0.07m; N=18.3s^{-1}; gas hold-up~10%, Fr=2.6; Re=1.18×10^5) with phosphate ore from Tapira-MG, provided by Fosfertil SA. The mineralogical composition of the ore was apatite=18%, carbonates=7%, micas=13%, quartz=20%, diopside=12%, and Ti,Fe-bearing oxides=30%. Flotation conditions were pulp with 12% of solids (in volume), 500g/t of cornstarch, 350g/t of collector (mixture composed of 80% rice bran oil + 20% KE-883® supplied by Cognis), and flotation pH=9.3. The froth was collected along cumulative time intervals (Δt) of 5s, 15s, 35s and 80s. The yielded masses were dried, weighed and sent to chemical analysis. The recovery of P_2O_5 was calculated based on the yielded mass and on the P_2O_5 content of the floated and sunk products.

Flotation rate and kinetic order was assessed by three different approaches: Method 1-Trial and Error (Foggler, 1999), Method 2-Ratio of Times (Vatcha, 1987) and Method 3-Fractional Time (Connors, 1990).

Pursuing Method 1, functions R(t) (recovery versus time) were resolved theoretically for different orders (zero, first, second and fractional). Points (t,R) obtained from experimental data were plotted in R *versus* t curves. The function R(t) that best fitted the plotted population of experimental points allowed the determination of order (n) and kinetic constant (k). The method of the least minimum squares ($-1<r^2<+1$) gave guidance to select the function R(t) that best fitted the experimental data, *id est*, the function whose r2 was closer to 1 (r^2 tending to 1).

The principle underlying Method 2 is that the ratio of the times at which two specific recoveries occur depends solely on the order of the "reaction." From experimental data (R versus t), one computes the ratio of the times corresponding to recoveries of 75% (3/4) and 50% (1/2) or 66% (2/3) and 33% (1/3). Therefore, equation 8 is valid for first order (n=1) and equation 9 is valid for other orders (n≠1).

$$(T_{3/4}/T_{1/2}) = \ln(1/4) / \ln(1/2) = 2 \quad (8)$$

$$(T_{3/4}/T_{1/2}) = (4^{n-1} - 1) / (2^{n-1} - 1) \quad (9)$$

According to Connors (Connors, 1990), for a chemical reaction, equation 10 is the integrated form of the rate equation $dc/dt = -kc^n$ (n≠1), where c_o is the initial concentration of a target substance in a reactor (apatite), c is the concentration of the same substance at a certain time t and p [$p = (1 - c/c_o)$] is the fraction of the substance reacted at time t.

The left side of equation 10 may be expanded in binomial series, which is truncated after the quadratic term, resulting in equation 11. Bringing the same concepts to our bench scale flotation system, it is possible to write equation 11 as equation 12, where R is flotation recovery and M_0 is the initial mass of apatite in the pulp. From the straight line defined by equation 12, it is possible to determine flotation order (n) and rate (k). This is Method 3 or the Fractional Time Method.

$$(1-p)^{1-n} = 1 + (n-1)\,k\,t\,c_o^{n-1} \quad (10)$$

$$t/p = (nt/2) + (1/\,kc_o^{n-1}) \quad (11)$$

$$t/R = (n/2)\,t + (1/\,kM_0^{n-1}) \quad (12)$$

RESULTS AND DISCUSSION

Recovery of P_2O_5 *versus* particle size obtained from bench scale flotation tests carried out with phosphate ore from Tapira-MG (Fosfertil SA) is depicted in Figure 1, where it is possible to observe that a maximum flotation response is attained when particle size lays in the range of 0.105mm to 0.037mm (–150 mesh +400 mesh). A markedly decrease in apatite recovery for particle size coarser than 0.105mm was observed under the strong hydrodynamic conditions that characterize a flotation experiment (Fr=2.6 and $Re = 1.2 \times 10^5$) carried out in Denver laboratory flotation cell. A slight decrease in apatite recovery for particle size lower than 0.037mm (-400 mesh Tyler) was also observed, but the behavior of fines is out of the scope of this work.

Since it is not possible to vary the hydrodynamic conditions in a wide range when handling with Denver laboratory flotation cell, microflotation tests were carried out with glass spheres versus apatite particles. The results are shown in Figure 2, Figure 3, and Figure 4, and they indicate that it is possible to obtain high recovery (R>90%) of coarse apatite particles under hydrodynamic conditions that are suitable to float them: $0.3 < Fr < 0.8$, $Re \sim 4 \times 10^3$, and $\varepsilon \sim 0.2$W/kg. Furthermore, flotation response of coarse apatite versus Fr, Re, and ε. (Figures 2-4) indicate that under more quiescent conditions ($Fr < 0.3$, $Re < 4 \times 10^3$, and $\varepsilon < 0.1$W/kg), there is a lack in particle suspension; whereas, under more turbulent conditions ($Fr > 0.8$, $Re > 4 \times 10^3$, and $\varepsilon > 1.0$W/kg), particle-bubble aggregate is destroyed (visual observation). To float coarse apatite particles successfully one should find a compromise between the forces that bring particle and bubble together and those forces that promote particle-bubble detachment.

Comparing flotation response of apatite versus glass spheres (Figures 2–4), it is possible to observe that under more quiescent conditions ($Fr < 0.1$, $Re < 2{,}000$, and $\varepsilon < 0.1$W/kg) the recovery of glass spheres is considerably higher than the recovery of apatite. As the specific gravity of apatite is considerably higher than glass spheres (3,080kg/m^3 versus 2,480kg/m^3), the adopted hydrodynamic conditions were able to promote particle suspension in a more effective way in the case of glass spheres than in the case of apatite particles. Since particle suspension could be considered as a result of transport phenomena caused mainly by the turbulence (macroturbulence) created by the impeller rotation (N), apatite particles demand more macroturbulence to be kept in suspension than glass spheres. Furthermore, under extremely quiescent conditions characterized by $Fr < 0.01$ (Figure 5), the

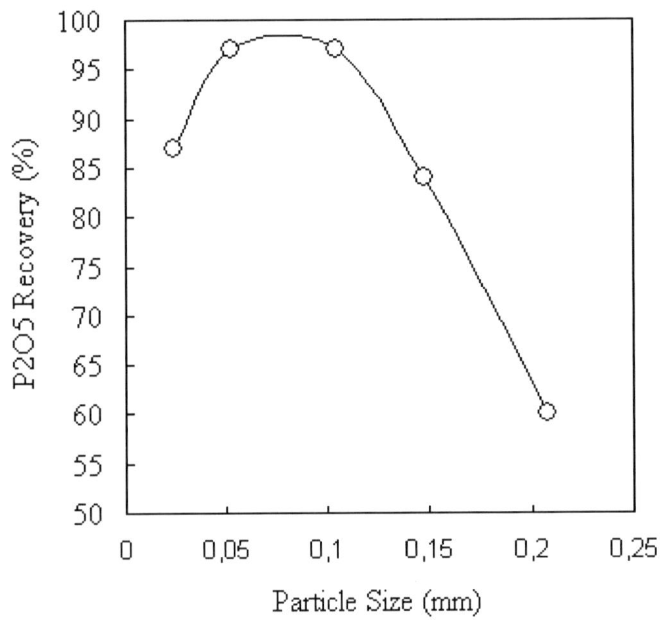

Figure 1 Recovery of apatite versus particle size from flotation experiment (rougher) carried out with phosphate ore in Denver laboratory cell

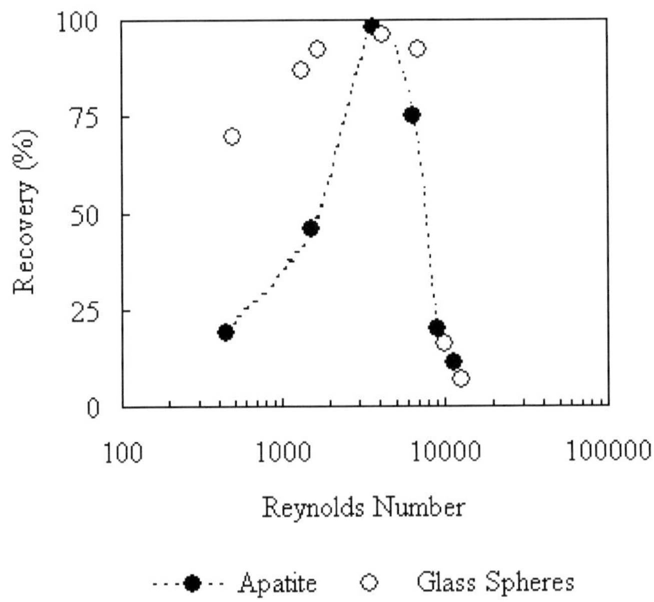

Figure 2 Recovery of apatite and glass spheres (d_p=0.248mm) versus Reynolds number

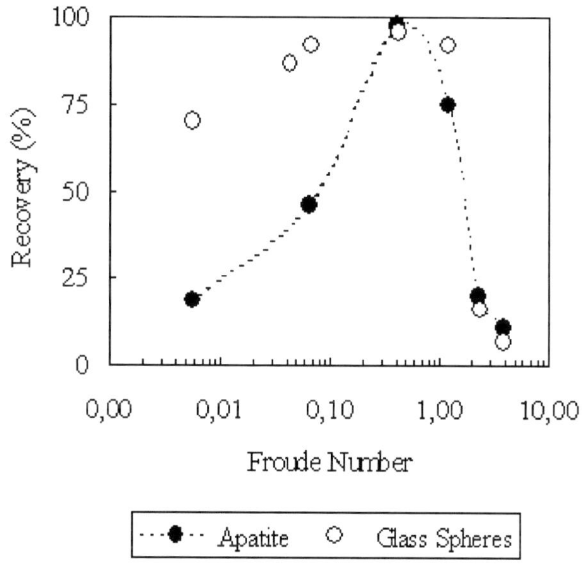

Figure 3 Recovery of apatite and glass spheres (d_p=0.248mm) versus Froude number

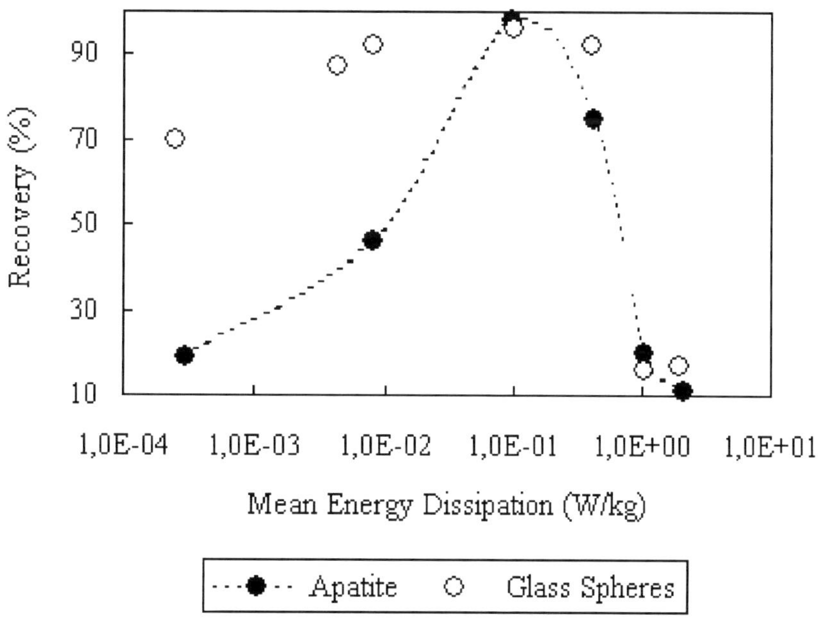

Figure 4 Recovery of apatite and glass spheres (d_p=0.248mm) versus mean energy dissipation (ε)

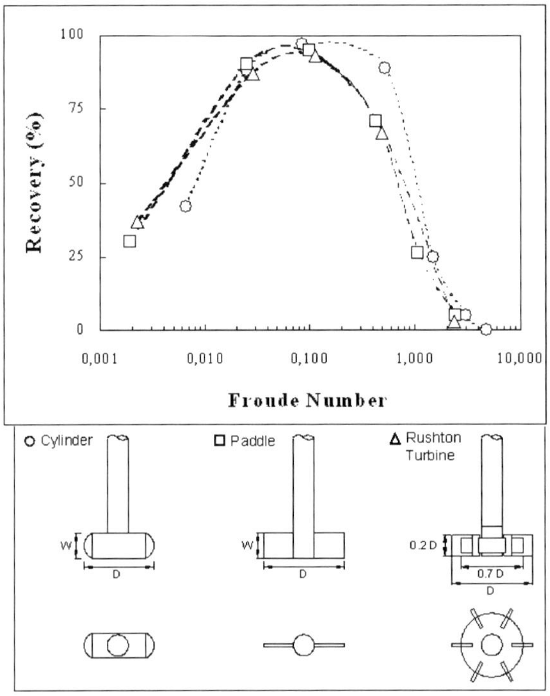

Figure 5 The influence of impeller geometry (D=15mm, W=6.52mm) on the recovery of glass spheres (d_p=0.248mm) versus Froude number

recovery of glass spheres is influenced by the geometry of the impeller: cylindroid-type impeller produced lower recovery than either paddle-type or Rushton turbine. Using Zwietering's one second-criterion, paddle-type impeller ($N_{critical}=7.5s^{-1}$) and Rushton turbine ($N_{critical}=7.0s^{-1}$) were far more effective than cylindroid-type impeller ($N_{critical}=9.5s^{-1}$) to promote full particle suspension.

Under the most turbulent conditions adopted in our experimental system (Fr>2.0, Re≥10^4, and ε≥1.0W/kg), the recovery of glass spheres and apatite particles were very low (Figures 2-4). Furthermore, according to Figure 5, under the same most turbulent conditions, the recovery of glass spheres was also very low, regardless of the type of the impeller used in the experiments. Since microturbulence is responsible for particle-bubble detachment, it is independent from the source that has generated it (Schulze, 1984).

Since particle-bubble detachment is likely to occur under very turbulent conditions, it is reasonable to expect that flotation rate of coarse particles may be lower than the finer ones. Under very turbulent conditions that characterize Denver laboratory flotation cell (Fr=2.6 and Re=1.18×10^5), flotation kinetics of a standard phosphate ore from Tapira Mine was studied. The results are shown in Table 2 and Figure 6. As depicted in Table 2, the magnitude of either flotation rate (k) or order (n) depends largely on the method used to determine them. However, a general trend could be outlined: the flotation rate (k) of coarse apatite (+65 mesh Tyler) was considerably lower than the flotation rate of the smaller particle sizes. The smaller the particle size, the higher the flotation rate of apatite within the experimental domain of this research.

Since the three methods used to assess flotation order (n) exhibited certain coherence, although yielding different magnitude, the geometrical average of the flotation order obtained by each method was calculated and plotted

Table 2 Flotation kinetics of apatite from a phosphate ore

Method	Particle Size	Total P_2O_5 Recovery (*)	Order (n)	Kinetic Constant (k)	Correlation Coefficient (r^2)
1	+65 mesh	60.4	10.0 (**)	$1.1 \times 10^{-13}\ g^{-9}\ s^{-1}$	0.9964
2			3.0	$2.2 \times 10^{-5}\ g^{-2}\ s^{-1}$	0.8691
3			3.1	$8.2 \times 10^{-5}\ g^{-2}\ s^{-1}$	1.000
1	-65+100 mesh	84.2	3.0	$2.3 \times 10^{-4}\ g^{-2}\ s^{-1}$	0.9999
2			2.6	$5.3 \times 10^{-4}\ g^{-1.6}\ s^{-1}$	0.9960
3			2.2	$1.5 \times 10^{-4}\ g^{-1.2}\ s^{-1}$	0.9998
1	-100+200 mesh	99.9	1.1	$8.3 \times 10^{-2}\ g^{-0.1}\ s^{-1}$	0.9996
2			1.6	$1.3 \times 10^{-1}\ g^{-0.6}\ s^{-1}$	0.9310
3			1.9	$1.7 \times 10^{-1}\ g^{-0.9}\ s^{-1}$	0.9992
1	-200+400 mesh	99.9	1.2	$9.7 \times 10^{-2}\ g^{-0.2}\ s^{-1}$	0.9967
2			0.8	$8.8 \times 10^{-2}\ g^{0.2}\ s^{-1}$	0.8955
3			1.9	$6.0 \times 10^{-2}\ g^{-0.9}\ s^{-1}$	0.9947
1	-400 mesh	87.3	1.0	$4.5 \times 10^{-2}\ s^{-1}$	0.9997
2			1.2	$3.7 \times 10^{-2}\ g^{-0.2}\ s^{-1}$	0.9972
3			Not valid for n=1	-	-

(*) After total exhaustion of mineralized froth (80 seconds)
(**) Value ignored in the calculation of the geometrical average

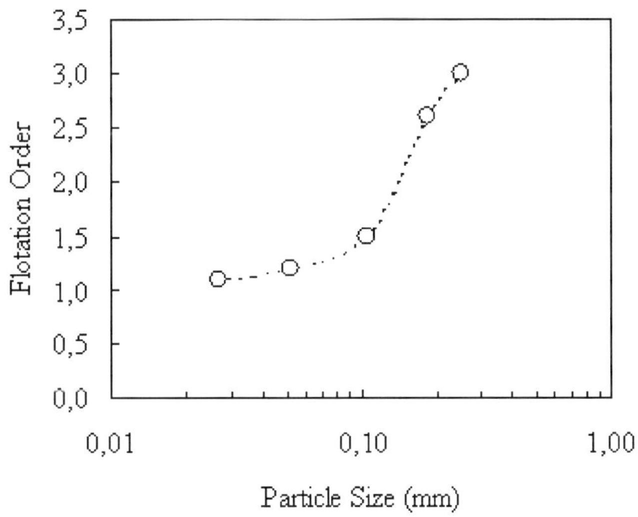

Figure 6 Flotation order versus apatite particle size in Denver laboratory cell (Fr=2.6, Re=1.18×10^5)

versus particle size (geometrical average of the aperture of the screens used to classify the ore). According to Figure 6, flotation order tends to first order (n=1) when the average particle size decreases from 0.248mm to 0.027mm. This finding could help those who are interested in understanding why coarse apatite particles (d_p>0.15mm) float preferentially along scavenger circuits in most Brazilian phosphate flotation mills inasmuch as finer ones float preferentially along the rougher stage. Actually, coarse particles of apatite (d_p>0.15mm) might demand much more time to float than the finer particles as a result of an inefficient collection process in which particle-bubble detachment might play a major role.

Information about hydrodynamics of rougher flotation cells used at Brazilian flotation mills is depicted in Table 3. Since all of the cells are supplied by Wemco, their impellers exhibit the same geometry. The value of the mean energy dissipation (1.3W/kg) determined at Catalao Mill (Copebras) suggests that particle-bubble detachment is likely to occur. Reynolds number and Froude number are in the range of other Brazilian industrial cells, except at mill Cajati Mill (Bunge).

Since Froude number at Cajati Mill (Fr=1.3) was considerably higher than Froude number at the other mills (Fr=0.6 and Fr=0.8), some modification was accomplished in a row of cells (row-1); whereas another row (row-4) was used as a reference. The impeller speed (N) of row-1 was slightly lowered in order to make it running with a lower Froude number (Fr~1.0). Results are presented in Table 4, where it is possible to see that the content of P_2O_5 of coarse fraction (+65 mesh Tyler) in tailings of row-1 is considerably lower than row-4.

Table 3 Hydrodynamics of mechanical cells of some Brazilian phosphate flotation mills

Company	Mill Location	Cell Type	Re (x10^6)	Fr	ε (*)
Fosfertil	Tapira-MG	Wemco-120	0.9	0.6	n.d.
Ultrafertil	Catalão-GO	Wemco-144	1.3	0.8	n.d.
Copebras	Catalão-GO	Wemco-120	1.0	0.8	1.3 W/kg
Bunge	Cajati-SP	Wemco-66	0.5	1.3	n.d.

(*) Power determined by current flow and voltage measurements
(**) n.d.=not determined

Table 4 Flotation performance of two rows of mechanical cells at Cajati Mill - Bunge

Characteristics of the cells and their performance	Row-1	Row-4
Impeller Diameter (m)	0.29	0.29
Impeller Rotation (min^{-1})	350	400
Froude Number	1.0	1.3
Content of P_2O_5 in the coarse fraction (+65 mesh Tyler) in the feed	(8.5±0.8)% (*)	(8.5±0.8)% (*)
Content of P_2O_5 in the coarse fraction (+65 mesh Tyler) in the tailings	(2.6±0.6)% (*)	(4.8±1.4)% (*)

(*) Average of P_2O_5 contents obtained during 9 hours of continuous operation

CONCLUDING REMARKS

Coarse particles of apatite (d_p=0.25mm, contact angle ~85°–88°) presented a maximum recovery (R>90%) in microflotation tests when 0.3<Fr <0.8, Re~4×10^3, and ε~0.1W/kg. For more quiescent hydrodynamic conditions (Fr<0.3, Re<4×10^3, and ε<0.1W/kg), the recovery of apatite decreased due to a lack in particle suspension; whereas, for the most severe hydrodynamic conditions (Fr>2, Re>10^4, and ε> 1.0W/kg), the recovery was almost nil due to particle-bubble detachment. Coarse glass spheres (d_p=0.25mm, contact angle ~85°–88°) presented flotation response similar to apatite except under more quiescent hydrodynamic conditions. In this case, glass spheres (specific gravity=2,480kg/m^3) floated better than apatite (specific gravity=3,080kg/m^3) because the impeller was able to promote better suspension of the lighter particles. These results show that high recovery of apatite (and glass spheres) is feasible under suitable hydrodynamic conditions.

Flotation kinetics of apatite was studied in laboratory scale with phosphate ore from Tapira-MG under strong hydrodynamic conditions that characterize Denver laboratory cell (Fr=2.6 and Re=1.2×10^5). The results indicated that coarse particles of apatite (d_p>0.15mm) float at a lower rate than finer particles of apatite (d_p>0.15mm) float at a lower rate than finer particles (0.15mm>d_p>0.037mm) and, further, the flotation kinetics of coarse particles does not fit a first order model. The smaller was particle size, the higher was flotation rate and flotation kinetics tended to first order. These results suggest that coarse particles demand more time to float than finer particles. Moreover, if flotation is carried out under hydrodynamic conditions that are suitable to float finer particles, those conditions may be not suitable to float the coarser ones.

To illustrate the role of hydrodynamics in coarse apatite flotation, a modification was accomplished in industrial scale: the impeller rotation (N) of a bank of cells was lowered in order to decrease Froude number from Fr=1.3 to Fr=1.0. The content of P_2O_5 of the fraction +65 mesh Tyler of the tailings decreased from 4.8% to 2.6%.

ACKNOWLEDGMENTS

Fapesp (Fundação de Amparo a Pesquisa do Estado de Sao Paulo) and CNPq (Conselho Nacional de Desenvolvimento Cientifico e Tecnologico) are acknowledged for supporting the research. The authors would like to extend the acknowledgments to phosphate mining companies Fosfertil, Bunge and Copebras for providing samples, chemical analysis and information.

REFERENCES

Connors, K.A., 1990. *Chemical kinetics—The study of reaction rates in solutions*. VCH Publishers, pp. 469.

Foggler, H.S., 1999. *Elements of chemical reaction engineering*, 3rd edition. Prentice-Hall, pp. 967.

Jewett, A., 1980. Formation and disruption of particle-bubble aggregates in flotation. In: *Fine Particles Processing*, Vol. 1. P. Somasundaran (Editor). AIME, New York, pp. 720–754.

Knudsen, J.G.; Kats, D.L., 1958. *Fluid dynamics and heat transfer*. McGraw Hill, New York, pp. 246–322.

Leal Filho, L.S.; Rodrigues, W.J.; Ralston, J., 2003. Importance of hydrodynamics in coarse particle flotation. *Proceedings of the International Conference Flotation and Flocculation—From Fundamentals to Applications*. University of South Australia, Mawson Lakes, pp. 203–212.

Manqiu, X.; Finch, J.A., 1988. Bubble diameter estimation in a mechanically agitated flotation cell. *Minerals and Metallurgical Processing*, 5, pp. 43–44.

Nagata, S., 1975. *Mixing: principles and applications*. John Wiley & Sons, New York, pp. 1–65.

Ralston, J.; Fornasiero, D.; Hayes, R., 1999. Bubble-particle attachment and detachment in flotation. *International Journal of Mineral Processing*, 56, pp. 133–164.

Rodrigues, W.J.; Leal Filho, L.S.; Masini, E.A., 2001. Dimensionless hydrodynamic parameters and their influence on flotation behavior of coarse glass spheres. *Minerals Engineering*, Vol. 14, pp. 1047–1054.

Roscoe, R., 1952. The viscosity of suspensions of rigid spheres. *British Journal of Applied Physics*, 3, pp. 267–269.

Schubert, H., 1999. On the turbulence control microprocesses in flotation machines. *International Journal of Mineral Processing*, 56 pp. 257–276.

Schulze, H.J., 1977. New theoretical and experimental investigations on stability of bubble/particle aggregates in flotation: a theory on the upper particle size of floatability. *International Journal of Mineral Processing*, 4, pp. 241–259.

Schulze, H.J., 1984. *Physico-chemical elementary processes in flotation*. Elsevier, Amsterdam, pp. 21–52.

Sissom, L.E.; Pitts, D.R., 1972. *Elements of transport phenomena*. McGraw Hill, New York, pp. 765.

Subramanyan, T.V.; Monte, M.B.M.; Middea, A.; Valdiviezo, E.; Lins, F.A.F., 1999. *Minerals Engineering*, 12, pp. 1347–1357.

Uhl, V.W.; Gray, J.B., 1966. *Mixing: theory and practice*, Vol. 1. Academic Press Inc., New York, pp. 123–124.

Vatcha, S.R., 1987. Directly determine reaction order. *Chemical Engineering*, Vol. 4, No. 2, pp. 195–196.

Yianatos, J.B.; Finch, J.A.; Dobby, G.S.; Manqiu, X., 1998. Bubble size estimation in a bubble swarm. *J. Coll. Interface Science*, 126, pp. 37–44.

Zwietering, N.T., 1958. Suspending of solid paticles in liquid by agitators. *Chemical Engineering Science*, 8, pp. 244–253.

CHAPTER 4

Mechanisms Involved in Reactive Flotation of Dolomite

H. El-Shall,* R. Stana,† and P. Somasundaran‡

ABSTRACT

Dolomite impurities are increasing in the currently mined phosphate deposits in Florida. Several research efforts have been conducted in various laboratories around the globe to find methods to economically remove the dolomite.

However, a universal and cost-effective process is not in commercial use. In this paper, a new process for dolomite/apatite separation is presented. Differential solubility of dolomite in slightly acidic solution is utilized to generate micro CO_2 bubbles at the dolomite/water interface. Particle (dolomite)/bubble aggregates float to the surface leaving phosphate concentrate with low MgO content. Interestingly, up to 6.0 mm particles can be floated by such micro bubble/particle aggregates. Different surface-active agents were tested to stabilize the bubbles. Results of these tests are presented and the mechanisms involved are discussed.

INTRODUCTION

Phosphate rock is one of the basic materials for fertilizers. Phosphoric acid is an important intermediate product for production of fertilizers. The phosphoric acid technology essentially comprises sulfuric acid attack and separation of the phosphoric acid produced from the calcium sulfate crystals resulting from the reaction. Both the attack and separation, which is typically filtration, are considerably affected by the nature of the ore and its impurities (Becker, 1989).

The presence of impurities can significantly affect the acid attack and separation and consequently the filtration process. The most important impurity in Florida phosphate ore is dolomite. The dolomite is (Ca, Mg) CO_3, is the main source for magnesium. As impurities, both CaO and MgO affect sulfuric acid consumption, In addition, the MgO stays with the acid phase, increasing viscosity significantly. Finally, it acts as a diluent, making it difficult to make on grade DAP (di ammonium phosphate). Therefore, developing an efficient and economical method for separating dolomite from phosphate is a one way to extend phosphate mine life (Geo et al., 2002).

Systematic research in carbonate phosphate ore beneficiation started in the early fifties. Some of the techniques employed include flotation (direct and reverse) (Fuerstenau et al., 1968; Ananthapadmanabhan and Somasundaran, 1985; Rao et al., 1988; Anon., 1989; Celetici et al., 1984; Ataly et al., 1985; Lawendy, 1993; Snow, 1979 and 1982; Lawver and Snow,

*Department of Materials Science and Engineering, University of Florida, Gainesville, Florida.
†Engineering Research Center for Particle Science and Technology, University of Florida, Gainesville, Florida.
‡Columbia University, New York, New York.

1980; El-Shall et al., 1996); physical methods such as gravity concentration, sizing, desliming, heavy media separation (Wiegel and Hwang, 1984; El-Shall and Bogan, 1990); and acid leaching (Rule et al., 1970; Abu El-shah, 1991).

Motivation for Current Study

The various beneficiation techniques mentioned above have their limitations or disadvantages. For example, in acid leaching, the strong acids can attack both the dolomite and phosphate and there are environmental problems associated with this process.

Also, in case of the flotation processes for high dolomitic phosphate, the MgO in the product rock is not less than 1% where it is required to be applied commercially. Even in the most recent process developed, the CLDRI fine flotation process (El-Shall and Zhang, 2004; Zhengxing et al., 1999; Zhizhong and Zhengxing, 1999), by which the MgO% could be reduced to less than 1%, there are some concerns such as handling of fine concentrate (dewatering and transport) capital cost (grinding and multiple process steps), and operating cost.

Heavy media separation (HMS) is the only process that has been applied commercially. However, it has its own shortcomings such as high product MgO (more than 1%) in the concentrate in addition to about 50% phosphate loss.

Current Study

This study is intended to address a process that can be applied to separate dolomite from phosphate pebbles to produce a phosphate concentrate with less than 1% MgO. The process idea stems from the well-known reaction between carbonates and acids producing carbon dioxide gas. Thus, dolomite, as a carbonate mineral, generates CO_2 when exposed to acidic solutions as shown in the following chemical reaction:

$$\text{Dolomite} + \text{Acid} \rightarrow (\text{Ca, Mg})\ \text{salt} + CO_2$$

Capturing the generated CO_2, at the dolomite particle surface, into bubbles using a coating agent, which is a kind of surfactant or polymer, can selectively float dolomite and separate it from phosphate rock. Since this process depends on the chemical reaction between the dolomite and the acidic medium to form a stable bubble at the particle surface and then to float it, this process is called "Reactive Flotation, RF" process as schematically depicted in Figure 1.

PROCESS DEVELOPMENT

A 100-pound sample of high dolomitic phosphate oversize pebble was obtained from an operating central Florida phosphate mine. Since most of the rock was relatively coarse, a portion of it was crushed and rescreened to obtain smaller fractions that could be used in the tests. The procedure used in these tests included dipping a single particle into a surfactant solution then adding the particle to the acidic solution. Several acids and surfactants were used at different concentrations as mentioned in the next sections.

Effect of Acid Type and Dosage

Different acids (H_2SO_4, HCl, HNO_3, and acetic acid) and the surfactant $SDSO_3$ were used in these experiments. Using inorganic acids (up to 2% by volume) such as HCl, HNO_3, and H_2SO_4 gave similar results (in terms of kinetics and number of bubble formation as shown in Table 1). On the other hand, the acetic acid (RCOOH) formed a very small number of bubbles. Increasing the concentration of acetic acid from 2 to 5% gave no significant variation in bubble formation, which means that the organic acids are too weak and would require large amounts compared with inorganic ones.

Since all the inorganic acids behave in the same manner, sulfuric acid was selected for the rest of the experiments since it is the least expensive and already widely used in the industry.

From Table 1, it can be clearly seen that the reaction between dolomite and acid to produce CO_2 decreases noticeably with decreasing acid concentration, and the flotation of dolomite

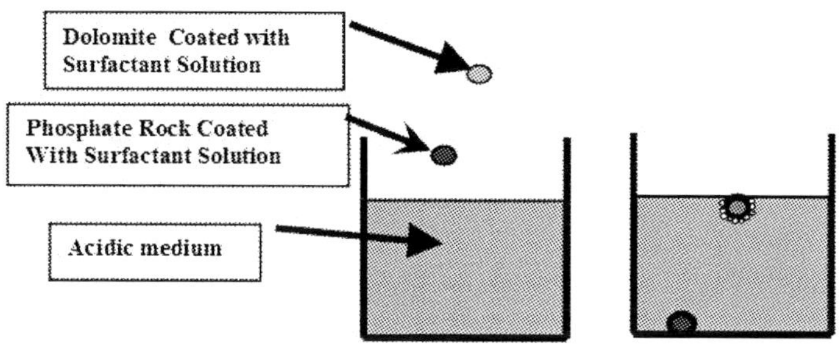

Figure 1 Schematic diagram depicting the reactive flotation process

Table 1 Effect of sulfuric acid concentration on the flotation of 1-2 mm dolomite particles coated, by dipping into 2% $SDSO_3$ solution

	Sulfuric acid concentration			
	0.25 %	0.5%	1%	2%
Bubble formation rate	Increasing →			
Number of bubbles formed	the interval between formation of two bubbles was about 30 seconds	Increasing number of bubbles → continuous bubble formation		
Flotation of dolomite particle	No flotation		Takes 5-10 minutes to float	floats after 1 minute

particle takes a longer time (5–10 min) in 1% acid, than the flotation in 2% acid. Consequently, the 2% sulfuric acid concentration was considered the most suitable acid concentration for bubble formation rate and quantity of bubbles formed.

Several types of surfactants were used as listed in Table 2. The experimental procedure involved coating the particle by dipping it into the surfactant solution (2%) and then adding it to 2% sulfuric acid solution. The observations from preliminary tests for different surfactants are shown in Table 3. Anionic surfactants formed stable "membranes" that captured CO_2 and persisted for up to 10 minutes. Of the anionic surfactants, $SDSO_3$ was the best. This may be due to its ability to work at the low pH. Therefore, $SDSO_3$ was used in the next experiments.

Effect of Surfactant ($SDSO_3$) Dosage

Different dosages (0.04-0.12%) of surfactant $SDSO_3$, as well as the role of coating particles with surfactant before transferring them to the acidic solution, were then studied. It was noted that particles without coating (without immersing in the surfactant solution before adding the particle to acidic surfactant solution) are highly reactive in comparison to coated ones. This may be due to the absence of a surfactant layer that prevents instantaneous contact between the particle surface and acidic solution.

Increasing the surfactant concentration increases the floatability of the particles, and large particles (3–4 mm) could be floated at the

Table 2 List of tested surfactants

Surfactant	Chemical Name	Type
Ethofat	Tall Oil	Anionic
$SDSO_3$	Sodium dodecyl-sulphonate	
$SDSO_4$	Sodium dodecyl-sulphate	
Oleic acid	Oleic acid	
Arlacel	Sorbitan sesquioleate	Non-ionic
Span-20	Sorbitan monolautrate	
G-1096	Polyoxyethelene-sorbitol hexaoleate	
Brij-97	Polyoxyethelene oleyl ether	
Arlatone T	Polyoxyethelene polyofatty acid ester	
Tween-21	Polyoxyethelene Sorbitan monolautrate	
Tween-85	Trioleate	
PVA	Poly-Vinyl-Alcohol	Non-ionic

Table 3 Preliminary tests of high dolomitic phosphate with different surfactants

Surfactant Name	Remarks
Ethofat (1%)	Formation of bubbles on the particle surface, then particle floats
$SDSO_3$ (1%)	Bubbles form and go up to solution surface, but particle doesn't float
$SDSO_4$ (1%)	Bubbles form and go up to solution surface, but particle doesn't float
$SDSO_3$ (2%)	Particle takes 2-3 sec to float
$SDSO_4$ (2%)	Flotation rate in case of $SDSO_4$ < Flot.rate in case of $SDSO_3$
Oleic acid	Added as it is, particle stays at surface of the solution (air-water interface) with formation of many small bubbles.
Arlacel Span-20 G-1096 Brij-97 Arlatone T Tween-21 Tween-85	**Common observations for non-ionic surfactants are:** No significant change in behaviour of different non-ionic surfactants used The flotation of high dolomitic particle after 3-5 seconds and stay at the surface, while the low dolomitic particles do not float. Formation of stable bubbles on the floated particle surface

surfactant concentration up to 0.08%. Increasing the surfactant concentration to 0.12% did not make any noticeable change in floatability.

Polyvinyl-Alcohol (PVA) as Surfactant

PVA is a polymer well known for its low permeability to gases, especially CO_2 (Finch, 1973). Thus, PVA was used as a viscous polymer to coat and to float the dolomite particles. The effect of acid concentration %, different particle sizes and different PVA concentrations were studied.

PVA forms a strong membrane on the particle. This membrane needs a high acid concentration to penetrate it and react with the particle surface, as it was observed that there was a very small number of generated bubbles. However, most of the bubbles form a cluster on the particle surface, which increases the volume of the bubbles at the surface and floats the particle. The floated particles, especially the large ones, are not stable at the beginning, teeter up and down and finally stay at the interface for a long time (sometimes more than 2 days). By increasing the PVA concentration, a longer time was needed to form bubbles and float the particle. This required a higher acid concentration (3% sulfuric), but in this case large particle sizes can float (up to 8 mm).

From exploratory tests, we found that fully hydrolyzed polyvinyl alcohol (PVA) with high molecular weight can be used as a coating agent before adding the particle to acidic medium. In acidic medium a membrane starts to form at the dolomite particle surface and consequently the dolomite particle can be floated.

Effect of Particle Size

As mentioned earlier, up to 8 mm particles could be floated after coating with PVA. Therefore, PVA was used to further investigate the role of particle size in the formation of bubbles and its efficient flotation. Figure 2 shows the flotation time needed to float different particle sizes.

The flotation times are found, as discussed below, to depend on the rate of CO_2 formation and the quantity needed to make the particle plus CO_2 bubbles buoyant in the aqueous solution. Therefore, the particle size effect on particle dissolution rate was investigated by testing different particle sizes in 3% sulfuric acid, H_2SO_4, solution (0.33 M). The results of reacting uncoated particles with acid at room temperature (23 ± 1°C) are shown in Figure 3 in terms of Mg^{++} ions dissolved. In all of the experiments, the solid loading in the reactor was 10%. Generally, magnesium ion concentration in the liquid phase first underwent an exponential increase and then slowly approached an asymptotic value. For the small particles, 0.2–0.3 mm, the reaction began very rapidly due to the higher surface area available for H^+ ions after which the dissolution reached a plateau

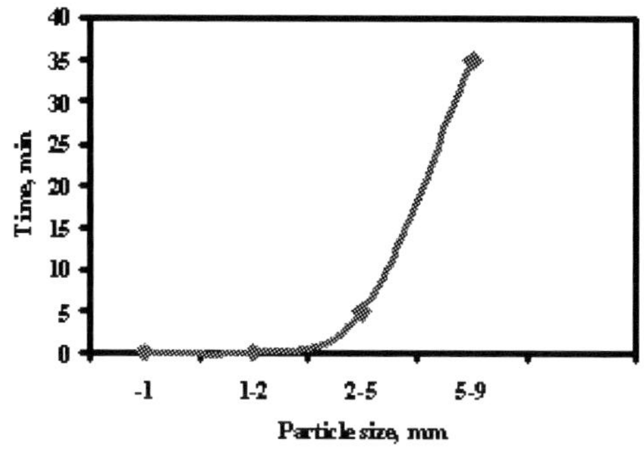

Figure 2 Effect of the particle size on the time needed to float the particle in 3% H_2SO_4 when the particle is precoated with 3% PVA

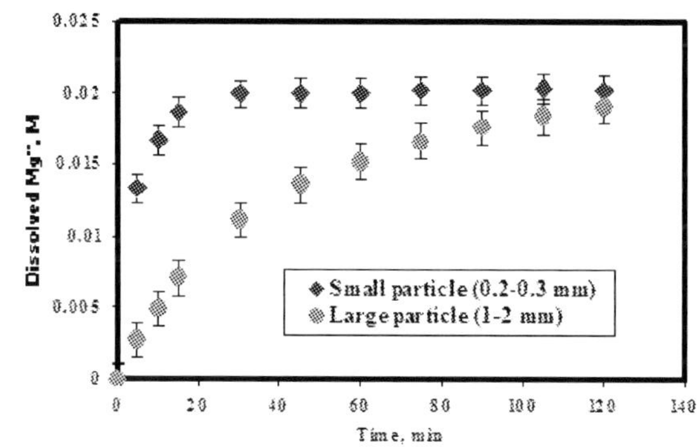

Figure 3 Dissolution kinetic of different particle sizes in terms of magnesium ions in 3% sulfuric acid solution at 23°C

(equilibrium) after 30 minutes. In the case of large particles, 1–2 mm, the reaction increased gradually as a result of lower surface area and did not reach equilibrium even after 2 hours.

The H^+ transport is perturbed by the presence of the reaction products and since the H^+ has a higher diffusivity, the reaction continued until it reached the point of equilibrium. At this point, the H^+ can no longer reach the surface due to the formation of an insulating layer of reaction products, most probably gypsum, $CaSO_4 \bullet 2H_2O$ and/or a CO_2 gas layer around the particle. This was especially true in the case of coated particles.

Estimation of CO_2 Required to Float the Dolomite Particle

Simple calculations were used to predict the size of the particle with its surrounding envelope of gas and to estimate the relation between the radius of the particle before and after the formation of the gas bubbles around the particles. For simplicity, the following conditions were assumed: spherical particle, uniform gas layer around the particle and at equilibrium condition. The quantity of the particle dissolved in the reaction was not considered in the calculations.

As a starting point, we assumed that the particle starts to float when the density of the particle and the gas layer around it is less than the density of acidic solution. The calculations are in equation 1.

In addition, the ratio between gas volume and particle volume and the ratio between the gas volume and particle surface area were calculated equations 2 and 3.

From previous calculations, we can deduce the following: The total radius is 1.4 of the original radius of the particle (without taking the dissolved part of the particle and the pores volume into account), the volume of gas around the particle is approximately twice the original particle radius and larger particle needs more volume of gas to float due to its lower surface area.

Flotation of Representative Samples

Instead of using single particles as in the previous tests, samples consisting of 10 grams were coated with 3% PVA solution using a dipping method and then added to 3% sulfuric acid solution in a 500mL beaker. No stirring or agitation was used. The floated and the non-floated (sink) fractions were separately collected after two days. The results are given in Table 4 in terms of wt % and chemical analyses of both floated and sink fractions. It was noticed that most of the particles (about 90%) floated within 30 minutes and about 10% in the rest of the two days of the experiment.

Although, the RF is a promising process to separate dolomite (as shown in Table 4), the results indicate that the MgO was still higher than 1%, which means that the process needs

$$\text{density (Dolomite} + CO_2 \text{ gas)} < \text{density (acidic solution)}$$

$$\left(\frac{M_d + M_g}{V_d + V_g}\right) < \frac{M_w}{V_w} = 1\ \text{gm/cm}^3 \quad (M_g \cong 0)$$

$$M_d < V_d + V_g$$

$$\frac{2.9 * 4\pi}{3} r^3 < \frac{4\pi}{3} r^3 + \frac{4\pi}{3}(r_1^3 - r^3)$$

$$\frac{2.9 * 4\pi}{3} r^3 < \frac{4\pi}{3} r_1^3$$

$$(\sqrt[3]{2.9})r < r_1$$

$$r_1 > 1.4\, r$$

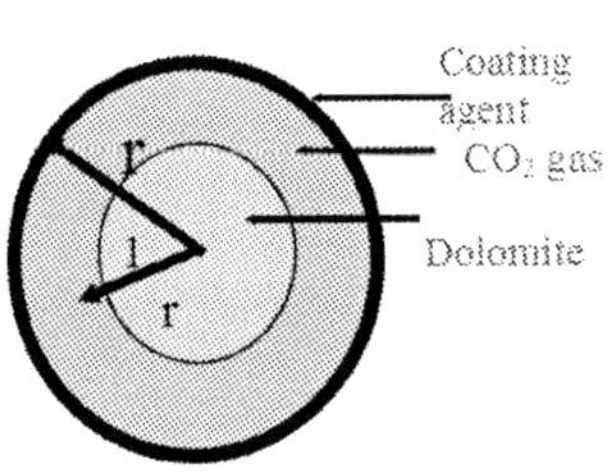

M_d : mass of dolomite particle
M_g : mass of CO_2 gas
M_w : mass of water
V_d : volume of particle
V_g : volume of CO_2 gas
V_w : volume of water
R: particle radius
R_1 : total radius (particle + gas)

Equation 1

(2)

$$\mu_p = \mu_w (1-\Phi)^{-2.5}$$

(3) *Gas volume : particle surface area*

$$\frac{V_g}{S_d} = \frac{\frac{4\pi}{3}(r_1^3 - r^3)}{4\pi r^2}$$

$$\frac{V_g}{S_d} = \frac{((1.4)^3 r^3 - r^3)}{3r^2}$$

$$\frac{V_g}{S_d} = \frac{((1.4)^3 - 1)r}{3} = 0.58r$$

Equations 2 and 3

to be optimized to produce a concentrate containing MgO < 1%. Therefore, a statistical design was used to screen the main factors in addition to determine the optimum conditions (El-Midany, 2004).

PROCESS OPTIMIZATION

Flotation of Dolomite

Detailed discussions of statistical designs as well as the results are presented in other publications (El-Midany, 2004). However, some of these data are presented in this section to illustrate the effect of the important parameters such as acid concentration, PVA dosage, particle size, etc. Figures 4 and 5 show the floatability of dolomite as a function of PVA and acid concentrations in terms of percent MgO removal % and percent MgO in the concentrate for different particle sizes.

It can be seen from these figures that the percent MgO removal and percent MgO in the concentrate are greatly affected by increased acid concentration and PVA dosage; for instance, MgO removal increases from 30% to more than 90% and the MgO concentration in the concentrate decreases from more than 2.1% to 0.6%, respectively. However, an increase in particle

Table 4 Separation flotation of dolomite from a representative sample (results of -4+1.19 mm particle size)

Sample	Wt%	P_2O_5	MgO
Float	20.5	4.00	16.00
Sink	79.5	24.79	1.25
Feed		20.53	4.27

size negatively affects the MgO removal and MgO concentration in the concentrate. At the maximum acid concentration and PVA dosage, MgO removal is as low as 75% and MgO concentration in the concentrate is higher than 1.1%. It should be remembered, however, that the flotation time is five minutes, which is the same for both sizes in these tests. In other words, more time is needed for enough CO_2 to form to float the coarse particles. This can be clear from the data presented in Figure 2 showing flotation time as a function of particle size.

The data shown in Figures 4 and 5 indicate that more than 3.0 lb/ton of PVA may be needed to achieve the best results. Such an amount is not economically feasible for the phosphate industry. Thus, further testing of PVA at different concentrations, and conditioning time, as well as method of conditioning (stirring vs. dipping), suggested that 1.0% PVA concentration and 1–2 lb/ton could be used to achieve concentrate of good quality (<1.0 MgO) and high removal of MgO (El-Midany, 2004).

PVA/Particle Interactions

The surface activity of a polymer mainly relies on its conformation in the aqueous solution; whereas, the conformation of macromolecule depends on its chemical structure and composition (Figure 6), etc. In the case of PVA, the hydrogen bonding and formation of a tight macromolecular film at high concentrations lead to an increase in its degree of crystallinity and increase in the local viscosity, which controls the droplet shape as well as its tendency to spread. This phenomenon is observed for PVA and also for some modified PVA end groups (Finch, 1992).

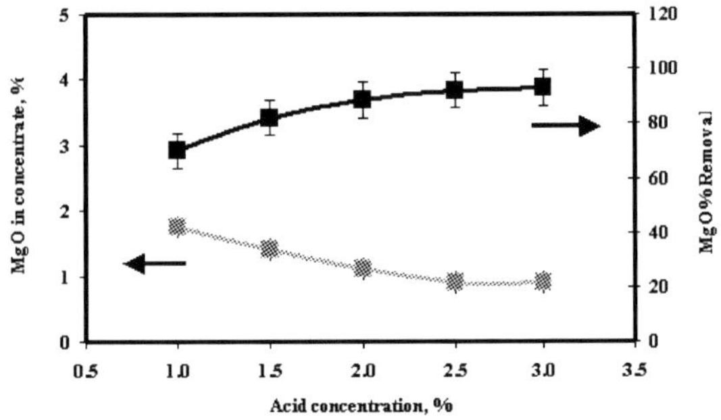

Figure 4 Flotation of dolomite particles (-0.5 mm) at 3 lb/t PVA dosage

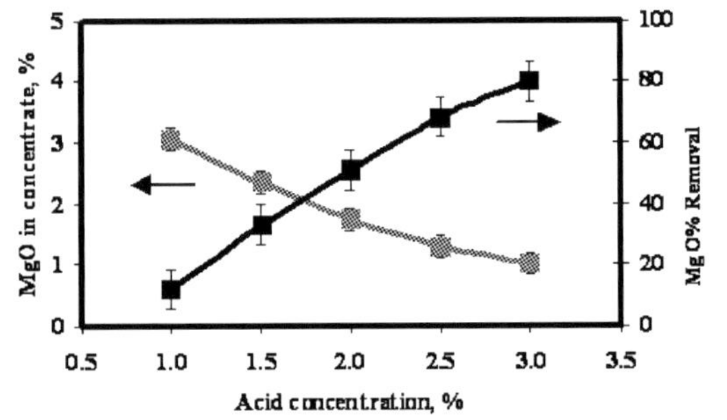

Figure 5 Flotation of dolomite (5–9 mm) at 3 lb/t PVA dosage

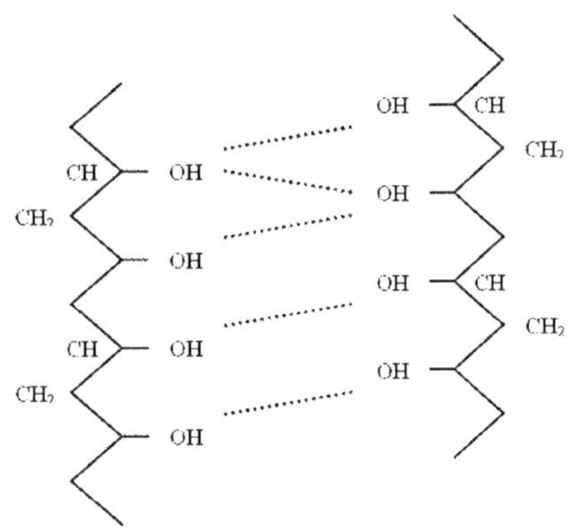

Figure 6 Polyvinyl alcohol crystal structure; the dotted lines indicate hydrogen bonds

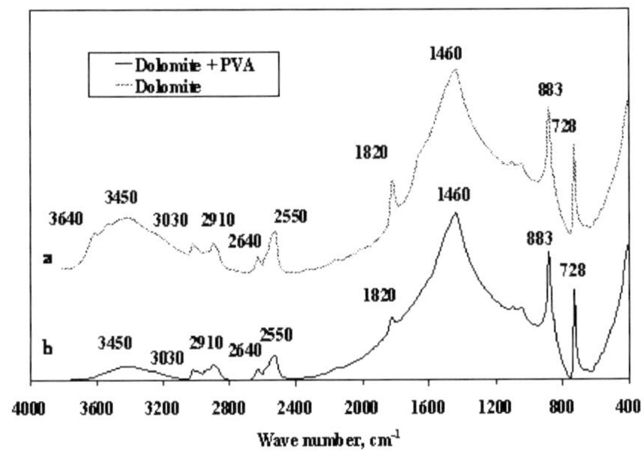

Figure 7 Infrared spectra (FTIR) of (a) dolomite alone, and (b) dolomite/PVA system

The PVA molecules form hydrogen bonding not only among themselves but also with the dolomite surface as verified by the FTIR spectra shown in Figure 7. In this figure, both coated and uncoated dolomite spectra are compared.

The isolated hydroxyl group peak at 3,640 cm^{-1} is present in dolomite powder without any polymer, but the spectra with polymer shows the disappearance of this peak. Moreover, the peak intensity of the hydrogen-bonded hydroxyl groups (3,450 cm^{-1}) is smaller in PVA-coated dolomite than the uncoated particles. These observations indicate that the adsorption of PVA on dolomite occurs by hydrogen bonding involving $^{-}$OH groups of the polymer. This finding is in concurrence with earlier studies on silica-PEO (Rubio and Kitchener, 1976; Cheng, 1985) and silica-polyacrylic acid PAA (Tadros, 1978), which showed that the presence of isolated OH– groups on the surface is responsible for the adsorption of polymer molecules by hydrogen bonding.

In addition, the FTIR spectrum of the apatite sample after having interacted with PVA was obtained by El-Midany (2004). However, unlike the case of dolomite, no change is shown in the bands' intensities or positions, either for bare apatite or apatite with polymer adsorbed on it (Chanchani, 1986; Ince, 1987). It is pertinent to mention that FTIR spectroscopic studies on the dolomite-PVA and apatite-PVA systems revealed the presence of hydrogen bonding only. This also has been confirmed by the zeta potential measurements of dolomite in presence and absence of PVA as shown in Figure 8. It can be seen that the isoelectric points (IEP) of dolomite is about 9. This is in agreement with the results reported in the literature (Amankonah and Somasundaran, 1985; Chander and Furestenau, 1982; Pugh and Stenius, 1985). The results of zeta potential in the presence of PVA solution show that the presence of PVA does not change the IEP of dolomite indicating physical adsorption. It does, however, cause a decrease in zeta potential values. This may be explained, as other investigators asserted (Ottewill, 1967; M'pandou and Siffert, 1984; Luckham, Vincent, and Tadros, 1983; M'pandou and Siffert, 1987), that the decrease in zeta potential in the presence of PVA is not due to a decrease in charge and surface potential, but rather to a shift of the shear plane. Further confirmation of the physical adsorption of PVA is evidenced by the shape of the adsorption isotherm as explained by El-Midany (2004), who conducted further studies such as polymer layer thickness, elasticity of the polymer membrane and rate of CO_2 evolution. In such studies, it was concluded that 1.0% PVA solution could form a membrane of required elasticity to capture CO_2 and float dolomite particles.

CONCLUSIONS

Dolomite can be separated from apatite by coating it with a surfactant and immersing it in a dilute acid solution where the dolomite will generate carbon dioxide gas that is trapped by the surfactant and floated to the surface.

Separation coefficients for the process are greater than 50 compared to about 4 for the heavy media separation process and about 20 for the CLDRI process.

This process will work on pebble size rock up to 8–10 mm in size.

Fully hydrolyzed polyvinyl alcohol (PVA) is the best surfactant for coating the mixture of phosphate rock and dolomite. It forms a membrane on dolomite surface entrapping carbon dioxide gas making dolomite buoyant.

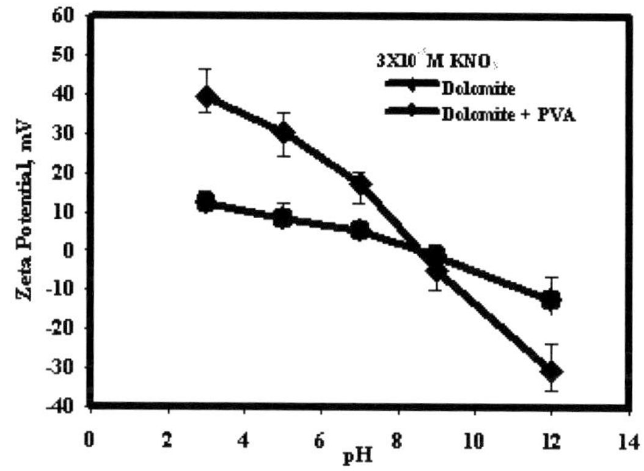

Figure 8 Zeta potential of dolomite in absence and presence of polymer

The volume of CO_2 required to float the dolomite particle was estimated based on the difference in density between the particle witha gas envelope and the acidic solution. The totalradius (particle + gas) is 1.4 of the original radius of the particle, the volume of gas around the particle is approximately twice the original particle radius, and a larger particle needs a larger volume of gas to float due to its lower surface area. PVA reduces the water surface tension and increases the contact angle leading to a lower spreading coefficient on the dolomite surface. Higher PVA concentrations than 1% do not significantly affect the spreading coefficient. This fundamental result can explain the flotation data showing 1% PVA concentration to be effective in flotation if PVA is stirred with the mineral to form a thin film.

The physical adsorption (hydrogen bonding) was confirmed using FTIR and zeta potential.

ACKNOWLEDGMENTS

The authors gratefully acknowledge the financial support received from the Florida Institute of Phosphate Research (FIPR). The authors also wish to acknowledge Particle Engineering Research Center (PERC), National Science Foundation, and IMC Phosphates for providing the rock samples used in this study.

Appreciation is also extended to A. El-Midany of the Department of Materials Science and Engineering at the University of Florida, Gainesville, for his valuable contribution to this paper.

REFERENCES

Abu El-shah, S.I., 1991. Beneficiation of calcareous phosphate rock using dilute acetic acid solution: optimization of operating conditions for Ruseifa (Jordan) phosphate. *International Journal of Mineral Processing* 31, 115–126.

Amankonah, J.O., Somasundaran, P., 1985. Effects of dissolved mineral species on the electrokinetic of calcite and apatite. *Colloids and Surfaces* 15, 335–353.

Ananthapadmanabhan, K.P., Somasundaran, P., 1985. Surface precipitation of inorganic and surfactants and its role in adsorption and flotation. *Colloids and Surfaces* 13, 151–167.

Anon. 1989. Evaluation of phosphatic clay disposal and reclamation methods, vol. 7: Engineering properties of flocculated phosphatic clays, FIPR Project #83-02-42.

Atalay, V., 1985. Beneficiation of low grade Tasit phosphate ore from Turkey. *World Congress on non-metallic Minerals Proceedings*, Belgrade, 389–396.

Becker, P., 1989. Phosphates and Phosphoric Acid: Raw Materials Technology and Economics of the Wet Process. *Fertilizer Science and Technical Series*, 2nd ed., 6, Marcel Dekker Inc.

Celetici, C., et al. 1984. Flotation of a phosphate rock with carbonate-quartz ganguc. Reagents in mineral industry, Jones, M.J. and Oblatt, R., eds, *IMM*, London, pp. 221–225.

Chanchani, R., 1986. Selective flotation of dolomite from apatite using sodium oleate as collector. Ph.D. Dissertation, University of Florida.

Chander, S., Furestenau, D., 1982. On the dissolution and interfacial properties of hydroxyapatite. *Colloids and Surfaces* 4, 101–120.

Cheng, Y.C., 1985. The effect of surface hydration on the adsorption and flocculation of a model silica suspension using polyethylene oxide. M.S. Thesis, University of Florida.

El-Midany, A. 2004. Separation of Dolomite from Phosphate Rock by Reactive Flotation. Ph.D. Thesis, University of Florida.

El-Shall, H., Zhang, P., Snow, R., 1996. Comparative analysis of dolomite/francolite flotation techniques. *Minerals and Metallurgical Processing* 7, 135–140.

El-Shall, H., Zhang, P., 2004. Beneficiation technology of phosphates: Challenges and solutions. *Minerals and Metallurgical Processing*, Vol. 21, No. 1, pp. 103–111.

Finch, C.A., 1973. *Polyvinyl alcohol: Properties and applications*. John Wiley and Sons, New York.

Finch, C.A., 1992. *PVA—Developments*. John Wiley and Sons, New York.

Fuerstenau, M.C., 1968. The influence of sodium silicate in non-metallic flotation systems. *Trans. AIME* 241, 319–323.

Geo, Z., Zheng, S., Gu, Z., 2002. Review of beneficiation technology for Florida high dolomite pebble. In: *Beneficiation of Phosphates: Fundamentals and Technology*. P. Zhang, H. El Shall, P. Somasundaran, and R. Stana (eds.), SME, USA, Chapter 21, 235–245.

Ince, D.E., 1987. Effect of sodium chloride on the selective flotation of dolomite from apatite. Ph.D. Dissertation, University of Florida.

Lawendy, T.A., 1993. Flotation of dolomitic and calcareous phosphate ores. Engineering Foundation Conference, Beneficiation of Phosphate: Theory and Practice, Dec 5–10, Palm Coast, Florida, Chapter 5, 60–72.

Lawver, L.E., Snow, R.E., 1980. Method of beneficiating phosphate ores. U.S. Patent 4,189,103.

Luckham, P., Vincent, B., Tadros, Th., 1983. The controlled flocculation of particulate dispersions using small particles of opposite charge. IV. Effect of surface coverage of adsorbed polymer on heteroflocculation. *Colloids and Surfaces* 6(2), 119–133.

M'pandou, A., Siffert, B., 1987. Polyethylene glycol adsorption at TiO_2–H_2O interface: Distortion of ionic structure and shear plane position. *Colloids and Surfaces* 24, 159–172.

M'pandou, A., Siffert, B., 1984. Sodium carboxylate adsorption onto TiO_2: Shortest chain length allowing hemimicelle formation and shear plane position in the electric double layer. Journal of *Colloids and Interface Science* 102(1), 138–145.

Ottewill, R., 1967. *Nonionic surfactant*. Martin J. Schick (ed.). Marcel Dekker, New York, 1, 627–682.

Pugh, R., Stenius, P., 1985. Solution chemistry studies and flotation of apatite, calcite and fluorite minerals with oleate collector. *International Journal of Mineral Processing* 15, 193–218.

Rubio, J., Kitchener, J.A., 1976. The mechanism of adsorption of poly(ethylene oxide) flocculant on silica. *Journal of Colloids and Interface Science* 57(1), 132–142.

Rule, A.R., 1970. Removal of magnesium impurities from phosphate rock concentrates. RI 7362, United States Bureau of Mines.

Snow, R.E., 1979. Beneficiation of phosphate ore. U.S. Patent 4,144,969.

Snow, R.E., 1982. Flotation of phosphate ore containing dolomite. U.S. Patent 4,364,824.

Tadros, Th.F., 1978. Adsorption of polyvinyl alcohol on silica at various pH values and its effect on the flocculation of the dispersion. *Journal of Colloids and Interface Science* 64 (1), 36–47.

Wiegel, R.L., Hwang, C.L., 1984. A predictive model for heavy media cyclone separation of phosphate pebble from dolomite. SME-AIME meeting, SME preprint #84-648.

Zhengxing, G., Zhizhong, G., Zheng, S., 1999. Beneficiation of Florida dolomitic phosphate pebble with a fine particle flotation process. In: *Beneficiation of Phosphates: Advances in Research and Practice*. Zhang, P., El-Shall, H., and Wiegel, R. (eds.). SME, pp. 155–162.

Zhizhong, G., and Zhengxing, G., 1999. Plant practices of phosphate beneficiation in China. In: *Beneficiation of Phosphates: Advances in Research and Practice*. P. Zhang, H. El-Shall, and R. Wiegel (eds.), SME, USA, Chapter 25, pp. 289–301.

SECTION 2

Waste Treatment/Utilization and the Environment

CHAPTER 5

Sustainable Mining: Trends and Opportunities

James L. Hendrix*

ABSTRACT

Sustainable development integrates economic, environmental and social considerations in order to improve the lives of the current generation and ensure that future generations will have adequate resources and opportunities. Sustainable mining is not a new approach or concept in resource consumption. It is a holistic approach for dealing with a complex, interlinked set of factors that determine the net societal worth of a project.

A holistic approach to phosphate beneficiation gives us opportunities for the improvement of economic results through reactant and waste minimization. It also gives direction for continued research thrusts to solve the problems associated with the clay ponds and the reduction of potentially hazardous wastes.

SUSTAINABLE DEVELOPMENT

Sustainable development is a growing concern expressed by many businesses, organizations and individuals. Yet, no workable quantifiable definition of sustainability is available for evaluation of specific projects or operations. The lack of such a definition inherently gives engineers difficulty. Sustainability of human activities (predominantly production and consumption) is a growing concern among businesses, customers, governments, international bodies and nongovernmental organizations. These concerns are often linked to energy efficiency, reduction of environmentally harmful emissions, ecosystem preservation and other "save the Earth" efforts (Hemanowicz, 2004). They are becoming a part of the "triple bottom line" for business accounting: financial, social and environmental (EBF, 2002). Despite its increasing importance, current definitions of "sustainability" are somewhat vacuous. The most commonly accepted description was provided by the World Commission on Environment and Development (1987) in the "Brundtland Report." According to this report, the goal of sustainability is to "meet the needs of the present generation without compromising the ability of future generations to meet their own needs."

SUSTAINABLE MINING

Obviously, sustainable mining is part of the more general notion of sustainable development. But what is it? The definitions and the perceived concepts associated with the term, "sustainable mining" are as diverse and numerous as the special interest groups that use the terminology.

*University of Nebraska-Lincoln, Lincoln, Nebraska.

How do we recognize sustainable development and mining? When the concepts of sustainable development are applied to mineral resources, there is little agreement on what is to be sustained, and by what means. Without such agreement, sustainable development will remain little more than a slogan of little practical value to public-policy makers. Part of what is missing from most discussions is an understanding of the nature of mineral resources and the dynamics of their development (National Academy of Sciences, 1996). Without this understanding, the use of the sustainable development and mining sustainability becomes phrases that often confuse sustainability with environmental protection and other lofty goals that, strictly speaking, are not required for sustainable operations.

In the search for an answer to the question of what is sustainable mining we will review two groups' perspectives of sustainable mining and development. The two largest United States public landholders, the US Bureau of Land Management and the USDA Forest Service, have produced a joint proclamation setting forth their applications and implementation. Contained in the bulletin, *Sustainable Development and Its Influence on Mining Operations Federal Lands—A Conversation in Plain Language* (Clarke and Bosworth, 2003), Kathleen Clarke, director of the Bureau of Land Management, and Dale Bosworth, chief of the USDA Forest Service, proclaim, "Mining, minerals and metals are important to the economic and social development of many countries. Minerals are essential for modern living. Enhancing the contribution of mining, minerals and metals to sustainable development includes actions at all levels to:

(a) Support efforts to address the environmental, economic, health, and social impacts and benefits of mining, minerals and metals throughout their life cycle, including workers' health and safety, and use a range of partnerships, furthering existing activities at the national and international levels among interested Governments, intergovernmental organizations, mining companies and workers and other stakeholders to promote transparency and accountability for sustainable mining and minerals development;

(b) Enhance the participation of stakeholders, including local and indigenous communities and women, to play an active role in minerals, metals and mining development throughout the life cycles of mining operations, including after closure for rehabilitation purposes, in accordance with national regulations and taking into account significant transboundary impacts;

(c) Foster sustainable mining practices through the provision of financial, technical and capacity-building support to developing countries and countries with economies in transition, for the mining and processing of minerals, including small-scale mining, and, where possible and appropriate, improve value-added processing, upgrade scientific and technological information and reclaim and rehabilitate degraded sites."

The corporate leaders of the minerals industry have responded to the call for sustainable development. The best source of their response is found in the report of the Mining, Minerals and Sustainable Development Project (MMSD, 2002). In the report they state, "One of the greatest challenges facing the world today is integrating economic activity with environmental integrity, social concerns, and effective governance systems. The goal of that integration can be seen as 'sustainable development.' In the context of the minerals sector, the goal should be to maximize the contribution to the well-being of the current generation in a way that ensures an equitable distribution of its costs and benefits, without reducing the potential for future generations to meet their own needs."

Sustainable development and mining sustainability are supported by four pillars: economic sphere, social sphere, environmental sphere and governance sphere. Each sphere has a set of guiding principles as indicated in Table 1.

The minerals industry stakeholders and leaders have described nine key challenges facing

Table 1 Sustainable development principles (from MMSD, 2002)

Sphere	Principles
Economic Sphere	• Maximize human well-being • Ensure efficient use of all resources by maximizing rents • Seek to identify and internalize environmental and social costs • Maintain and enhance the conditions for viable enterprise
Social Sphere	• Ensure a fair distribution of the costs and benefits of development for all those alive today • Respect and reinforce the fundamental rights of human beings, including civil and political liberties, cultural autonomy, social and economic freedoms, and personal security • See to sustain improvements over time; ensure that depletion of natural resources will not deprive future generations through replacement with other forms of capital
Environmental Sphere	• Promote responsible stewardship of natural resources and the environment, including remediation of past damage • Minimize waste and environmental damage along the supply chain • Exercise prudence where impacts are unknown or uncertain • Operate within ecological limits and protect critical natural capital
Governance Sphere	• Support representative democracy, including participatory decision-making • Encourage free enterprise within a system of fair rules and incentives • Avoid excessive concentration of power through checks and balances • Ensure transparency through providing all stakeholders with access to relevant and accurate information • Ensure Accountability for decisions and actions, which are based on comprehensive and reliable analysis • Encourage cooperation to build trust and shared goals and values • Ensure that decisions are made at the appropriate level, adhering to the principle of subsidiarity where possible.

the minerals industry in the quest for mining sustainability. These challenges are

1. Visibility of the minerals industry
2. The control, use and management of land
3. Minerals and economic development
4. Local communities and mines
5. Mining, minerals, and the environment
6. An integrated approach to using minerals
7. Access to information
8. Artisanal and small-scale mining
9. Sector governancc: roles, responsibilities, and instruments for change

Indeed, the mining industry is making more than a good gesture effort in embracing the sustainability initiatives. The leaders of the mining industry believe they will benefit by at least one or more of the following:

- Enhancing public reputation
- Establishing a business leadership position

- Building competitive advantage
- Securing public license to operate
- Attracting and retaining quality employees
- Attracting investment capital
- Driving internal behavior change
- Initiating efficiency improvement
- Attracting better quality business partners
- Maintaining strong shareholder support
- Minimizing business risk
- Avoiding, or at least delaying, onerous statutory regulation;
- Increasing capacity for innovation
- Maintaining access to land
- Believing it is simply the right thing to do

The principles of sustainable mining are well-defined by various groups, but the challenges for meeting the principles are great and complex. Many of the challenges facing sustainable mining are challenges that public opinion makers and government officials will have major roles in overcoming them. As technical leaders and educators we do have influence on how the challenge of mining and minerals beneficiation can be less threatening to the local environment and ecology. We can certainly aid in meeting the challenge of developing an integrated approach to using minerals.

One of the challenges is the large amounts of material involved in large-scale mining and minerals extraction. Although Korte and others (2000) have written about the effects of the cyanide leaching gold recovery process on the environment and the ramifications of it in terms of sustainable mining, some of their conclusions can be easily extrapolated to all large mining operations. Without question their point that the problems arising from the change of the chemistry of million of tons of natural ore during the grinding procedure, producing changes in the bioavailabilities of metals and other substances are not well understood. In the case of phosphate production not only are there large quantities of ore, but large quantities of mineral product that is chemically treated. Their conclusion is that when living in a time where the word "sustainability" has a certain weight, can such operations ever be considered sustainable.

Another is challenge of the significant altering of material by the use of chemicals in the treatment of the ores. The American Chemical Society (ACS) supports the use of the twelve Green Chemistry Principles (Anastas, P. T., Warner, J. C., 1998). Bucknam (2004) points out that four of these principles: prevention, atom economy, safer auxiliaries, and real-time analysis for pollution prevention provide guidance for sustainable use of chemicals in metallurgical processing.

Mining produces very large volumes of waste, so decisions about where and how to dispose of it are often virtually irreversible. Because decisions about waste handling and other aspects of operations are often so difficult and expensive to reverse, they need to be made correctly in the first place through mine closure planning.

A third challenge is the environmental legacy left by mining. The environmental issues of current mining operations are daunting enough. But in many ways far more troubling are some of the continuing effects of past mining and smelting. The development of an environmental management system (EMS) in which an environmental impact assessment (EIA) is an integral part would permit the mining company and the regulating authority a structure method for having an awareness and control of the performance of a project that could be applied at all stages of the mining life cycle. In the times of sustainable development, the EIA is insufficient. Now and in the future there is a greater need for integrated impact assessments. The minimum loss of biodiversity is the other great challenge of mining sustainability. The loss of biodiversity is an irreversible loss. Conservation practices that guarantee a minimum impact on biodiversity must be adopted and implemented.

As important as the methods of mining and beneficiation is how the minerals are used in efforts to develop sustainable development. An integrated approach for the use of minerals must be utilized. The use and downstream supply of mineral products has implications for sustainable development and must be considered along with mining and processing of minerals. Current patterns of minerals use raise concerns about efficiency and the need for more equitable access to resources world-wide. The environmental and health impacts of different mineral products in use need to be carefully managed. Where the risks associated with use are deemed unacceptable or are not known, the costs associated with using certain minerals may outweigh the benefits.

PHOSPHATE MINING

Each mineral/metals commodity that is produced through mining and metallurgical activities has a nearly unique set of conditions that need to be considered when studying sustainable mining issues. As mentioned before, three challenges, in one form or another, accompany any mining and beneficiation activity if it is large. Certainly, phosphate production has sustainability challenges including large amounts of material, use of chemicals to aid beneficiation and the improvement of the use and downstream supply.

The large scale of the phosphate produced and the location of the production facilities create one of the greater challenges for achieving sustainable phosphate rock production. The scale of the phosphate operations makes some of the ideals of sustainable mining difficult to approach. Jasinski (2004) reports that the U.S. production of marketable phosphate rock increased slightly to 36.1 million metric tons (Mt) in Crop Year 2004 compared with 35.6 Mt in Crop Year 2003. The manufacturing wet-process phosphoric acid for fertilizers and animal feed supplements was estimated to have accounted for more than 95% of the phosphate rock consumption.

Millions of tons of material are produced within the United States. But only four states have production and two states produce the bulk of it: Florida and North Carolina. The production of phosphate rock in the United States, by region, is given in Table 2 (Jasinski 2004).

Florida, alone, accounts for approximately 75% of domestic production and approximately 20% of world production, which is greater than any other country in the world (Florida Phosphate Council, 2003).

The relative magnitude of the worldwide mining and mineral processing of phosphate rock is illustrated by the compilation of the world production of some bulk minerals in 1998/1999 that was prepared by the United Nations Environment Programme (2001). The compilation is given in Table 3.

The United Nations and the International Fertilizer Commission have sponsored an excellent report (UNEP, 2001) on the mining sustainability of the minerals used for fertilizers. In the report it is claimed that the phosphate mining industry has had a significant improvement in the environmental performance. Improved performance has been caused by the following:

- Growing public awareness of environmental issues
- Greater appreciation by companies of the environmental issues
- Increasing regulation
- Scientific and technical progress that permits resolution of some of the issues
- Technological developments that improve environmental performance as a by-product of better efficiency

A number of emerging issues and trends may affect the industry in coming years, including the following:

- Consumers as a political and economic force
- The public's demand for accountability and transparency across the board, for both businesses and government
- Sensitivities about globalization
- A multidisciplinary approach to resolving issues

Table 2 Phosphate rock production (×1,000 metric tons) in the United States, by region

Period	Florida and North Carolina				Idaho and Utah			
	Crude Ore		Marketable Rock		Crude Ore		Marketable Rock	
	Tonnage	P_2O_5	Tonnage	P_2O_5	Tonnage	P_2O_5	Tonnage	P_2O_5
Crop year 2003	149,000	13,600	30,400	9,130	6,850	1,489	5,150	1,430
Crop year 2004	144,200	13,430	31,100	9,190	6,840	1,487	4,980	1,292

Table 3 World production of some bulk in minerals in 1998/1999

Product	Tonnage
Coal	4,655,000,000
Iron	1,020,000,000
Salt	186,000,000
Phosphate Rock	144,000,000
Bauxite	126,000,000
Gypsum	107,000,000
Potach Ore	45,000,000

- Ongoing public debate on what constitutes sustainability and good environmental performance
- Increasingly stringent expectations of environmental performance
- Scrutiny of business at a local and international level

In the future, community concerns may shift away from a focus on environmental damage at the mine site to the need to balance competing demands for limited natural resources such as fresh water and agricultural land. The mining industry, as a consumer of natural resources, will not be isolated from these pressures. Foresight and the adoption of adequate and effective solutions to arising issues will assist the industry's response.

Because of the magnitude of the operations used for the production of phosphate, the impacts are significant for the local area. The potential environmental effects can be delineated by the various phases of the production. In the mining phase in which there is overburden removal and the mining of the ore the potential environmental effects are land surface disturbance, water contamination, water table lowering, air emissions, topsoil degradation, vegetation and wildlife disruption, and possible noise and vibration problems. The second phase is the material handling from the mine to the beneficiation plant. Potential environmental effects in handling include air emissions, water contamination and noise.

Mineral beneficiation certainly is not without potential adverse environmental effects. Beneficiation of the phosphate rock includes the traditional steps of size reduction, separation, concentration, contaminant removal, drying, compaction and other product preparation steps. Waste generation is potentially one of the greatest adverse environmental effects. A variety of wastes is generated from the beneficiation of phosphate rock and may include following:

- Oversize low-grade rock from screening
- Coarse tailings composed of sand
- Fine tailings composed of clays and similar size materials
- Process water contaminated with fines and process reagents

The beneficiation of phosphate produces large volumes of waste that may cause a variety of adverse environmental effects if not managed and disposed of properly. The major environmental concerns related to waste disposal and management include the following:

- Land surface disturbance from the construction and operation of large waste disposal impoundments such as dams, ponds and stacks
- Surface and ground water contamination by wastes such as fines, tailings effluents and brines
- The safety and stability of the storage facility

Failure can result in extensive and widespread offsite effects.

Mine closure is now an integral part of any mine plan. During the past decade more and more public and government attention has been devoted to the potentially adverse environmental effects developed after the cessation of active mining operations. The goals and objectives of the attention and the appropriate legislation are to leave the mine site in a stable and safe condition. The concepts of mine sustainability and sustainable development have made mine closure more complicated than it has been in the past. It now involves the following objectives:

- Finalizing rehabilitation that commenced earlier in the mine life
- Rehabilitation and other activities such as sealing shafts and removing plant and equipment that could not be removed until after mining and beneficiation were completed
- Monitoring the keep of rehabilitation and closure activities in the long term

Rehabilitation is generally conducted progressively throughout the mine life. Social impacts on the workforce and on the community, associated with the closure of an operation, may be complex. The obviation of these impacts is certainly within the spirit of mining sustainability.

How much has the concepts of mining sustainability been accepted by the public and private sectors? One needs to study no more than the 2003–2008 Strategic Plan of the Florida Institute of Phosphate Research (FPIR, 2003). The FIPR research priorities mirror the potential environmental and sociological effects of phosphate mining and processing described in the United Nations Report, a report co-sponsored by the International Fertilizer Industry Association. The goals for chemical processing are directly aimed at solving the mine sustainability issues created by large processing wastes. They are (1) to develop procedures for reducing the magnitude of the process water problem, and (2) to reduce the accumulation of phosphogypsum in the stacks.

The goals in the strategic plan for mining and beneficiation also are crafted to fit the needs of sustainable mining. And once again, the issue of large quantities of material drives the goals, which are listed below:

- Develop methods for reducing or eliminating clay settling ponds
- Find environmentally acceptable uses for phosphatic clays
- Find environmentally acceptable uses for clay settling ponds
- Develop technologies for solving the dolomite problem
- Improve mining and transportation efficiency
- Improve flotation efficiency

One important aspect of sustainable mining is taking into account the whole mining venture. During and after the life of the mining operation environmental considerations must be evaluated along with sound reclamation practices. The strategic plan addresses goals for the environment and reclamations:

- Evaluate the effects of phosphate mining, processing and reclamation on the environment and develop methods for minimizing and ameliorating impacts
- Develop reclamation technology
- Increase knowledge of the functioning of hydrologic systems in mining areas and develop methods for enhancing them

The last thrust of the strategic plan is public and environmental health. This focus area also fits nicely into the goals of mining sustainability, which include the following:

- Evaluate the occupational, public, and environmental health aspects of phosphate industry emissions and by-products
- Evaluate the occupational, public, and environmental health aspects of exposure to technologically enhanced naturally occurring radioactive materials (TENORM)
- Develop procedures or technology to reduce risks of occupational-related illness or injury to persons employed within the Florida phosphate industry
- Evaluate the occupational, public, and environmental health aspects of technologies, procedures, and practices developed by FIPR through the mining and beneficiation, chemical processing, reclamation, and public and environmental health research programs
- Respond to new occupational, public, and environmental health concerns raised by the public, the phosphate industry, and governmental agencies

The Florida phosphate industry is an example of why the philosophy of sustainable mining must be adopted and implemented. After review of some of the articles published by local press agencies and transmitted on action groups web sites, it is obvious the industry has some political problems. For instance, one of the latest reports is the coverage of the failed retention dam that held an estimated 150 million gallons of phosphogypsum wastewater (MSNBC, 2004).

The true test of how the phosphate mining industry has done relative to mining sustainability is how potential new mining activities are received by the local public and government officials. The questions asked indicate what significant issues must be addressed. Nearly all are centered on local environment and natural habitat impacts based on the size of the operations. In the forefront is probably the use of ground water and its eventual return to the watershed for human and wildlife use. Because of the scale of the mines and their beneficiation plants the magnitude of the waste problem is enormous. One waste, phosphogypsum is a particularly difficult one to dispose or store without some habitat impact (Call, 2003).

Another possible long-term impact is the possible destruction of topsoil over a large area. It has been alleged that the long-term effect of overburden removal and the subsequent replacement with mine and mill wastes will be the long-term effect of the reduction in the quality and quantity of plants and animals on reclaimed mine sites. Questions have been raised regarding the reclamation and restoration plans for proposed new mining sites (Erwin, 2004). It is argued that the cost to meet the Florida Department of Environmental Protection's measures of success is prohibitive. Those measures of success include the following:

- All applicable water quality standards are met
- The mitigation area achieves viable and sustainable ecological and hydrological functions
- Specific success criteria contained in the permit are met

Most of the criticism leveled at the phosphate industry stems with the problems generated by the large-scale mining and processing required for the phosphate production needed for agriculture. Of course having the deposits in close proximity to both fragile ecosystems and large urban population centers doesn't help.

THE AGRICULTURE CONNECTION

Phosphate mining and beneficiation is not only impacted by the concerns regarding mining sustainability, but it is also impacted by those of sustainable agriculture. Any trend in phosphate usage in agriculture has a direct impact on phosphate rock production. Ninetyfive percent of the national production of phosphate rock is consumed in the manufacture of phosphoric acid, the basic material for producing most phosphatic fertilizers. The most common phosphatic fertilizers are diammonium phosphate (DAP) and monoammonium phosphate (MAP). The United States supplies most of the phosphate fertilizers in the world. Overall, more than 50 percent of the phosphoric acid produced in the United States is exported as finished fertilizers or commercial acid.

One misconception is that sustainable agriculture will have most of its impacts in the developing nations of the world. The acceptance of sustainable agriculture and development practices will have major impacts in the Corn Belt of the United States and it will be felt well beyond the Belt's boundaries. Two significant trends will impact the use of inorganic fertilizer.

The Florida Institute of Phosphate Research (FPIR, 2002) has given an answer to the question of why use chemical fertilizer instead of organic fertilizers. The response to the question includes, "The key to growing crops that are plentiful and that contain the nutrients we need is assuring that the local soil has the nutrients it needs. Manure and compost for example, typically have, and provide, relatively low nutrient content in comparison to commercial fertilizer. If enough is spread, it may provide adequate nitrogen, but likely will not provide enough phosphate."

Two relatively new variables may alter the phosphate balance between organic phosphorous and phosphorous from chemical fertilizers. The first is the response to the regulations, which go into effect in 2007, regarding Concentrated Animal Feeding Operations (CAFO). The second is the production of ethanol from corn use as a motor fuel. The major byproduct of the production of ethanol from corn is wet distillers grain (WDG). The major use of WDG is cattle feed. Nutritionally, the WDG is a better feed to cattle than corn. One of the positive aspects of WDG is the phosphorous content is significantly greater than that of corn. Fortunately, since the ethanol plants essentially process the starch in the corn into ethanol, the waste and byproducts have higher concentrations of phosphate than the original corn. The WDG supplies more phosphorous than the cow requires so manure produced has a higher content of phosphorous than manure originating from a corn feed.

Animal manure contains more phosphorous than nitrogen relative to plant needs, meaning that less manure can be spread on a given acre under the phosphorous limit than a nitrogen limit (Mullins, 2000). With the adopted phosphorous rules for a CAFO, approximately 2.4 acres of farmland are required for manure spreading for each cow contained in a CAFO. If WDG is used as a feed the amount of land needed for manure disposal is approximately five acres. If the integrated ethanol-cattle production facilities grow as anticipated, the need for inorganic phosphorous fertilizer will be significantly negatively impacted.

Phosphate mining is nearly unique because of the potential impacts of sustainable agriculture on a mining industry and, therefore, sustainable mining. Truly, this is an overarching case of sustainable development.

CONCLUSION

The cost/benefit analysis of phosphate mining, and therefore, the sustainability, is dependent on the geographic system used for the calculation. Because of the magnitude of the material involved in the mining and beneficiation efforts the local impacts can only be minimized, not eliminated. Any potential research effort must be evaluated on the potential for minimization of the impacts. The technical expert's triple bottom line is now technical feasibility, financial feasibility and environmental feasibility. Government and industry demand no less.

REFERENCES

Anastas, P.T., Warner, J.C., 1998. *Green Chemistry: Theory and Practice*. Oxford University Press, New York, p. 30.

Bucknam, C.H., 2004. A Green Chemistry Approach to Mercury Control during Cyanide Leaching of Gold. Preprint 04-101, SME Annual Meeting, Feb. 24–26, Denver, Colorado.

Call, C.L., 2003. Phosphate Company Give Presentation, http://www.sun-herald.com/newsarchive2/092603/ch11.htm?date=092603&story=ch11.htm, September 26.

Clarke, K., and Bosworth, D., 2003. Sustainable Development Mineral Applications, http://www.fs.fed.us/geology/fs_blm_statement_of_support.pdf.

EBF, 2002. CFO Europe: Investors Demand Environmental and Social Reporting, http://www.ebusinessforum.com/index.asp?layout=printer_friendly&doc_id=5392.

Erwin, 2004. The Cost of Reclamation, The Price of Failure, The Tampa Tribune, http://tampatrib.com/News/MGACSFIYOTD.html.

Florida Institute of Phosphate Research, 2002. Why Use Chemical Fertilizer Instead of Organic Fertilizers, http://www.fipr.state.fl.us/southo_organic_fertilizers.htm.

Florida Institute of Phosphate Research, 2003. Florida Institute of Phosphate Research 2003–2008 Strategic Plan, Bartow, FL, November.

Florida Phosphate Council, 2003. 2002 Florida Phosphate Facts: Tallahassee, FL, Florida Phosphate Council, 1–8.

Hermanowicz, S.W., 2004. Entropy and Energy: Toward a Definition of Physical Sustainability, Presentation at the International Eco-Efficiency Conference: Quantified Methods for Decision Making, Leiden, the Netherlands, April 1–3. 14 pp.

Jasinski, S.M., 2004. Marketable Phosphate Rock—Crop Year 2004. Mineral Industry Surveys, U.S. Geological Survey, U.S. Department of the Interior, September.

Korte, F., Spiteller, M., and Coulston, F., 2000. The Cyanide Leaching Gold Recovery Process Is a Nonsustainable Technology with Unacceptable Impacts on Ecosystems and Humans: The Disaster in Romania. Ecotoxicology and Environmental Safety, 46: 241–245.

MSNBC, 2004. Cargill Scrambles to Mitigate Wastewater's Effect on Creek. Tampa Bay Online, http://www.msnbc.msn.com/id/5923648/.

Mullins, G.G., 2000. Nutrient Management Plans—Poultry. In *Natural Resource, Agriculture, and Engineering Service, Managing Nutrients and Pathogens from Animal Agriculture*, NRAES-130, Ithaca, NY.

National Academy of Sciences, 1996. *Mineral Resources and Sustainability: Challenges for Earth Scientists, Commission on Geosciences, Environment, and Resources*, National Academy Press, Washington, D.C.

United Nations Division of Sustainable Development, 2002. World Summit on Sustainable Development Plan for Implementation, Johannesburg, September, Paragraph 46.

United Nations Environment Programme, 2001. Environmental Aspects of Phosphate and Potash Mining, Paris.

World Commission on Environment and Development, 1987. *Our Common Future*. Oxford University Press, Oxford.

CHAPTER 6

Plasma Vitrification of Sewage Sludge Containing Phosphorus Compounds

Krystyna Cedzynska*

ABSTRACT

The purpose of this study was to find the best method for destroying sewage sludge. Thermal destruction was chosen due to a high content of heavy metals and other compounds, including phosphorous compounds. These sludges were destroyed by a thermal plasma pyrolysis. It uses high temperature plasma for the destruction of all organic compounds with final vitrification of inorganic residues. During this study the structure of vitrified product was analyzed. The influence of inorganic and phosphorous compounds on the process of vitrification and the effectivity of immobilizing the metals was investigated. The properties of pyrolytic products were established with computer simulation to optimize the plasma treatment process.

INTRODUCTION

According to Polish law (Dz. U. z 1997 r. Nr 96 poz. 592), sewage sludge is classified as a waste. Utilization of sewage sludge is a particularly important problem and the final destination of the sludge in the EU is disposal to landfills and agriculture or decomposition via a thermal treatment.[1,2] The utilization of sewage sludge as a fertilizer seems to be limited because of the harmful components in the sewage sludge, such as heavy metals or toxic organic compounds. There are certain actions necessary prior to depositing sewage sludge in landfills in order to avoid any emission. Thermal conversion processes such as combustion offer interesting alternatives for the stabilization or elimination of these wastes. It can be burned in incinerators, but since the sludge has a high heavy metal content the use of incineration alone does not eliminate the metal hazard since these metals can show up in even a greater concentration in the post-incineration residues (ash and slag) and be emitted to the environment.[3] An alternative method to incineration can be the destruction of sewage sludge in plasma treatment. The thermal plasma method can transform toxic waste into a chemically stable and mechanically resistant product. Heavy metals and toxic components of sewage sludge can be converted into a non-leachable slag and the final product (vitrified material) is environmentally friendly.[4–6]

Thermal plasma is one of the methods especially suited to the conversion of waste in a pyrolytic process. The pyrolysis process involves heating the sludge in an inert atmosphere, which leads to the production of carbon, tar and gases. Pyrolysis of the sewage sludges gives the

*Technical University of Lodz, Institute of General Food Chemistry, Lodz, Poland.

advantage of concentrating the heavy metals in the final residue.[7,8] The leaching of these metals is small in the case of vitrified material from plasma pyrolysis, in contrast to that of ashes from incineration.[3]

The aim of this study was the application of a plasma vitrification process to transform toxic sewage sludge from a wastewater treatment plant into a chemically stable material with possibly the smallest negative effect on the environment, i.e., with the smallest leachability of vitrificates.

The objective of this paper was the study of plasma pyrolysis decomposition for two types of sewage sludge with different origins (metal and phosphorus compounds concentration) and behavior in the thermal decomposition.

To optimize the pyrolysis process, a thermochemical calculation was performed before the experimental part of study.

EXPERIMENTAL

Material

Two types of sewage sludge have been studied corresponding to various waste waters in the urban community. They were a typical sewage sludge from the Group Waste-Water Treatment Plant in Lodz (sludge I), and second, the sediment from a water purification plant from a municipal center with a number of industrial plants including a plant producing fertilizers and protection agents (sludge II). They had the following parameters: dry organic mass 55–64%, pH 7–7.4, and 60–72% moisture. Before testing the sewage sludges ware dried, crumbled and piled.

The composition and content of particular clements (mainly metals) were determined for various lots of sewage sludge by elemental analysis and by spectrophotometric methods (e.g., AAS). Table 1 shows the contents of both sewage sludges chosen for the investigation (I and II) and the average content of components for various lots of sewage sludge from the Wastewater Treatment Plant in Lodz (for dry mass basis).

The leachability of metals was accepted as the most important characteristic parameter for the chemical resistance. Leaching tests were performed for dry and mixed pyrolytic products and vitrificates with the grain size below 4 mm according to the European Standards pr EN 12457. The concentrations of metals in water solutions were investigated using an atomic absorption spectrometer with graphite furnace–GFAAS (GBC 932plus, GF 3000) according to EPA Tests Methods Series 7000.

Thermal Plasma Treatment

The samples of the sewage sludge were vitrified to obtain a better quality product in the same processes of vitrification.

The vitrification processes were run using a "Little JetArc" system, designed and built in the Institute of Electrical Apparatus at Lodz Technical University, to obtain quick testing of different feed samples in controlled repetitive working conditions. An arc furnace (up to 20kW) was placed inside a stainless steel chamber connected to a vacuum pump. This allowed at first the evacuation of air and then continuous cleaning of the furnace interior from the exhaust gases using argon as a purging gas. A detailed description of the plasma reactor was reported in an earlier paper.[9]

The measurement system was equipped with an optical temperature recorder Minolta/Land-Cyclops 152 and off-gas analyzer Madur GA-40T. The process and measurements were computer controlled with a self-made program based on DasyLab software.

All the vitrification processes were run under similar conditions in an inert argon atmosphere. For all samples the mass was about 10g; time, 300s; and power, 1400J.

RESULTS AND DISCUSSION

Thermochemical Calculation

The experimental part of the process study was made after first performing thermochemical calculations. For the calculation we used a ChemSage program with a database covering organic

Table 1 Contents of some elements in sewage sludge

Element	Sewage sludge average content	Sewage sludge I	Sewage sludge II
N_{total}	7%	9.5%	5.7%
P	3%	4.1%	9.1%
Ca	3%	4.0%	7.9%
Mg	0.8%	1.1%	1.8%
Pb	24mg/kg	35mg/kg	92mg/kg
Cd	4mg/kg	5mg/kg	85mg/kg
Cr	163mg/kg	85mg/kg	125mg/kg
Cu	234mg/kg	92mg/kg	173mg/kg
Ni	45mg/kg	58mg/kg	72mg/kg
Fe	80mg/kg	35mg/kg	81mg/kg
Zn	1052mg/kg	828mg/kg	1105mg/kg

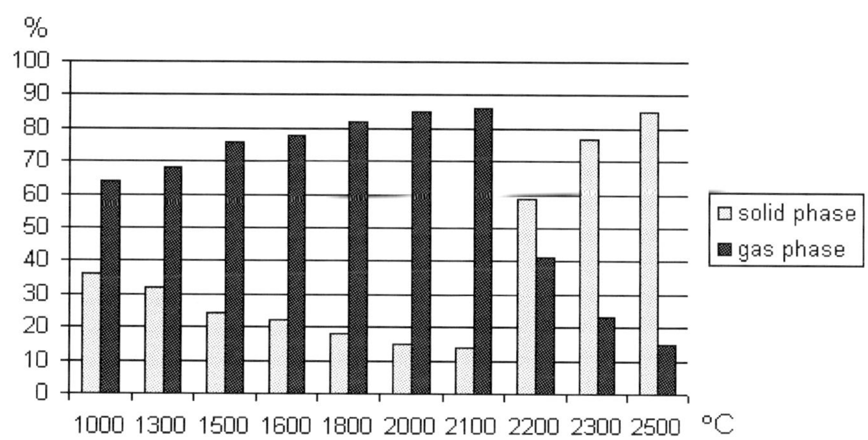

Figure 1 The percent ratio for gaseous and solid products

and inorganic compounds undergoing a thermal destruction.

The calculations were made for the process of pyrolytic decomposition of sewage sludge with the average content of chemical elements (shown in Table 1) within the range of temperatures between 1,000–2,500°C and under a pressure of 1 atm.

The composition of products obtained during the process was determined from simulation of the sludge vitrification. Figure 1 shows the relation between the process temperature and the percentage ratio between the gaseous and solid products formed during vitrification.

As the temperature rises, the percent of gaseous products grows to a maximum level at

2,100°C when the quantity of gaseous products is six times greater than the solid products. CO and H_2 are the main products with a calorific value of 13–13.5 MJ/Nm^3. The high concentration of organic matter in sewage sludge is the reason for such a high concentration of fuel gases. The quantity of gaseous products rapidly falls above 2,100°C. This study shows that 2,200°C is the optimum process temperature since at higher temperatures toxic compounds are formed, e,g., CuCN, C_2Cr_3.

Table 2 shows the concentrations of the main solid components obtained in the pyrolysis process. Carbon is the main element and its concentration is almost independent of temperature. This table shows that up to 1,000°C, solid products are mainly carbon and a mixture of phosphates, Ca and Mg. Above 1,200°C they undergo a slow decomposition to oxides and then to gaseous elements (from 1,400°C for Mg and from 1,600°C for Ca and P). The concentration of the phosphates is proportional to the P and Ca concentration in the initial sewage sludge.

Computer simulations were made to select the best conditions for vitrification. They were run as a function of the process temperature. If the pyrolytic process takes place at 1,000–1,200°C then apart from carbon, there is also phosphate present as a solid product and the gases will have a reasonable calorific value. When the process is run above 2,200°C, then solid products prevail in molten forms. In both cases the heavy metals concentration in the solid product is higher than in the feed sludge.

Description of Obtained Pyrolysis Products and Vitrificates

Plasma pyrolysis of the sewage sludge I at 1,000–1,200°C and vitrification of sewage sludge I and II were performed to estimate the rate of bonding of heavy metals in the product. Both the solid product and vitrified products of dry sewage sludge I and II were obtained in a plasma thermal treatment and tested for physical and chemical (leaching) properties.

Plasma pyrolysis of dry sewage sludge at the temperature 1,000–1,200°C gave products of a gravel crystalline structure (Figure 2). This was caused by the large volume of gas formed during the pyrolysis process (over 50% organic compounds). The products were very porous and had a tendency to crumble easily.

The vitrificates (plasma vitrification at 2,100–2,200°C) of sewage sludge I and II have a uniform structure (Figures 3a and 3b). They have a smooth, black surface with a pearly shine when broken and contain small gas bubbles inside.

Table 2 The concentration of the main solid components

Temp.[°C] / Compound	1000	1300	1500	1600	1800
C	57.9%	60.7%	69.1%	73.6%	71.4%
$Mg_3(PO_4)_2$	10.5%				
$Ca_3(PO_4)_2$	29.3%	32.9%			
MgO		5.9%			
CaO			15.7%	25.8%	
$CaMgO_2$			14.6%		
CaC_2					20.3%

Figure 2 Solid product after plasma pyrolysis of sewage sludge I at 1,000–1,200°C

Figure 3a Vitrified sewage sludge I

Figure 3b Vitrified sewage sludge II

Table 3 Metal concentration in an aqueous extract from pyrolysis product of sewage sludge I and extract of vitrificates from sewage sludge I and II and standard acceptable metal contents in sewage water and in I class water (mg/dm^3)

Element	Pyrolysis product I	Vitrificate I	Vitrificate II	Sewage water	I class water
Zn	3.48	0.15	0.14	2.0	0.2
Fe	2.52	0.45	0.32	10	1.0
Ni	1.15	0.38	0.33	2.0	1.0
Cu	0.92	0.20	0.18	0.5	0.05
Cr	0.85	0.02	0.01	0.2	0.05
Cd	0.28	0.005	0.005	0.1	0.005
Pb	0.65	0.03	0.03	0.5	0.05

The results of leaching tests for vitrificates of dry sewage sludge (I and II) and for dry pyrolysis product of sewage sludge I obtained at 1,000–1200°C and standard acceptable metal contents in I class water and in sewage water (mg/dm^3) are shown in Table 3.

Contrary to results of dry pyrolysis of sewage sludge I obtained at 1,000–1,200°C, vitrificates made from this sludge had a very low metal leachability. For some metals, the leachability was decreased over 20–50 times (Zn, Cd, Cr and Pb). The concentration of all metals leached from vitrificates did not exceed any values allowed by the standard for sewage waters and in most cases also by standard values for surface water of I class.

The main reason for such a low metal leachability from vitrificates is its internal structure, completely different than in pyrolysis product of sewage sludge in which the metals are mostly easily leached as various salts. Vitrificates are minerals, and they can absorb into their structure ions of other metals through the exchange of calcium and magnesium atoms for other cations, e.g., chromium. Metals in the vitrificate structure have strong bonds, giving with high resistance to leaching. Even if some part of salt or metal oxide were not built into the vitrificate structure, the vitrificate would still physically confine them inside, inhibiting the leachability[9]. In contrast to the products of pyrolysis, the vitrificates have a very compact structure with a much smaller area of contact with water significantly reducing the effect on the leaching of metals. An addition of the vitrificate of sludge II gives even better results. Despite the higher content of some heavy metals in sludge II, their leachability from vitrificates was comparable and even lower due to higher content of Ca and P in this sludge.

CONCLUSIONS

The results of analyses confirm the possibility of bonding heavy metals in the vitrified material during thermal treatment in a plasma reactor. A significant decrease of heavy metals leaching could be achieved by sludge vitrification. Such vitrificates are safe for the environment since the metals leach in quantities not exceeding standards for water of I class. The vitrificates have a compact structure without trapped gas bubbles and do not crumble easily. The results obtained from this study allow us to conclude that the thermal treatment utilization of even toxic sewage sludge can give an environmentally acceptable product. Good results of leaching tests, which do not exceed standards, confirm this conclusion. Mechanical tests made for this material also give good results. Increased content of Ca and P in the investigated sludge also improves the vitrificate structure and decreases the leachability.

In summary, plasma vitrifiction can be used for the disposal of toxic wastes, including those which are generated by sewage water plants, and can be an alternative for methods presently used, i.e., dumping in special landfills.

REFERENCES

1. CEN, 1999. Recommendations to preserve and extend sludge utilisation and disposal routes. Comite Europeen de Coordination des Normes (European Committee for Co-ordination of Standards); CEN TC 308.
2. EU, 2000. Working document of sludge; 3rd Draft; European Union, Brussels. En V.E.3/LM.
3. Cedzynska, K., Kolacinski, Z., Izydorczyk, M., and Sroczynski, W., 1999. *Thermal Plasma Treatment of Waste Incinerator Ash. Progress in Plasma Processing of Materials*. P. Fauchais and J. Amouroux, eds. Begell House Inc., 707–711.
4. Cedzynska, K., and Kolacinski, Z., 1998. Plasma Pyrolisys—A Novel Method of Waste Destruction. *Polish Journal of Applied Chemistry*, Vol. 2, 143–149.
5. Calaminus, B., and Stahlberg, R., 1998. Continuous In Line Gasification/Vitrification Process for Thermal Waste Treatment: Process Technology and Current Status of Projects. *Waste Management*, Vol. 18, 547–556.
6. Kolacinski, Z., Cedzynska, K., and Szymanski, L., 2000. Thermal Plasma Systems for Waste Treatment. *Proc. 5th Conference on Environment and Mineral Processing*, Ostrava, Czech Republic, Vol. 2, 663–668.
7. Font, R., Fullana, A., Conesa, J.A., and Liavador F., 2001. Analysis of the pyrolysis and combustion of different sewage sludges by TG. *Journal of Analytical and Applied Pyrolysis*, 58–59, 927–941.
8. Cedzynska, K., and Kolacinski Z., 2004. Plasma Destruction of Toxic Chloroorganic Wastes Toward Zero Residues. *Journal of Advanced Oxidation Technologies*, 7/1, 74–78.
9. Cedzynska, K., Winsche, M., Kolacinski, and Z., Szymanski L., 2004. Chromium rich sewage sludge vitrified in thermal plasma. *Proc. Conf. European Commission—CANMET Joint Workshop on Clean Production Technologies*. A. Adjemian, J. Dutrizac, P. Negre, and L. Yuramendi, eds. Madrid, Spain, pp. 153–162.

CHAPTER 7

Uranium From Phosphoric Acid: Is It Time (Again)?

Thomas C. Pool*

INTRODUCTION

Uranium as a by-product of phosphoric acid was an important component of world uranium supply from 1980 through the1990s. Production ceased entirely by 1999 and the production facilities have been completely decommissioned. This decline was in direct correlation with decreasing prices that ultimately plummeted to record lows (see Figure 1). Uranium prices have increased rapidly over the past two years. Is it time for a reincarnation of this once thriving industry? Let's examine some of the factors that bear on this issue.

Uranium Market

Uranium is currently sold only for use as nuclear fuel. This is in contrast to most other metals which have, typically, multiple uses and multiple markets. Uranium's market is small; a total of perhaps 100 buyers worldwide. Yet it is a very important market as nuclear power accounts for approximately 20 percent of the world's electricity supply.

In characterizing the uranium market, it is necessary to review the balance between demand and supply which, in turn, dictates prices which, in turn, provide the impetus for exploration and production.

Demand

Demand for uranium comprises the fuel needs of approximately 440 nuclear reactors worldwide. A 1,000 megawatt reactor, for example, will require some 500,000 pounds of natural uranium per year. Current world uranium requirements are approximately 175 million pounds U_3O_8 per year.

Consumption of uranium is increasing both because of the increasing number of reactors and an increasing reactor utilization factor (see Figure 2). Such utilization factors were typically in the range of 60 to 80 percent just a decade ago, but the incentive of profit provided by deregulation has pushed most utilization factors into the 85 to 95 percent range.

Supply

Uranium supply has progressed in just a few years from a simple, country-by-country oriented primary production and inventory basis to a complex system of international trade in interrelated fuel components. This transition derived mainly from the dissolution of the former Soviet Union and a subsequent agreement between Russia and the U.S. for down-blending Russia nuclear weapons into nuclear fuel. Figure 3 sets

*International Nuclear, Inc., Golden, Colorado.

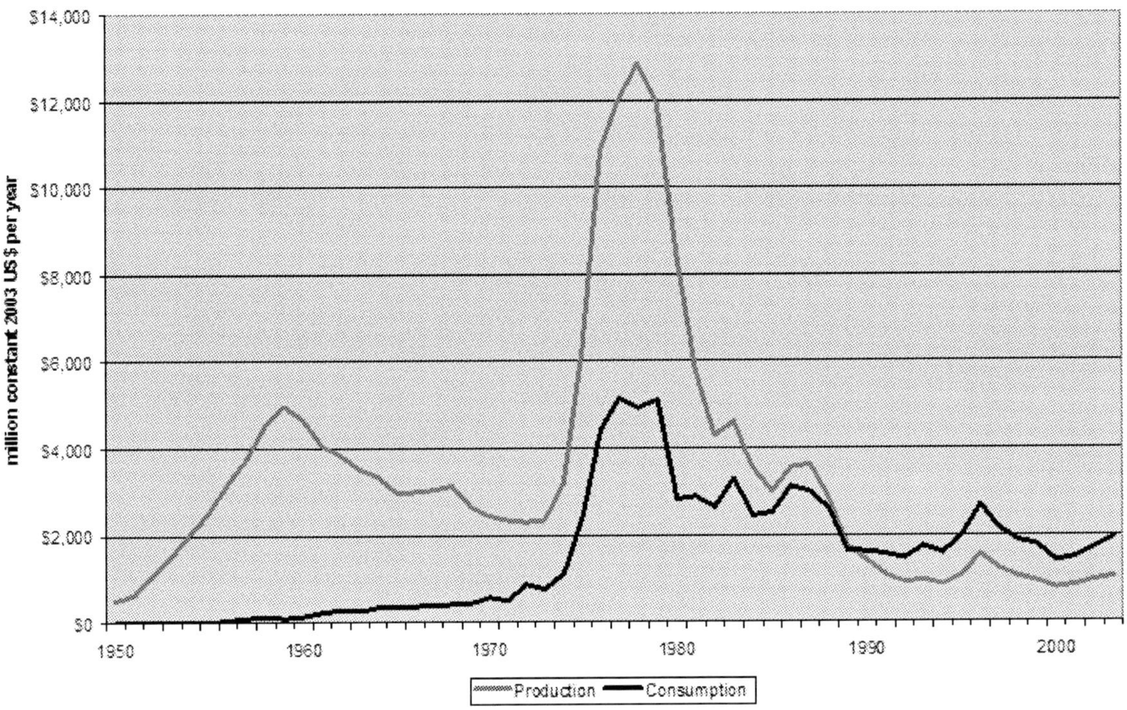

Figure 1 Uranium market value

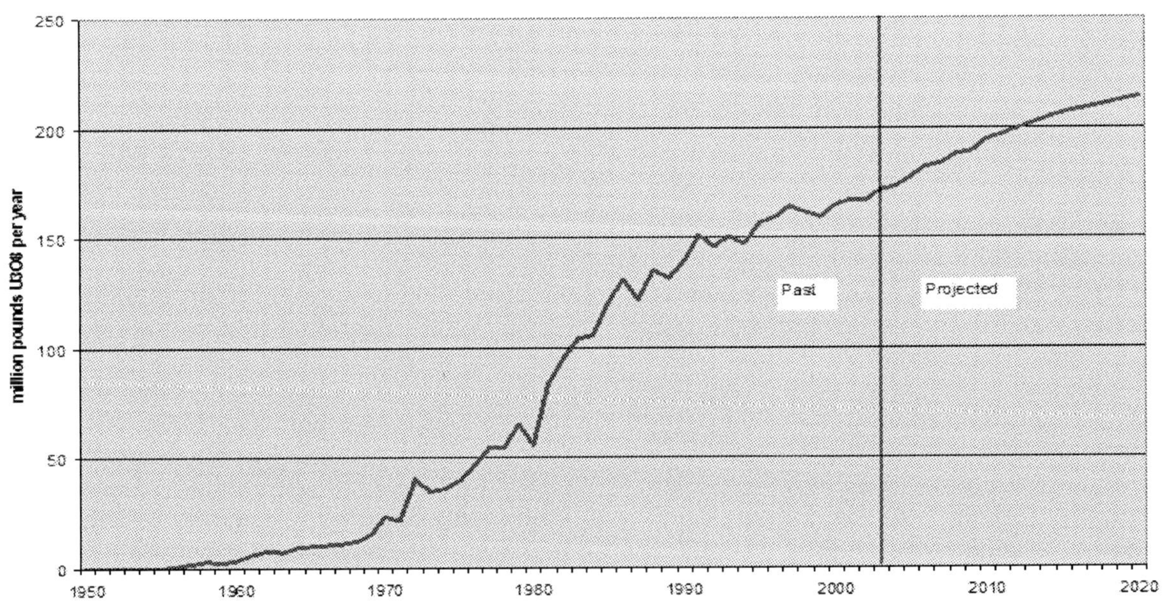

Figure 2 Uranium consumption—past and projected future

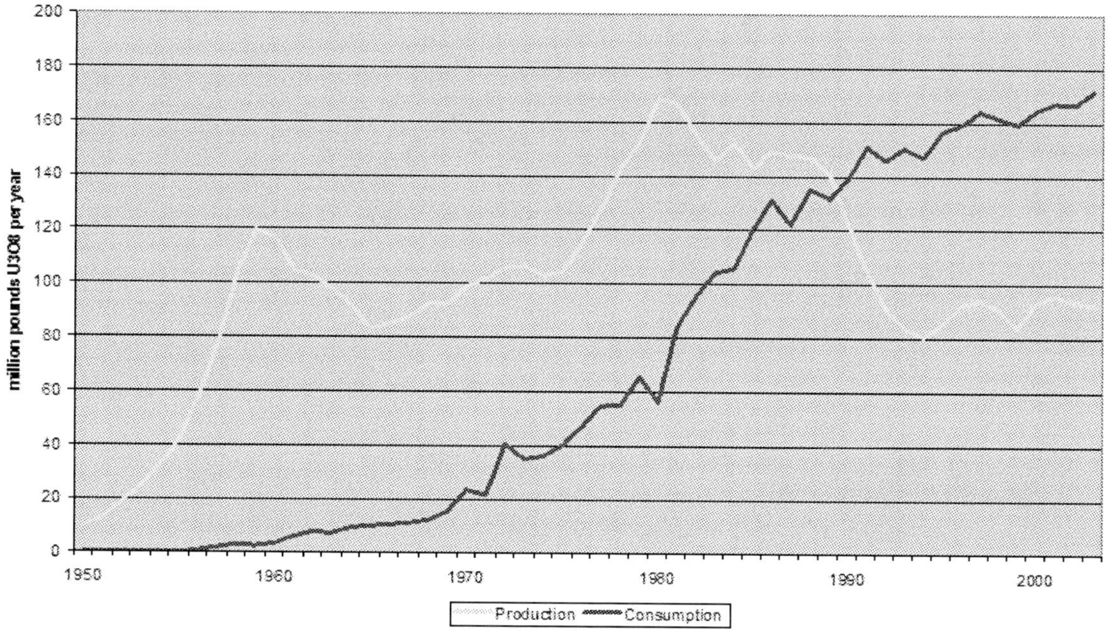

Figure 3 Uranium production and consumption—historical

forth historical uranium production in comparison to consumption.

Overproduction in the early years of the atomic age produced a very large inventory of material that is still being drawn down. This inventory constitutes the basis for most current secondary supply.

Primary Production

Primary production of uranium currently includes 42 production facilities in 18 countries. Of these facilities, 20 are conventional mine-mill operations, 18 produce uranium by means of in-situ leaching and 4 produce uranium as a byproduct. Current primary production is 94 million pounds U_3O_8 per year; just slightly more than 50 percent of the world's total current requirements.

Difficulties attend the process of bringing a new uranium mine into production. Financing has been difficult to obtain not only because of recent low prices, but also because of the controversy surrounding any endeavor in which radioactivity is involved. Environmental controls continue to become more stringent. Anti-nuclear activism remains a serious impediment to uranium development, but the increasingly stellar record of nuclear power in general and its ability to reduce greenhouse gases in particular are making it more difficult for activists to achieve credibility.

Secondary Supply

Primary uranium producers have faced significant competition from secondary supplies for more than two decades. Government, utility and supplier inventories of previously produced uranium have been a major feature of the market in recent years, but are rapidly waning. Current secondary supply features nuclear fuel blended down from Russian nuclear weapons. In 1993, the Russian and U.S. governments reached agreement for the conversion of approximately 40 percent of Russia's nuclear weapons arsenal to be converted to nuclear fuel. By 2004, this process had removed some 5,000 warheads from the arsenal and was supplying the world with over 20 million pounds U_3O_8 equivalent per year, some 12 percent of the world's fuel needs. The agreement expires in 2013, but it is anticipated that the process will continue if only for the fuel needs of Russia itself. U.S. progress in this regard is slow, but it is to be hoped that the example of Russian disarmament will prove to be compelling. Other notable sources of secondary supply include reprocessing of enrichment tails, reprocessing of spent fuel, and utilization of the plutonium component of spent fuel in mixed oxide (MOX) fuel.

Most nuclear reactors require an enriched uranium fuel of 3 to 5% ^{235}U. Natural uranium contains only 0.711% ^{235}U. Current enrichment processes can efficiently recover only about 60% of the ^{235}U contained in natural uranium. When reprocessing costs are quite low as is the case currently in Russia, a portion of the remainder can be recovered economically. Spent fuel contains both unutilized uranium and a fission product of potentially usable plutonium. While reprocessing of spent fuel is costly, it does avoid the substantial costs associated with disposal.

Through time, the balance between primary production and secondary supply has changed markedly and has had a major impact on prices. Figure 4 sets forth this balance.

Uranium Prices

Uranium is typically priced and sold on a US$ per pound U_3O_8 basis. Historical prices have been relatively volatile as shown in Figure 5.

Early prices in the late 1940s and the 1950s were driven by cold war needs for nuclear weapons. As that need was more than fulfilled by the early 1960s, prices declined until rapidly escalating needs for nuclear fuel pushed prices up once more. Note, however, that US government policy regarding enrichment scheduling in the mid- to late-1970s was a major factor in the record high prices during that era.

Most uranium is sold under term contracts of three to five years. Prices, however, are set mainly by spot market sales of which there may be only a few per month and of which not all may be reported accurately. Thus, uranium prices are not fully transparent, a disconcerting element to many market participants. These prices, nevertheless, are closely followed and are the controlling factor in most plans for exploration and production.

Uranium prices have increased substantially in the past 18 months from about US$10 per pound U_3O_8 in May 2003 to over US$20 per pound in late 2004 (Figure 6). This increase is driven by a number of factors including a weak US dollar, production interruptions at producing facilities, depletion of excess inventories and withdrawal from the uranium concentrate market by Russia.

URANIUM PRODUCTION FROM PHOSPHORIC ACID

History of Process Development

A process for extracting uranium from phosphoric acid was first developed during an intensive research and development program sponsored by the US Atomic Energy Commission in 1947. Four US producers of phosphoric acid participated in the program and were instrumental in developing a recovery process which became known as the pyro solvent process, or OPPA process.[1] Three of the four companies recovered uranium on a limited basis at their respective phosphoric acid plants producing a total of approximately 1,000,000 pounds U_3O_8 during the period 1954 to 1962. Uranium production from phosphoric acid ceased in 1962 when it became noncompetitive with increased production by conventional methods from uranium mines in Wyoming and New Mexico.

Continuing research, however, at Oak Ridge National Laboratory in Tennessee throughout the 1960s and early 1970s resulted in the development of two new recovery processes utilizing improved solvents and new technology. The DEPA-TOPO[2] process was developed in the late 1960s and the OPAP[3] process in 1974.

All commercial uranium recovery from phosphoric acid has utilized one of the three above processes, or modifications thereof.

Commercial Production Facilities

Increasing prices for uranium in the mid-1970s spurred a renewed interest by private industry in uranium recovery from phosphoric acid.

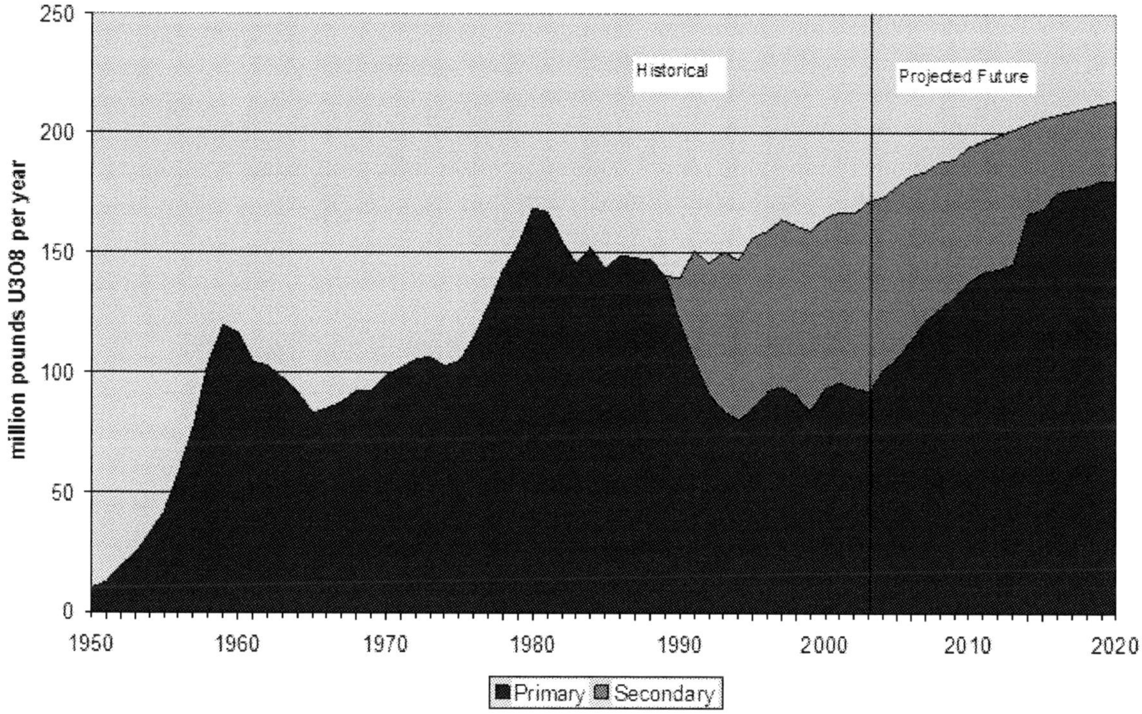

Figure 4 Primary and secondary uranium supply

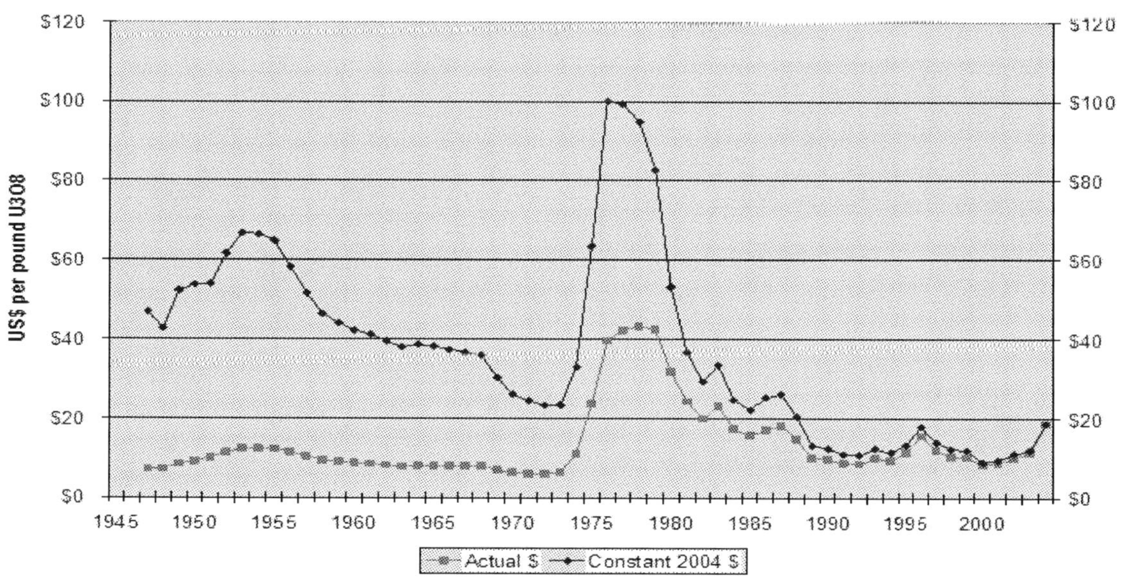

Figure 5 Historical uranium prices

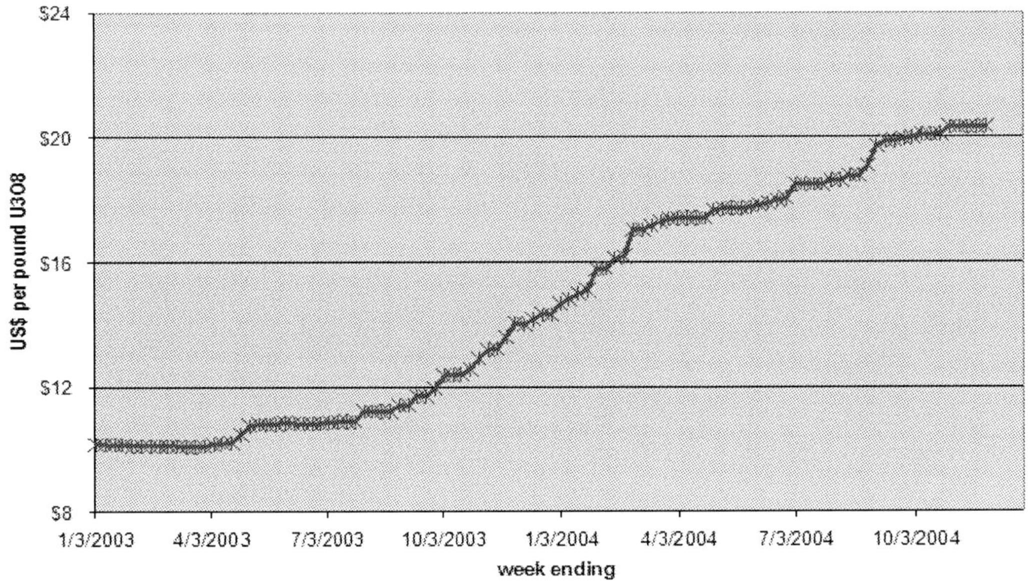

Figure 6 Recent uranium spot market prices

UNC Recovery Corporation, a subsidiary of United Nuclear Corporation, commenced uranium recovery at the W. R. Grace phosphoric acid plant near Bartow, Florida, in 1975. This was the first of a series of new-generation plants to recover uranium from phosphoric acid. Wyoming Minerals Corporation, a subsidiary of Westinghouse Electric Corporation, opened a uranium recovery plant at Pierce, Florida, in 1978. It was closely followed by Freeport Minerals Company which initiated commercial operations at its Uncle Sam plant in Louisiana in January 1979.

International Minerals & Chemical Corporation (IMC) has been involved in the recovery of by-product uranium from phosphoric acid since participating in the USAEC research and development program in the 1950s. After years of additional work, IMC built a pilot plant adjacent to its new phosphoric acid facilities at New Wales, Florida, in 1977 and initiated construction of the world's largest plant for recovery of uranium from phosphoric acid at the same site in 1978. Commercial operation of the New Wales plant commenced in May 1980. This plant also operated as a central recovery facility for two satellite plants at Plant City and Bartow. Table 1 sets forth a series of data on uranium-from-phosphoric acid facilities.

As noted above, most of the world's uranium from phosphoric acid production facilities were located in the USA. Historical production from these facilities is illustrated in Figure 7.

In relating uranium production from phosphoric acid to uranium prices, as set forth in Figure 8, we note that prices peaked prior to start-up of most production facilities. Most uranium sales from these facilities were made under long-term contracts, however, and the full impact of declining prices was not felt for some years.

Even at spot market prices, the industry generated substantial revenue for almost a decade (Figure 9).

Outlook for the Future

In considering whether or not uranium might once again be extracted profitably from phosphoric acid, several factors come into play, including the production of phosphoric acid, advances in technology, regulatory regimes, and, not the least, future prices for uranium.

Table 1 Uranium recovery from phosphoric acid facilities

Acid Producer	Uranium Producer	Location	Cpacity P_2O_5 (tons/year)	Capacity U_3O_8 (lbs/year)	Process	Operating Period
Farmland	Wyoming Minerals Corp.	Pierce, FL	450,000	400,000	WMC, DEPA-TOPO	1978 – 1981
Freeport	Freeport Minerals Co.	Uncle Sam, LA	675,000	690,000	FMC, DEPA-TOPO	1978 - 1999
Agrico	Freeport Minerals Co.	Donaldsonville, LA	360,000	420,000	FMC, DEPA-TOPO	1981 - 1998
IMC	IMC	New Wales, FL	1,000,000	800,000	IMC, DEPA-TOPO	1980 - 1992
CF Industries	IMC	Bartow, FL	720,000	600,000	IMC, DEPA-TOPO	1981 - 1985
CF Industries	IMC	Plant City, FL	680,000	600,000	IMC, DEPA-TOPO	1980 - 1992
W. R. Grace	Uranium Recovery Corp.	Bartow, FL	na	330,000	URC, OPAP	1976 – 1980
Gardinier	Gardinier	East Tampa, FL	500,000	420,000	Gardinier, OPPA	1979 – 1982
Western Coop. Fertilizer	Earth Sciences, Inc.	Calgary, Alberta, Canada	144,000	120,000	ESI, OPAP	1981 – 1987
Chemie Rupel	Umipray	Purrs, Belgium	100,000	150,000	IMC – Prayon, DEPA-TOPO	1980 - 1998

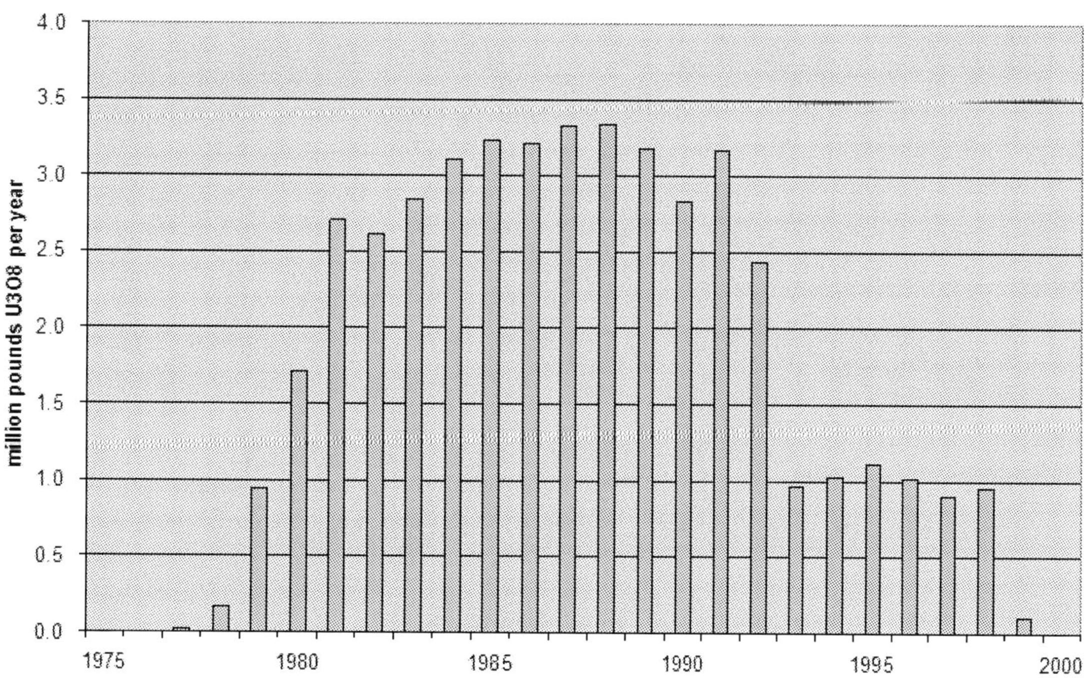

Figure 7 U.S. uranium production from phosphoric acid

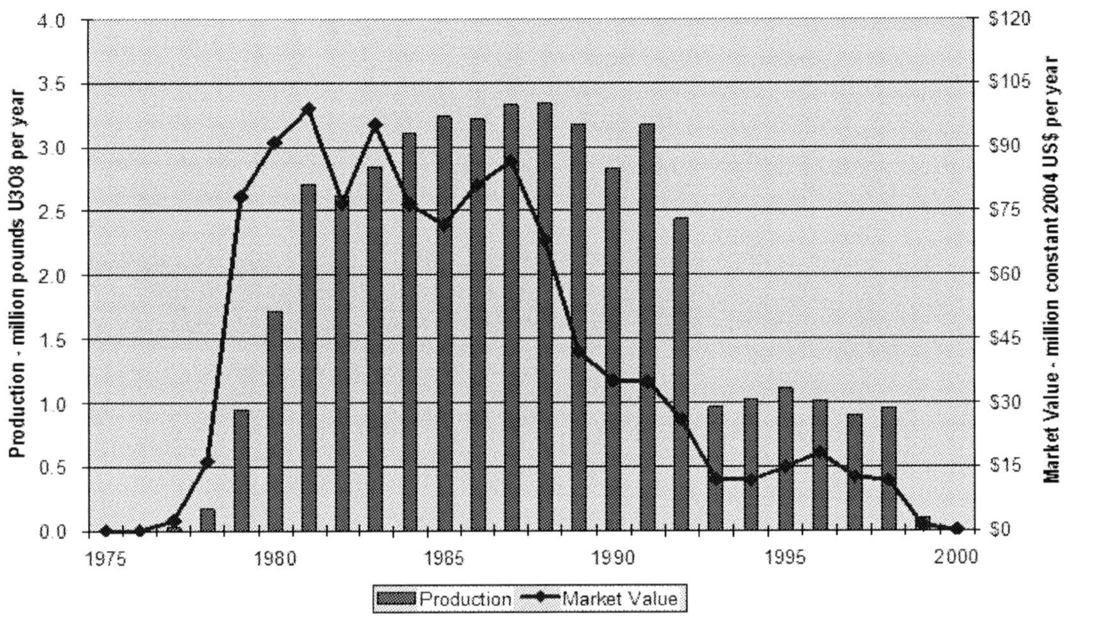

Figure 8 Uranium price and production

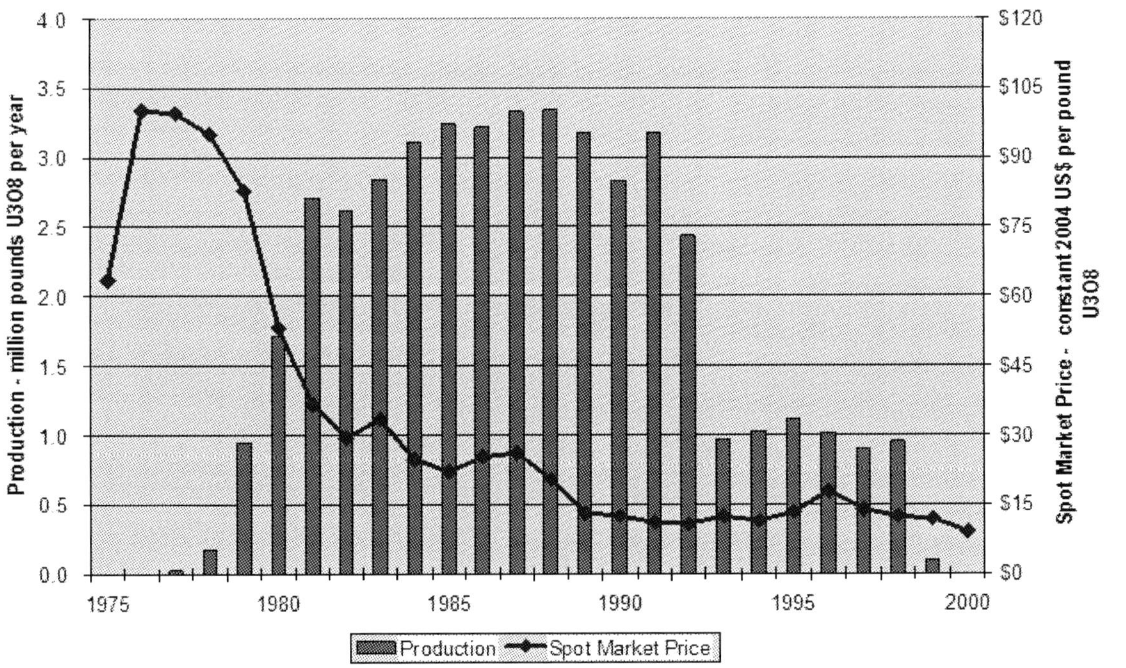

Figure 9 Value of uranium from phosphoric acid

P_2O_5 Production

While U.S. production of phosphate products has declined somewhat over the past few years, the industry still produces more than 10 million tons equivalent P_2O_5 per year (Figure 10). The proportion of this material which might pass through a phosphoric acid stage is not known with accuracy, but is estimated to be more than 90 percent.

Resources

Phosphate resources are large both in the U.S. and the world. A comprehensive study of uraniferous phosphate resources was compiled by Earth Sciences[4] in 1979. This study provided the following estimates:

- Phosphate resources: 1,305,136 million tons of phosphate rock
- Potentially mineable phosphate product: 687,832 million tons (approx. 30% P_2O_5)
- Estimated recoverable phosphate product: 292,727 million tons (30% P_2O_5)
- Average uranium concentration of phosphate product: approx. 100 ppm U
- Uranium contained in potentially mineable phosphate: 67,816,000 tons
- Estimated recoverable uranium: 29,532,000 tons

Considering current consumption of about 175 million pounds U_3O_8 per year, this resource would provide over 400 years of fuel.

Technology

Little work has been done directly on the recovery of uranium from phosphoric acid in the past 20 years, but it is probably fair to say that the field of process control has advanced substantially. We certainly see this in the uranium industry where staffing of processing facilities has been reduced significantly. I certainly wouldn't rule out the application of other technology that has been developed since the advent of uranium from phosphoric acid some 30 years ago.

Regulation

I had the pleasure during much of last year to work closely with the Australian regulatory authorities on a series of issues involving a uranium in situ leaching project. A few minor spills of uranium-bearing solutions with a few to

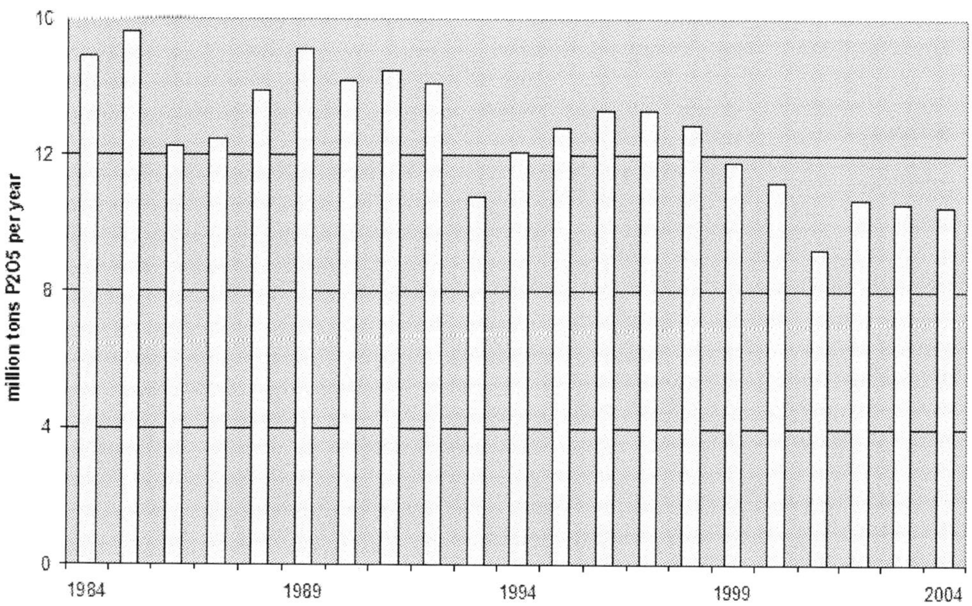

Figure 10 U.S. P_2O_5 production

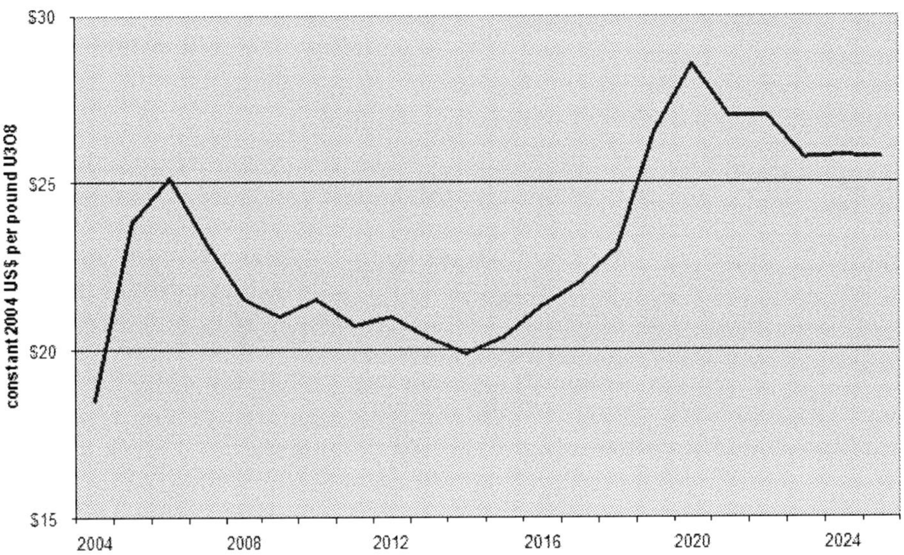

Figure 11 Uranium price forecast

perhaps as much as 20 ppm uranium were greeted with newspaper headlines and Senate hearings. This is in spite of the fact that the Australians too use phosphate fertilizer containing as much as 100 ppm uranium. I find it very interesting that, to date, uranium in fertilizer seems to be totally overlooked from a regulatory standpoint. Will there come a time when the USEPA mandates uranium removal from fertilizer?

Uranium Price Outlook

International Nuclear prepares an annual price forecast which assesses the balance between supply and demand and which relates future prices to the production cost of the marginal producer. The results of our latest effort are set forth in Figure 11.

In comparison with the recent past, this forecast indicates significant strength in uranium prices; a strength that will certainly translate into additional conventional uranium production facilities. We can also opine that these prices coupled with substantial future phosphate fertilizer production, new technology and increasing regulatory concern are likely to lead at least some new production of uranium from phosphoric acid.

REFERENCES

1. The fundamental OPPA process consists of reduction of uranium contained in phosphoric acid by the addition of iron; solvent extraction utilizing octyl pyrophosphoric acid diluted with kerosene; clarification of the loaded solvent; stripping, precipitation, purification and drying of the uranium.
2. The two-stage DEPA-TOPO process uses a solvent 1 mixture of di(2-ethylhexy) phosphoric acid and trioctyl phosphine oxide to extract, concentrate and purify uranium contained in phosphoric acid.
3. The OPAP process uses the solvent octylphenyl acid phosphate to concentrate contained uranium in the first stage of the process. DEPA-TOPO solvent is used in the second stage for extraction and purification.
4. *Uraniferous Phosphate Resources and Technology and Economics of Uranium Recovery from Phosphate Resources, United States and Free World*; Earth Sciences, Inc.; June, 1979.

SECTION 3

Technology Innovations: Reagents

CHAPTER 8

Hidroveg Chemicals and Their Efficiency in American Phosphate Rock Flotation

S.C.A. França,* J.W. Schlanz,† and S.M.R. de Lima*

ABSTRACT

Research has been conducted at the North Carolina State University Minerals Research Laboratory, in conjunction with Brazil's Center for Mineral Technology, to test a recently developed series of collectors for phosphate rock flotation on United States ore. The Hidrocol collectors (produced by Hidroveg Inc., a Brazilian company) consist of a biodegradable fatty acid produced by hydrogenation of vegetable oils. Because the collectors have several inherent advantages, among which include enhanced biodegradability, low residual oil content and the potential to improve flotation performance, the collectors may benefit US producers, who face challenges related to ore grades and environmental concerns, as well as production costs and water usage. The Hidrocol series could lead to lower reagent consumption and operating costs, flotation circuit streamlining and improved concentrate grades and recoveries while lessening the environmental impact of current practices.

INTRODUCTION

Hidrocol is a biodegradable product manufactured by fatty acid hydrogenation in hydrogen cells. The chemical principle is based on improving the molecular structure by increasing the hydrogen chain lengths. This product is produced with combinations of fatty acids obtained from vegetable oils and, with specific catalysts, subjected to the hydrogenation. Research using the Hidrocol collectors focused on increasing the efficiency of phosphate rock flotation, which was not satisfactory using traditional fatty acid based chemicals.

Current ores have different characteristics from the first phosphate rocks studied for the development of flotation collectors. After several years of exploration at a particular mine, the raw ore became lower grade (P_2O_5 content) and presented liberation challenges, jeopardizing the productivity and quality of the phosphate concentrate.

Today, Hidroveg is the supplier to Fosfertil/Ultrafertil, the largest industrial group in Brazil producing phosphate rock, phosphoric acid and fertilizers. This group produces 2,500,000 mt/year of phosphate rock concentrate, accounting for 43% of Brazilian production (Souza 2004, Jasisnky 2004) and requires 6,200 mt/year of Hidrocol (2.48 kg/mt). It offers several advantages compared to other chemicals used in phosphate rock flotation, such as enhanced biodegradability and low residual oil content,

*CETEM—Center for Mineral Technology, Ilha da Cidade University, Rio de Janeiro, Brazil.
†Minerals Research Laboratory/North Carolina State University, Asheville, North Carolina.

both of which lessen environmental impacts while enhancing flotation performance.

OBJECTIVE

Because bench-scale flotation tests have traditionally been used for evaluating flotation reagents (Agar and Barrett 1983), a test series was designed to produce comparative data between the reagents in question, allowing Hidroveg Chemicals to study the efficiency of Hidrocol in US phosphate rock flotation. Research objectives were defined as producing equivalent or better grades and recoveries while supplying cleaner chemicals that can increase production and product quality and lessen the environmental impacts.

Major differences exist between Brazilian and American phosphate rock, specifically those related to geological origin (Harben 1996). Brazilian phosphate rock was formed by volcanic action while American deposits are sedimentary, resulting in different grain size distributions and P_2O_5 content (Bartels and Gurr 1994). These typological differences may require a chemical restructuring of the Hidrocol family of reagents to achieve optimum surface interaction given the rock characteristics.

METHODS AND MATERIAL STUDIED

Three different phosphate rock samples were tested, originating in North Carolina (sample 1) and Florida (samples 2, 3-C and 3-F) (França et. al. 2002).

Sampling

The samples consist of four different flotation feeds. They were dried and split into 500 g representative samples for the bench testing, all of which were conducted in a Denver laboratory flotation cell.

Reagents

Flotation tests were performed with Hidroveg Chemicals' fatty acid collectors Hidrocol A, B and C, with degree of hydrogenation being the distinguishing difference. This characteristic of the collector customizes its effectiveness based on rock typology. The concentration used was 5% (w/w) for all three collectors.

Cornstarch and sodium silicate were employed as depressants in concentrations of 2 and 5% (w/w), respectively. Cornstarch was selected as the primary for the test program as it has shown good depressing characteristics on Brazilian ores when used in conjunction with Hidrocol collectors.

Bench Flotation Tests

The bench flotation tests were run in a 4.5 L Denver lab cell. General conditions in the flotation tests are presented in Table 1. These tests were run in January-September, 2002, at the North Carolina State University Minerals Research Laboratory (MRL/NCSU) in Asheville, N.C., with follow-up tests conducted at CETEM's facilities in Rio de Janeiro, Brazil. All tests consisted of single stage flotation with rougher and cleaner steps (no second stage amine flotation).

RESULTS AND ANALYSIS

Test results and procedures on the different samples are presented in the following tables and figures. Batch flotation tests were completed using the three Hidrocol collectors (A, B, C) with cornstarch or sodium silicate as depressants. Results are presented in terms of grade and recovery of P_2O_5, SiO_2, and CaO.

Sample 1

Sample 1 data is presented in Table 2 with 14 batch flotation tests completed using the three Hidrocol collectors (A, B, C) with cornstarch or sodium silicate as depressants. Results are presented in the following figures in terms of grade and recovery of P_2O_5, SiO_2, and CaO.

Figure 1 shows poor flotation grades were produced with collector Hidrocol B (Tests 6 and 7), indicating poor selectivity between SiO_2 and P_2O_5. Tests employing collectors A and C showed significant improvements in flotation performance in both grade and recovery. Figures 1 and 2 show tests 9, 10, and 11 produced the best results (Hidrocol C), with grades over 30% P_2O_5 and plus 90% recoveries.

Table 1 Experimental conditions in the batch flotation tests

Conditioning	
Percent Solids – 70%	
Retention time (pulp/depressant) – 4 min	pH = 9.5
Retention time (pulp/depressant/collector) – 1 min	pH = 9.5 - 10
Flotation	
Percent Solids – 20 - 25%	
Rougher - pH = 9.5 – 10	
Cleaner – pH = 8.0 – 8.5	

Table 2 Batch flotation tests—sample 1

Experiment numbers	Collector (mL)			Depressant (mL)		Comments
	A	B	C	Corn starch	Sodium silicate	
1	11	-	-	25	-	-
2	11	-	-	-	25	-
3	11	-	-	25	-	add attrition scrub
4	11	-	-	-	25	add attrition scrub
5	6	-	-	25	-	decrease Hid. A
6	-	11	-	25	-	-
7	-	11	-	-	25	-
8	-	-	11	25	-	-
9	-	-	11	-	25	-
10	-	-	11	-	-	no depressant
11	-	-	8	-	-	no depressant
12	12.5	-	-	-	25	increase Hid. A
13	14	-	-	-	25	increase Hid. A
14	11	-	-	-	30	increase sodium silicate

Cornstarch was not effective on US ore as a silica depressant because of the typology of this phosphate rock. The use of the cornstarch as a silica depressant was based on Brazilian phosphate rock flotation, where it is used with Hidrocol and exhibits high flotation efficiency in terms of P_2O_5 grade and recovery with low SiO_2.

Overall, the best results were obtained using Hidrocol and sodium silicate, producing enhanced selectivity between SiO_2 and P_2O_5. Sodium silicate is often used as a silica depressant by phosphate companies in the United States.

Data presented in Figures 3 and 4 confirms the efficiency of Hidrocol A and C in the phosphate rock flotation and the ability to attain satisfactory P_2O_5 grades and recoveries in the float products and tails. The best results were obtained with Hidrocol C.

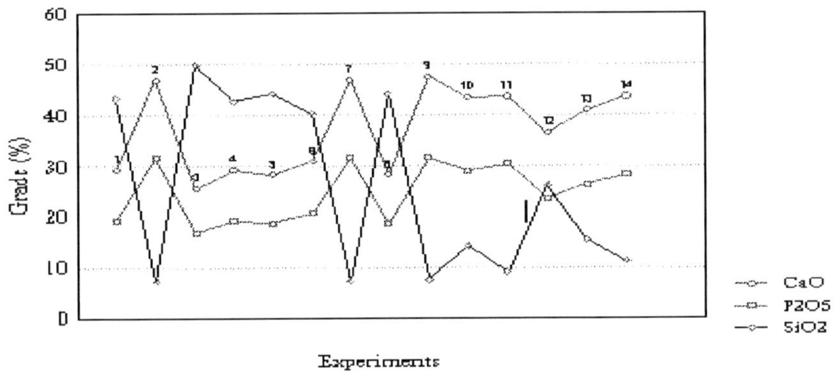

Figure 1 Grades of phosphate rock concentrates (cleaner float products) from the flotation of sample 1

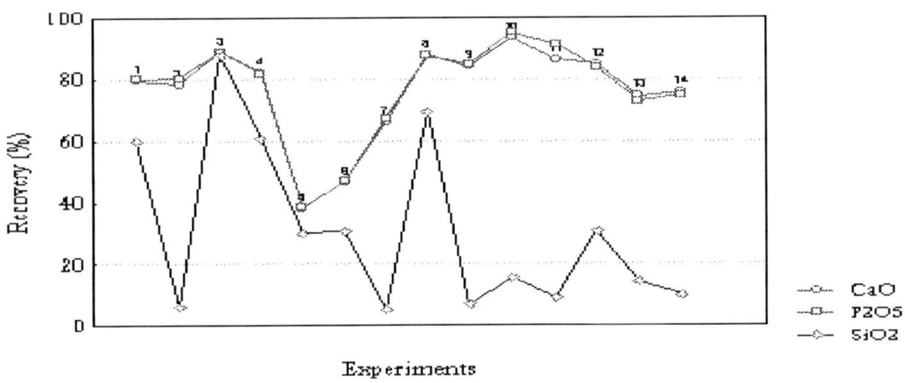

Figure 2 Recovery of phosphate rock concentrates (cleaner float products) from the flotation of sample 1

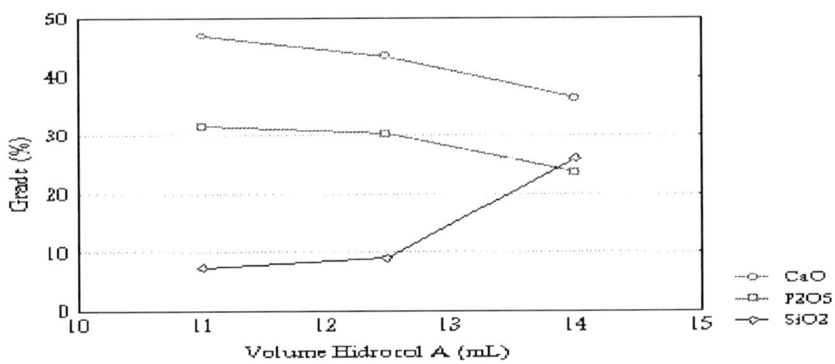

Figure 3 Variation between volume of Hidrocol A and grade for the main compounds from batch flotation of sample 1

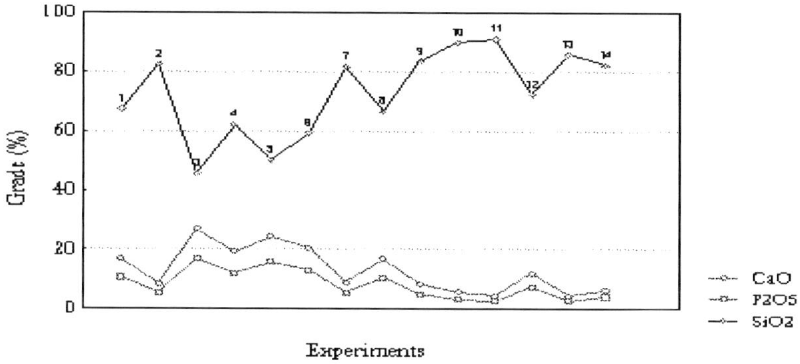

Figure 4 Grade of rougher tails from sample 1

Table 3 Batch flotation results from sample 2

Experiment numbers	Collector (mL)			Depressant (mL)		Comments
	A	B	C	Corn starch	Sodium silicate	
1	11	-	-	25	-	-
2	11	-	-	-	25	-
3	-	11	-	25	-	-
4	-	11	-	-	25	-
5	-	-	11	25	-	-
6	-	-	11	-	25	-
7	-	-	14	-	25	increase Hid. C
8	12.5	-	-	-	25	increase Hid. A
9	14	-	-	-	25	increase Hid. A
10	12.5	-	-	-	25	increase Hid. A
11	-	-	11	-	25	
12	11	-	-	-	25	

Sample 2

Process parameters on the 12 batch flotation tests conducted on sample 2 are summarized in Table 3. These tests utilized Hidrocol collectors A, B, and C with cornstarch or sodium silicate as depressants (results in Figures 5–7).

Hidrocol B again produced poor flotation results on sample 2, with Hidrocol A and C showing superior performance. Figures 5 and 6 show the best flotation selectivity was obtained using the sodium silicate depressant, with float products assaying 32.5% P_2O_5 or better with corresponding recoveries as high as 87%.

Although concentrate grades were very consistent, recoveries show some variation. However, improved recoveries can be anticipated with the Hidrocol + sodium silicate by small adjustments in the collector dosage.

Sample 3

In this case flotation takes place in two different circuits: fine and coarse flotation. The plant feed ratio is 6:1 between fine (91.6% -35 mesh) and coarse flotation feed (96.1% -16 mesh). Two flotation circuits are required due to challenges floating coarse and fine particles together because of competitiveness for the reagents.

Data collected on sample 3-C (coarse) and sample 3-F (fine) are presented in Table 4. Eighteen tests were completed using Hidrocol

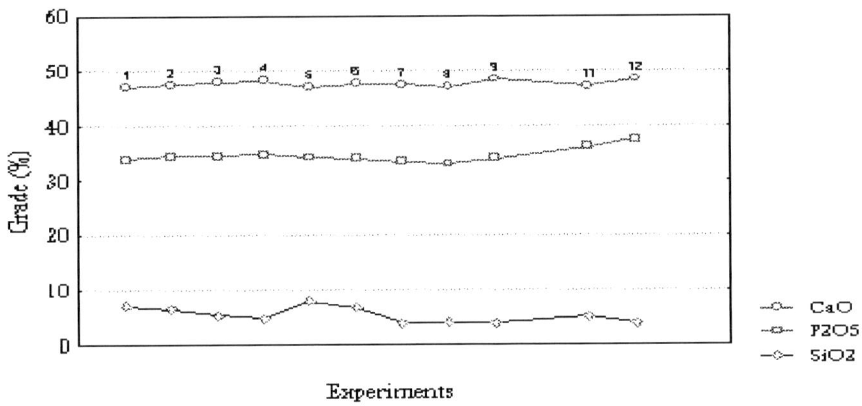

Figure 5 Grade of cleaner float product to the phosphate rock flotation of sample 2

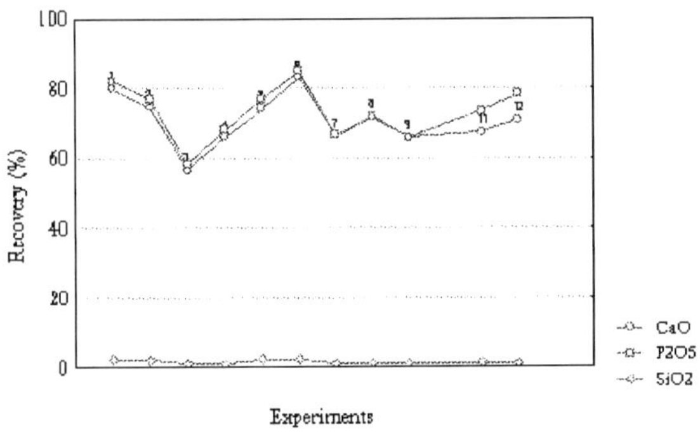

Figure 6 Recovery of cleaner float product from sample 2

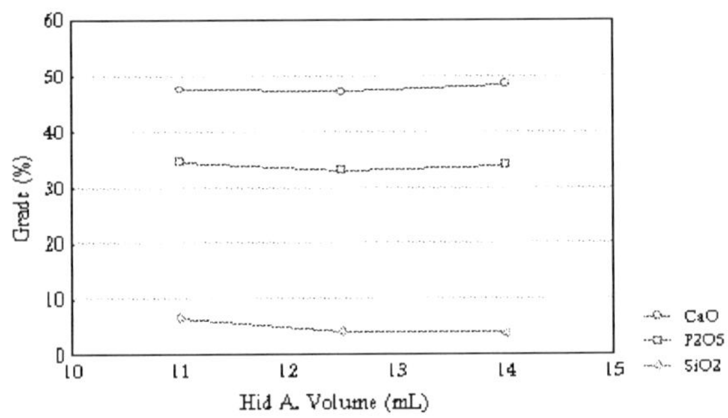

Figure 7 Concentrate grade versus dosage of Hidrocol A—sample 2

collectors A, B, and C with cornstarch or sodium silicate as depressants.

Figures 8–11 present the results in terms of grade and recovery of P_2O_5, SiO_2, and CaO. Data are presented separately for the two flotation feeds.

Coarse Feed Flotation

The collector's performance will be expressed in CaO, P_2O_5, and SiO_2 grade and recovery in the phosphate rock flotation from sample 3-C. Results from tests using Hidrocol A and C are shown in Figures 8 and 9.

Figures 8 and 9 show the measured values to judge the flotation efficiency on P_2O_5 grade and recovery. Figure 8 shows excellent concentrate grades ranging from 33 to 35% P_2O_5, while Figure 9 indicates variable recoveries. Restructuring of the Hidrocol may be required to improve the recoveries.

In the Brazilian process, this problem was solved by blending both feeds (coarse and fine) and is referred to as *carrier flotation*. Similar tests can be run with samples 3-C and 3-F if the concentrate's final size distribution is not critical.

Fine Flotation Feed

Figures 10 and 11 indicate reasonable performance of the Hidrocol A and C on the fine flotation feed. Consistent grades were produced ranging from 33 to 35% P_2O_5, and recoveries

Table 4 Batch flotation tests to phosphate rock from sample 3

Experiment numbers	Collector (mL)			Depressant (mL)		Comments
	A	B	C	Corn starch	Sodium silicate	
Sample 3-C (coarse feed flotation)						
1	11	-	-	25	-	-
2	11	-	-	-	25	-
3	-	-	11	25	-	-
4	-	-	11	-	25	-
5	-	-	12.5	-	25	increase Hid. C
6	-	-	14.5	-	25	increase Hid. C
7	14	-	-	-	25	-
8	-	-	11	-	25	
9	11	-	-	-	25	
10	-	-	14	-	25	
Sample 3-F (fine feed flotation)						
1	11	-	-	25	-	-
2	11	-	-	-	25	-
3	-	-	11	25	-	-
4	-	-	11	-	25	-
5	-	-	10	-	25	decrease Hid. C
6	-	-	9	-	25	decrease Hid. C
7	-	-	11	-	25	
8	11	-	-	-	25	

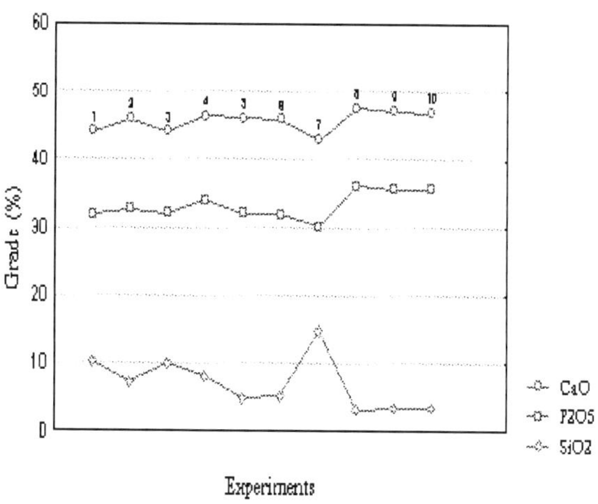

Figure 8 Grade of cleaner float product from sample 3-C

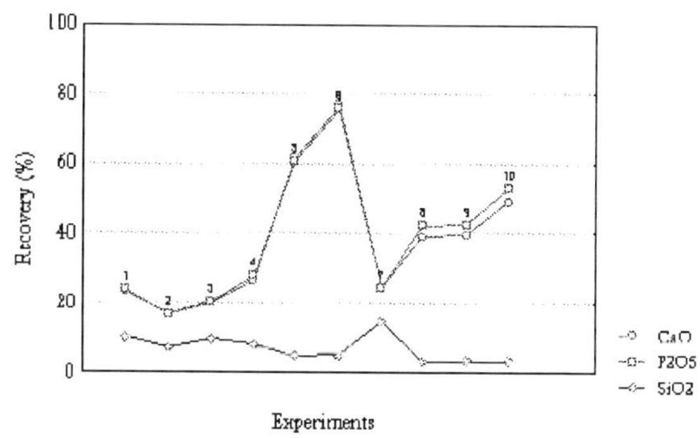

Figure 9 Recovery of cleaner float product from sample 3-C

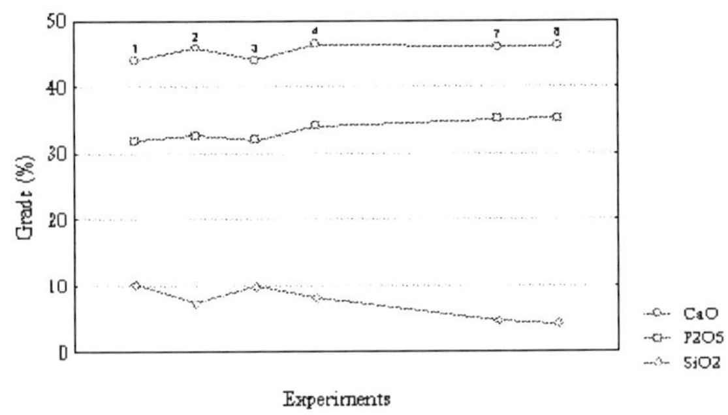

Figure 10 Grade of cleaner float products in the sample 3-F

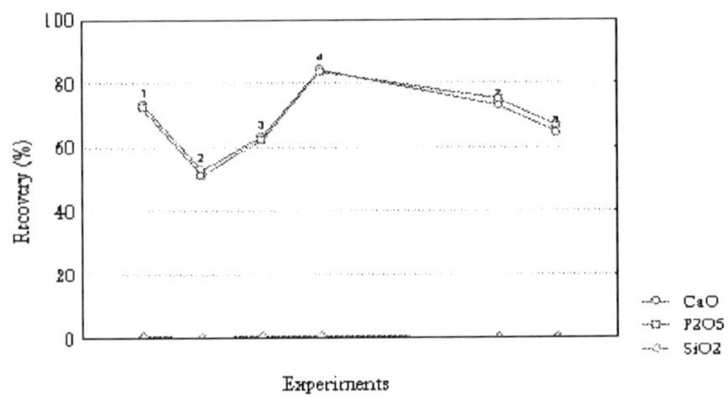

Figure 11 Recovery of cleaner float products to sample 3-F

improved significantly over the coarse feed, increasing to 70 to 85%.

Based on the results of these preliminary tests, data for reagent consumption and production costs for the different feeds can be estimated as shown in Table 5, for single stage flotation/rougher and cleaner steps.

DISCUSSION AND CONCLUSIONS

Preliminary results obtained from US phosphate rock batch flotation tests Hidroveg collectors were satisfactory. Hidrocol A and C produced good P_2O_5 grades and recoveries. However, the process can be optimized and improved results anticipated through structural adjustments of the reagents (hydrogenation) to create a better match with the typology of each phosphate rock. Hidrocol B proved inefficient in the flotation of all studied ores, exhibiting poor selectivity between the SiO_2 and P_2O_5. Two different silica depressants were used in the flotation tests, and results allow the conclusion that sodium silicate performed better than cornstarch, probably due to the rock typology.

Although Hidrocol A and C have different degrees of hydrogenation in the chemical structure, both presented very good and similar results in the flotation tests. However, results obtained in the flotation tests with Hidrocol C show better grades and recoveries. This trend was observed in all three rock types.

It is important to emphasize that all results documented in this paper were obtained using a single flotation stage, with rougher and cleaner steps (no amine/silica float). In comparison to the industrial plants, the use of Hidrocol reagents shows a strong potential to improve the flotation circuits in terms of reagent additions and consumptions, plant equipment, operating costs and environmental concerns.

RECOMMENDATIONS

Although the bench testing of the Hidrocol reagents was limited, the data generated indicates a true potential for improvement in the phosphate flotation circuits. Therefore, the following plans are strongly recommended to conclusively determine the full potential of the Hidrocol family of collectors on American phosphate rock flotation:

1. Follow-up bench testing using current plant reagents to establish baseline parameters for direct comparison to the Hidrocol performance
2. Follow-up bench testing to determine benefits of attrition scrubbing and its effects on reagent consumption, product quality, and recovery
3. A preliminary economic evaluation once baselines are established to justify additional testing
4. If economics warrant, pilot plant testing should be conducted to optimize all process

Table 5 U.S. phosphate rock beneficiation process reagents consumption and production costs

Feed	Mass recovery (%)	Grade (%)			Recovery (%)	Reagents consumption (lb/ton feed)	Production costs (US$/ton feed)	Reagents consumption (lb/ton product)	Production costs (US$/ton product)
		CaO	P_2O_5	SiO_2	P_2O_5				
Sample 1	50	43.5	30.7	9.0	91.2	2.2	0.80	4.4	1.58
Sample 2	24	47.7	34.1	6.9	84.0	2.2	0.80	9.2	3.30
Sample 3-C	22	46.0	31.9	4.9	72.1	2.2	0.80	10.0	3.60
Sample 3-F	8	46.5	34.0	6.1	66.0	2.2	0.80	27.5	9.90

parameters and supply data for projected plant performance and scale-up

5. Final economic evaluation to determine if plant changes are warranted

ACKNOWLEDGMENTS

Hidroveg Chemicals Inc. and CETEM—Center for Mineral Technology would like to thank the Minerals Research Laboratory Team, with a special regard to Dr. Bruce Tippin (ex-director), John Schlanz (interim director/senior engineer) and Robert Baldwin (mineral technician). We thank you for support in the preliminary flotation tests and technical discussions.

Appreciation is also extended to Robert Baldwin for his contribution to this paper.

REFERENCES

Agar, G.E. and Barrett, J.J., 1983. The use of flotation rate data to evaluate reagents. *CIM Bulletin*, 76, 851: 157–162.

Bartels, J.J. and Gurr, T.M., 1994. Phosphate Rock. In: *Industrial Minerals and Rocks*. Breese, R.O.Y., and Carr, D.D. (Editors). SME, Littleton, Colorado, pp. 751–764.

Harben, P.N., 1996. Phosphate Rock. In: *Industrial Minerals—A Global Geology*, pp. 289–303.

França, S.C.A., Schlanz, J.W., and Lima, S.M.R., 2002. Hidroveg chemicals and their efficiency on the flotation of the American phosphate rock flotation. Technical Report–RT 2002-065-01, CETEM, Rio de Janeiro, Brasil.

Jasisnky, S.M., 2004. Phosphate Rock. In: *US Geological Survey—Mineral Commodity Summaries*, January, pp. 122–123.

Souza, A.E., 2004. Fosfato. In: Nery, M.A.C. Sumário Mineral (electronic version). DNPM Departamento Nacional da Produção Mineral, Brasília (Brasil).

Technical reports, flowsheets and data from the U.S. Phosphate companies.

Technical reports from Minerals Research Laboratory/North Carolina State University.

CHAPTER 9

Pilot-Plant Evaluation of a Water Insoluble Hydroxamic Acid Collector for Single Stage Flotation of Florida Phosphate Rock

X. Wang* and J.D. Miller*

ABSTRACT

Recent laboratory results have demonstrated that selective flotation of phosphate rock can be achieved in a single stage with improved separation efficiency when compared to the standard double float process. The new reagent chemistry is based on the use of a water insoluble alcoholic solution of alkyl hydroxamic acid as collector following a conditioning procedure similar to that currently used by the Florida phosphate industry. The effectiveness of the new collector chemistry has been demonstrated by the results from single stage bench scale flotation experiments for feed material from plants in Florida, North Carolina, and Utah. Now pilot plant testing at a feed rate of 700–900 lb/hr has been completed in order to evaluate the hydroxamic acid collector on a continuous basis. Results from the pilot-plant campaign confirm the findings from laboratory experiments. Single stage flotation recovery of 95% was reached with a concentrate grade of 32% P_2O_5 for coarse feed material (16×35 mesh).

Initial economic analysis of the new flotation chemistry with AERO 6493 shows a strong dependence on the reagent price. If the cost of AERO 6493 could be reduced to $1.80/lb, the single stage flotation of coarse phosphate with the alkyl hydroxamic acid/alcohol collector would be cost competitive with the traditional double float process using fatty acid and fuel oil.

INTRODUCTION

Recent research results (Miller et al. 2000; Miller et al. 2002a; Miller et al. 2002b) demonstrate that water-insoluble alcoholic solutions of alkyl hydroxamic acids can be used as selective collectors for phosphate flotation that can increase separation efficiency especially for coarse feed material; in this way the standard double float process can be reduced to a single step. The experimental results indicate that high separation efficiencies can be achieved in single stage flotation with these new collectors (Miller et al. 2000). The effectiveness of the new collector chemistry was demonstrated by single stage bench scale flotation experiments of feed material from plants in Central Florida, North Carolina, and Utah (Miller et al. 2002a). Particular effectiveness was achieved for coarse feed material from these phosphate operations. In some cases, the single stage concentrate grade reached

*Department of Metallurgical Engineering, University of Utah, Salt Lake City, Utah.

31% P_2O_5 with 95% recovery from low grade (5% P_2O_5) coarse feed (16×35 mesh). Such separation efficiency could not be achieved using the traditional fatty acid/fuel oil collector. High solids conditioning (70–75% solids by weight) was found to be necessary to achieve the desired separation efficiency. Significantly, the flotation response is essentially independent of pH for the hydroxamic acid collector, and the natural pH of the system was found to be satisfactory for flotation in most cases.

Subsequently, the new collector has now been evaluated for use by the Florida phosphate industry in pilot-plant studies and the experimental results, which confirm the findings from laboratory experiments, show an improved separation efficiency especially for coarse phosphate flotation feed from the Cargill South Fort Meade (SFM) plant.

PILOT-PLANT TESTING AT JACOBS ENGINEERING

Pilot-plant testing was carried out at the Jacobs Engineering pilot plant facility in Lakeland, Florida. The flowsheet for the pilot flotation system is shown in Figure 1. The feed material was loaded by a front end loader into a screw feeder and pulped by tap water. The slurry was pumped into a screw classifier to dewater to 70–75% solids. Then the slurry was discharged into a vertical conditioning tank with impeller and the new collector (AERO 6493) was added. The conditioning time was about 2–3 minutes. After conditioning, the slurry was diluted to 20% solids and fed to the flotation circuit which consisted of two flotation cells (DECO Flotation machine, total active volume 5 ft^3). Flotation products and feed samples were taken every 20 minutes. Three samples were combined and the composite taken for analysis. Shown in Figure 2 are photographs of operations at the pilot plant.

The flotation variables considered were the collector dosage and percent solids during conditioning. The pilot-plant testing used feed material from the Cargill SFM plant. Two flotation feeds, fine flotation feed and coarse flotation feed, were tested. The particle size distributions for fine feed and coarse feed are shown in Figures 3 and 4 together with the particle size distributions for concentrate and tailing products. The feed rate was about 700–900 lb/hr.

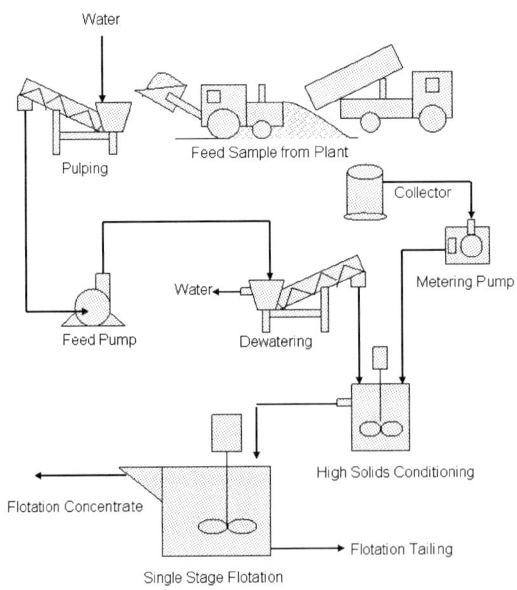

Figure 1 Flowsheet for pilot-plant testing

Loading, pulping, and pumping of feed

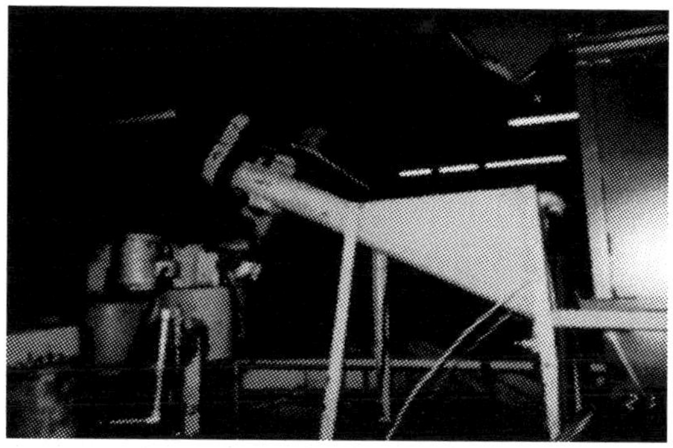

Dewatering and high solids conditioning

Single stage flotation

Figure 2 Photographs of pilot-plant operation

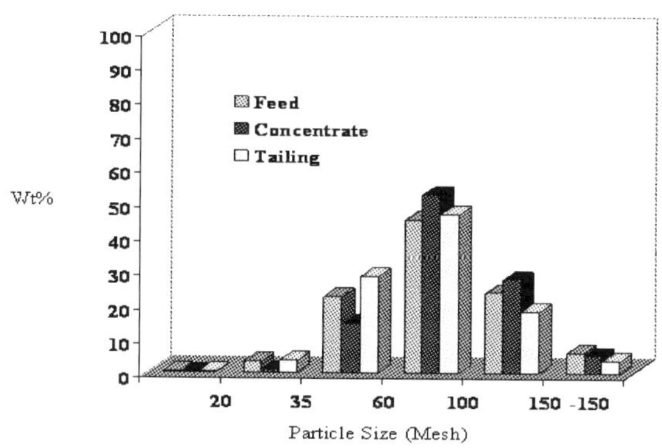

Figure 3 Particle size analysis for fine flotation feed, concentrate, and tailing from pilot testing at a feed rate of 900 lb/hr and conditioning at 74% solids

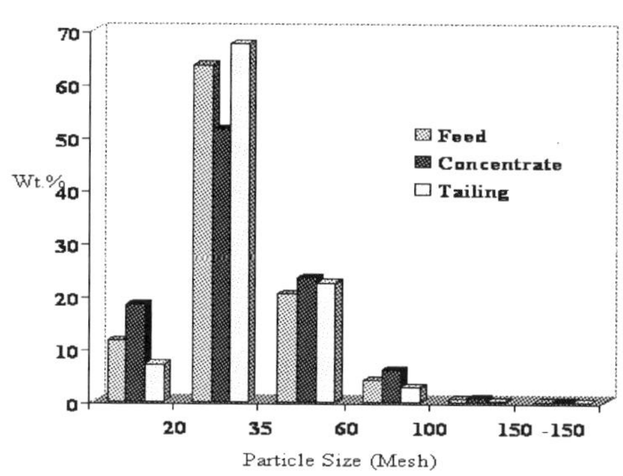

Figure 4 Particle size analysis for coarse flotation feed, concentrate, and tailing from pilot testing at a feed rate of 700 lb/hr and conditioning at 71% solids

RESULTS AND DISCUSSION

Flotation With Cargill SFM Fine Feed

The new collector (AERO 6493) was tested with Cargill SFM fine flotation feed. The results in Figure 5 are from collector addition experiments. Conditioning was at about 74% solids and a natural pH of 7. The feed rate for the pilot plant testing was about 900 lb/hr for fine feed. The results show that a recovery of 86% with a concentrate grade of 30% P_2O_5 was achieved in single stage with an addition of 850 g/t of the new collector.

Flotation with oily, water-insoluble collectors is frequently governed by selective wetting/ spreading phenomena and the formation of aeroflocs, which are complex aggregates of air, water, the liquid collector phase, and mineral particles. In this case, high solids conditioning is preferred to achieve the creation of stable aeroflocs of sufficient buoyancy for flotation. Shown in Figure 6 is the X-ray CT spectroscopy image of a cross-section of aeroflocs formed with francolite particles, air bubbles, and the image of a cross-section of aeroflocs formed with francolite particles, air bubbles, and the new collector. It can be observed that the rate of agglomeration and size of aeroflocs increase as the amount of small air bubbles increases during the flotation experiment. The hydrophobic particles are triggered by the small air bubbles to form large aeroflocs. It is evident that the separation efficiency does not vary significantly for a collector addition.

Flotation With Cargill SFM Coarse Feed

Pilot-plant flotation results for the coarse feed at different collector additions are shown in Figure 7. The conditioning was done at 71% solids and a natural pH of 7. It can be seen from the results presented in Figure 7 that with 1,200 g/t collector dosage a concentrate with a grade of 32% P_2O_5 AT 95% recovery can be achieved in single stage pilot scale flotation.

During the coarse feed flotation, the feed rate was reduced from 900 lb/hr to 700 lb/hr because the conditioning solids were difficult to control and the flotation time appeared to be insufficient. The results from pilot testing were consistent with the results from bench scale flotation experiments. The best results from the pilot plant campaign are summarized in Table 1.

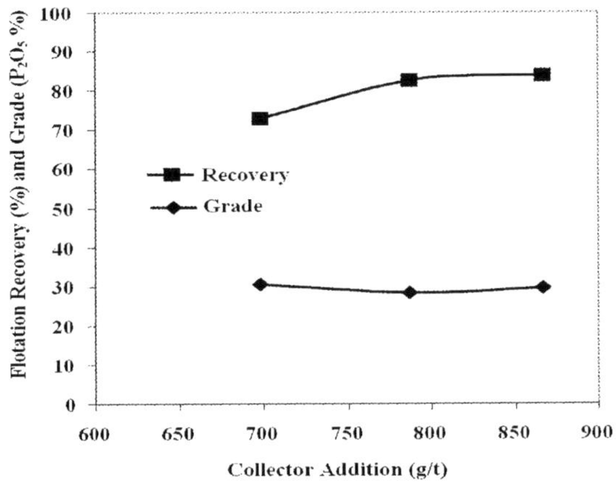

Figure 5 Effect of collector addition on single stage pilot-plant flotation of Cargill fine feed at a feed rate of 900 lb/hr and conditioning at 74% solids

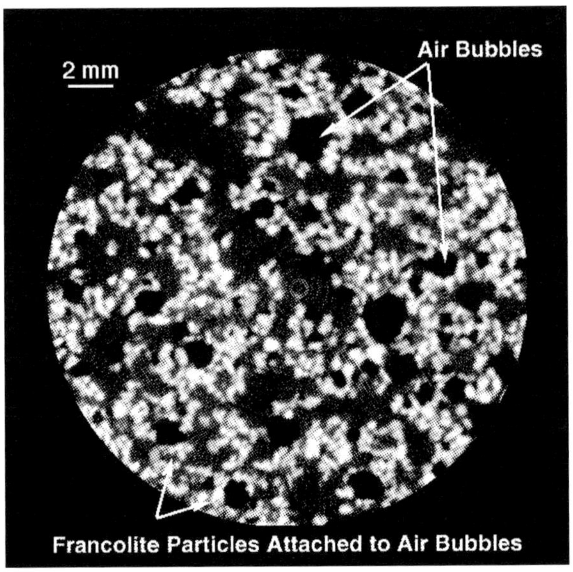

Figure 6 X-ray micro CT image of aeroflocs formed by francolite particles, air bubbles during flotation with the hydroxamic acid collector

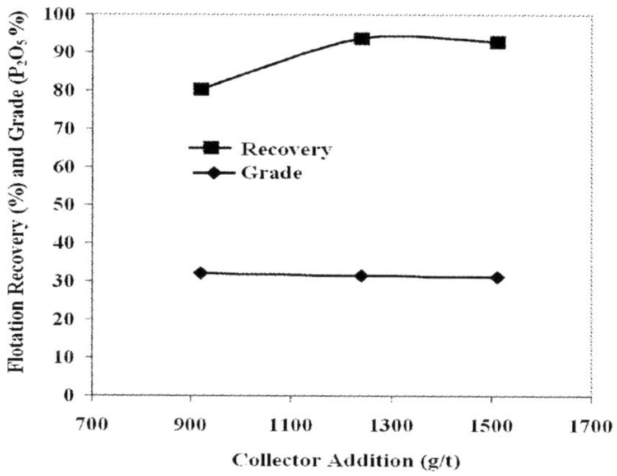

Figure 7 Effect of collector addition on single stage pilot-plant flotation of Cargill coarse feed at a feed rate of 700 lb/hr and conditioning at 71% solids

Flotation Water Quality

Tap water was used in these flotation experiments. The feed was conditioned with collector, AERO 6493, at a dosage of 800 g/t at 75% solids for 3 minutes. Then the slurry was transferred into the flotation cell and diluted to 20% solids with tap water. After flotation the water from filtration of the tailing and concentrate products was collected and analyzed. The Whatman #4 coarse fast filter papers were used for filtration. The flotation results and water analyses are presented in Table 2. These data are used as a baseline. Notice that the TOC content is 6.50 mg/L, of which 35% or 2.28 mg/L would be contributed from the collector if none of the collector adsorbed/attached to the particles.

Figure 8 shows the locked cycle flotation procedure. In this test the tap water was used for

Table 1 Results from single stage pilot-plant flotation of phosphate feed from the Cargill SFM plant using the new hydroxamic acid collector (AERO 6493)

Feed	Condition	Product	Wt %	P_2O_5 %	Recovery %
	Dosage 850 g/t	Concentrate	27.96	29.6	85.94
Fine Feed	Conditioning Solids	Tailing	72.04	1.88	14.06
	74 %	Feed	100.00	9.63	100.00
	Dosage 1200 g/t	Concentrate	38.80	31.55	94.88
Coarse Feed	Conditioning Solids	Tailing	61.20	1.08	5.12
	71%	Feed	100.00	12.9	100.00

Table 2 Analytical results for open cycle flotation of Cargill feed (12×80 mesh) and discharge water quality without water recycle using AERO 6493 as collector; flotation with tap water at pH 7.41

Grade P_2O_5%	Recovery %	pH	Hardness(1) mg/L	TSS(2) mg/L	TDS(3) mg/L	Turbidity(4) NTU
30.98	93.36	8.07	650	140	820	150
TOC(5) mg/L	**COD(6) mg/L**	**P(7) mg/L**	**Ca(8) mg/L**	**Mg(9) mg/L**	**F(10) mg/L**	**Dissolved Oxygen(11) mg/L**
6.5	240	15	170	54	9.1	10.82

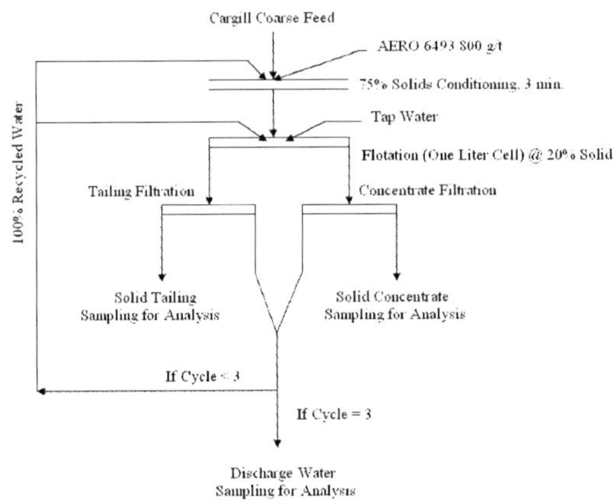

Figure 8 Locked cycle single stage flotation procedure with recycled water

Table 3 Flotation results for locked cycle experiment for Cargill feed (12×80 mesh) with 100% water recycle using AERO 6493 as collector at pH 7.4

Cycle	Products	Wt. %	Grade P_2O_5%	Recovery %
	Concentrate	42.57	32.41	89.36
1	Tailing	57.43	2.86	10.64
	Feed	100.00	15.44	100.00
	Concentrate	45.94	30.07	95.12
2	Tailing	54.06	1.31	4.88
	Feed	100.00	14.52	100.00
	Concentrate	46.13	30.07	95.30
3	Tailing	53.87	1.27	4.70
	Feed	100.00	14.56	100.00

Table 4 Water analyses from cycle 3 locked cycle flotation experiments with Cargill feed (12×80 mesh) water recycled using AERO 6493 as collector at pH 7.4

Grade P_2O_5%	Recovery %	pH	Hardness(1) mg/L	TSS(2) mg/L	TDS(3) mg/L	Turbidity(4) NTU
30.07	95.30	7.97	590	110	820	110
TOC(5) mg/L	COD(6) mg/L	P(7) mg/L	Ca(8) mg/L	Mg(9) mg/L	F(10) mg/L	Dissolved Oxygen(11) mg/L
22	440	13	150	52	16	10.81

the first cycle of flotation. In the first cycle of flotation, a feed sample of 250 grams was conditioned with the new collector, AERO 6493, at a dosage of 800 g/t and 75% solids for 3 minutes. Then the slurry was transferred to the flotation cell and diluted to 20% solids with tap water. After the first cycle of flotation, water from the filtration of tailing and concentrate was collected for the second cycle of flotation. In the second cycle of flotation a second 250 gram feed sample was conditioned and diluted with the recycled water from the first cycle flotation at the same reagent schedule, conditioning time, and flotation conditions. After the second cycle of flotation the water from filtration of tailing and concentrate was collected for the third cycle of flotation. After completing the same procedure as used for the second cycle, the filtration water from the third cycle was collected and analyzed. The Whatman #4 coarse fast filter papers were used for filtration.

Flotation results and the water analyses (essentially the third cycle water) are listed in Tables 3 and 4. The results indicate that the recovery and grade of the phosphate concentrate have stabilized in the third cycle. It also can be seen that there was no significant change in water quality (most analytical results) when the open cycle results are compared to the closed cycle. Notice that the total organic carbon (TOC) in closed cycle increased from 6.50 mg/L to 22.0 mg/L. The chemical oxygen demand (COD) also increased from 240 to 440 mg/L. This change may be due to a build up of the reagent in the closed cycle. Usually in the closed cycle the reagent dosage would be reduced.

Shown in Table 5 is the average analyses of 11 samples of plant water taken from flotation operations in Central Florida, as reported by Jacobs Engineering Group, Inc. (1998) in a survey of phosphate beneficiation plants including the IMC plants and the Cargill plants. Since the individual plant water data is not available, a complete comparison cannot be made. In general, all the analyses listed in Table 2 and Table 5, including total suspended solids (TSS), hardness (as Ca_2CO_3), pH, and P (total P), have higher values for the discharge water from AERO 6493 bench flotation experiments than those for plant water samples, especially for TSS and hardness. Apparently, the difference in water quality is significant. However, the final conclusion on the environmental impact of the new flotation chemistry using AERO 6493 still cannot be made based only on these results since many incomparable variables are involved, such as the time for water sampling after discharge, filtration conditions for concentrate, etc. The analytical results from Table 2 and Table 5 suggest that the hydroxamic acid collector, AERO 6439, addition could be reduced in a locked cycle flotation which would be expected to reduce the TOC content in discharged water without having a negative influence on the separation efficiency.

Further environmental evaluation regarding water quality should be made in plant testing.

ECONOMIC CONSIDERATIONS

It has been documented in a recent workshop (Zhang, 1991) that beneficiation of phosphate is a significant portion of the overall cost of production and at the flotation process is the most costly of all the beneficiation steps including washing, flotation preparation, flotation, in-process storage, and hydraulic station operation. A matter of particular concern in the development of the hydroxamic acid collector

Table 5 Average analyses of 11 plant water samples from central Florida (Jacobs Engineering Group, Inc.)

TSS[1]	Hardness[2]	pH[3]	P[4]
15 ppm	129 ppm	7.6	2.7 ppm

for phosphate flotation is the cost of the new collector chemistry relative to current reagent costs for typical plant operations.

Sources of Data

Since AERO 6493 generally showed significant improvement in flotation efficiency for coarse feed, flotation results used for this preliminary economic evaluation were those from testing of the coarse feed from the Cargill South Fort Meade Plant, Central Florida. Specifically, data for the cost comparison of the new collector chemistry with the traditional collector chemistry were from two sources:

1. The single stage pilot plant flotation results obtained using the hydroxamic acid collector (AERO 6493).
2. The typical flotation results using the traditional double float process with fatty acid/fuel oil (FA/FO) and amine as collectors. Data for the traditional process were obtained from several different references (Wang 1999; Wiegel 1999; Zhang et al. 1997b).

Flotation results and reagent costs are presented in Table 6 for single stage flotation with AERO 6493 and Table 7 for double stage flotation (Crago process) with fatty acid/fuel oil (FA/FO) and amine chemistry.

Economic Analysis

This preliminary economic analysis was carried out considering separation efficiency and, in this regard, includes recovery, grade, reagent cosumption, and the reagent prices. The analysis is based on the economic evaluation model developed by Jacobs Engineering (El-Shall 2001) for phosphate mineral processing.

In this model, a scheme for penalizing lower grade rock has been developed and the economic performance measure is defined as:

$$M = C_pP - C_fF - FC_r$$

Where

M – The economic performance measure, representing dollars earned per year ($/year)

C_p - Sales value of product: C_p = Price of 66% BPL rock x $(B_L/66)^{1.5}$ (B_L = % BPL)

Price of 66% BPL rock (30% P_2O_5): $30/ton (Zhang et al. 1997a)

C_f - Sales value of feed: C_f = Price of 66% BPL rock x $(B_L/66)^{1.5}$

C_r - Total reagent cost: C_r ($/ton feed)

P - Product flowrate: $P = F(\alpha/100)(\varepsilon/100)(100/\beta)$ (ton/year)

F - Feed solid flowrate: F = 4,000,000 ton/year (assumed)

α - Feed P_2O_5 grade (see Tables 6, 7)

β - Concentrate P_2O_5 grade (see Tables 6, 7)

ε - Product recovery (see Tables 6, 7)

The preliminary results comparing the AERO 6493 collector with the traditional double float process using fatty acid/fuel oil (FA/FO) are presented in Table 8 using Jacobs Engineering's model for economic performance. The results indicate that when the price for AERO 6493 is less than $1.80/lb the economic performance for AERO 6493 should be competitive with the traditional double float process.

SUMMARY AND CONCLUSION

It has been found that excellent flotation selectivity is achieved when pure mineral samples are conditioned at high percent solids with water insoluble alcoholic solutions of alkyl hydroxamic acids as collectors. The results from contact angle and high-speed video studies indicate that the hydroxamic acid collector has different spreading characteristics at the surfaces of fluorapatite, dolomite, and quartz. It is evident that colcoposition is an important variable which influences the flotation response. Long branched chain alcohols increase the hydrophobicity and reduce reagent consumption. Understanding of the mechanism of collector attachment and spreading is still unclear and surface chemistry research should be continued to further understand the phenomena involved in the flotation of

Table 6 Pilot-plant flotation results and reagent schedule for single stage flotation with AERO 6493

Process	Feed Grade P_2O_5 % (α)	Concentrate Grade P_2O_5 % (β)	P_2O_5 Recovery % (ε)
Cargill South Fort Meade Fine Feed , Single Stage, pilot Scale Flotation at AERO 6493 Consumption of 1.87 lb/t feed	9.6	29.6	86
Cargill South Fort Meade Coarse Feed, Single Stage, pilot Scale Flotation at AERO 6493 Consumption of 2.65 lb/t feed	12.9	31.55	95

Table 7 Typical flotation results and reagent costs for coarse feed using traditional double float process (FA/FO and amine)

Process	Feed Grade P_2O_5 % (α)	Concentrate Grade P_2O_5 % (β)	P_2O_5 Recovery % (ε)
Typical Double Float Plant Scale Flotation with FA/FO	8.00	31.5	85
Reagents	Reagent (lb/t feed)	Price ($/lb)	Cost ($/t feed)
Na_2SiO_3	0.1-0.4	0.05	0.01~0.02
Na_2CO_3	0.8-1.4 (pH 9)	0.08	0.07~0.11
Fatty Acid	1.4 ~ 2.1	0.15-0.20	0.21 ~ 0.42
Fuel Oil	0.6 ~ 0.9	0.07	0.042 ~ 0.063
H_2SO_4	2 (pH 3.3)	0.03	0.06
Amine	1	0.25-0.30	0.25 - 0.30
Total			0.64~ 0.97

Table 8 Typical results from the economic performance analysis (4 million tons per year of capacity)

Process	Reagent Price $/lb	Reagent Cost $/t feed	Product Flowrate Ton/year	Economic Performance Million $/Year
Single Stage AERO 6493	1.6 ~ 1.8	1.93 ~ 2.17	1,445,324	7.55 ~ 8.51
Double Float FA/FO Amine	0.63 ~ 0.73	0.64 ~ 0.97	1,295,238	7.41 ~ 8.73

phosphate minerals with insoluble alcoholic solutions of hydroxamic acids.

Based on these bench scale flotation results with a water insoluble alcoholic solution of hydroxamic acid (AERO 6493), it seems that a single stage phosphate recovery of 90–95% with a concentrate grade of 31% P_2O_5 is possible for coarse feed from the IMC-Agrico Four Corners Plant. As is the case for the traditional phosphate flotation, high solids conditioning is necessary with the hydroxamic acid collector. Conventional plant practice using traditional fatty acid/fuel oil collector (1,200 g/t) results in only 75–80% recovery at a grade of 31% P_2O_5 after multiple flotation stages in different flotation circuits (double flotation). In the case of coarse feed from the Cargill Plant, the concentrate grade reached 34% P_2O_5 with 93% recovery in single stage bench scale flotation with the new collector. Also significant improvements in flotation were achieved with plant samples from North Carolina and Utah. The results indicate that the hydroxamic acid collector is more effective for the flotation of coarse feed than for the flotation of fine feed. Pilot-plant testing at a feed rate of 700–900 lb/hr was carried out in order to evaluate the new collector in a larger scale continuous flotation experiment. The results from the pilot plant with single stage flotation are summarized in Table 6. For the coarse feed, in the best case a concentrate grade of 31.5% P_2O_5 and 95.0% recovery was achieved. The average concentrate grade is about 31.4% P_2O_5 with 92.8% recovery and 6.96% insol. The results from the pilot-plant testing confirmed the results from bench scale flotation experiments.

The preliminary economic analysis of the new flotation chemistry with the alcoholic solution of the hydroxamic acid collector, AERO 6493, indicates that the economic performance is sensitive to the feed grade and reagent cost. Significant economic effectiveness could be achieved under certain circumstances. For example, at a plant capacity of 4 million tons per year, a marginal economic performance of 7.55~8.51 million dollars per year can be reached for a price of AERO 6493 at $1.6~1.8/lb, in contrast to a price of 0.63~$0.93/lb for the existing traditional double float process with FA/FO and amine. A price lower than $1.60/lb for AERO 6493 would give a much more significant improvement in economic performance. The value of using AERO 6493 would be much more significant if other aspects involved in the entire flotation process were counted, such as energy reduction, water quality, water consumption, equipment investment, the cost of equipment maintenance, etc. Analysis of water from bench scale flotation experiments and review of the literature indicate that there is no significant environmental impact from utilization of the new collector. In fact the AERO 6493 reagent is currently used in the kaolin industry and is not detected in discharged water from these operations.

ACKNOWLEDGMENTS

Financial support from the University of Utah, the Florida Institute of Phosphate Research, and a DOE basic science grant (FG-03-93-ER 14315) is gratefully acknowledged. Appreciation is given to Patrick Zhang, Florida Institute of Phosphate Research, for providing samples of francolite and dolomite. The IMC, Cargill, PCS, SF Phosphate, Cytec, ARR-MAZ, and Jacobs Engineering are recognized with gratitude for their participation.

REFERENCES

Jacobs Engineering Group, Inc. (1998). Understanding the Basics of Anionic Conditioning in Phosphate Flotation. No. 04 031-068, FIPR, Bartow, Florida.

El-Shall, H. (2001). Bubble Generation, Design Modeling and Optimization of Novel Flotation Columns for Phosphate Beneficiation. Final Report: Volume 2, FIPR Publication #02-111-175, Florida Institute of Phosphate Research.

Miller, J.D., Li, M., and Wang, X. (2000). Selective flotation of phosphate minerals with hydroxamate collectors. U.S. Patent 6341697.

Miller, J.D., Wang, X., and Li, M. (2002a). Bench Scale Flotation of Sedimentary Phosphate Rock with Hydroxamic Acid Collectors. *Beneficiation of Phosphates: Advances in Research and Practice*. P. Zhang, H. El-Shall, and R. Wiegel, eds., SME, Littleton, Colorado, 93–101.

Miller, J.D., Wang, X., and Li, M. (2002b). A Selective Collector for Phosphate Flotation. FIPR Publication No. 02-142-187, Florida Institute of Phosphate Research, Bartow, Florida.

Wang, X. (1999). Flotation of Phosphate Rock from Florida with the Air Sparged Hydrocyclone. MS Thesis, University of Utah, Salt Lake City.

Wiegel, R.L. (1999). *Phosphate Rock Beneficiation Practice in Florida*, SME, Littleton, Colorado.

Zhang, P., Yu, Y., and Bogan, M. (1997a). Challenging the "Crago" Double Float Process II, Amine-fatty Acid Flotation of Silcieous Phosphates. *Minerals Engineering*, 10(9), 983–994.

Zhang, P., Yu, Y., and Bogan, M. (1997b). Challenging the "Crago" Double Float Process II. Amine-Fatty Acid Flotation of Siliceous Phosphates. *Minerals Engineering*, 10(9), 983–994.

CHAPTER 10

Updating the Knowledge of Amine Flotation in Phosphate Processing II: Effect of Particle Size

Francisco Sotillo,* Jinrong "Patrick" Zhang,† and Robert Snow†

ABSTRACT

The effect of particle size on the performance of six types of amines in the cleaning step of the Crago process was evaluated by screen assays of the concentrates and tailings of the best batch flotation tests performed on unsized flotation feed. For this purpose, the Insol and P_2O_5 weight frequency and distribution as a function of particle size were used. In addition, the amine consumption necessary to drop the Insol in the concentrate to about 2% (related to the coarse particles, 20×28 mesh) and the flotation results on coarse and fine amine flotation feed of Plant II were also considered to evaluate this particle size effect. Two indices were designed to compare the screen assay data, R and D. R corresponds to the ratio of Insol distribution at a given particle size to the P_2O_5 distribution at the same particle size in the tailings (specificity of the amine for the Insol). D corresponds to the difference between the Insol weight frequency to that of plain weight frequency for each size fraction in the tailings (selectivity to reject Insol). $P_{Insol\ 50}$ parameter was obtained from the Insol cumulative distribution data (particle size at which an Insol particle has 50% probability to report in the concentrate or the tailings) and used for the evaluation of the effect of particle size on the amine flotation. Furthermore, the amine consumption to achieve about 2% Insol grade in the concentrate as well as the ratio of Insol rejection in the tailings for the flotation tests of Plant II coarse and fine feed were considered. The flotation ranking from coarser to finer Insol (taking into consideration all these evaluations) indicated the following trend:

Primary > Ether > Condensate » Tertiary > Quaternary

INTRODUCTION

This paper corresponds to the second part of an investigation aimed at filling the gaps in the understanding of amine flotation of silica from phosphate. For this purpose, the areas of interest were define as (i) the selectivity of different types of amines for silica, (ii) the tolerance of these amines tested to different levels of slimes content, and (iii) the effect of particle size on the performance of the amine tested. Part I of this paper deals with the selectivity and the tolerance of slimes of different amines (Zhang et al. 2002). It was concluded that the selectivity rank from most selective to least selective changed slightly depending on the evaluation method, which in-turn depends on the sensitivity

*PerUsa EnviroMet Inc., Lakeland, Florida.

†Florida Institute of Phosphate Research, Bartow, Florida.

of the amines on the operating conditions of the flotation methods used. By using batch flotation tests, the following rank was determined:

Quaternary > Primary ≈ Tertiary ≈ Condensate > Ether > Secondary (?)

However, using locked cycle tests, the selectivity rank obtained was

Quaternary > Primary ≈ Tertiary > Ether > Condensate > Secondary (?)

The first part also determined the amine collecting power using batch tests. In order to obtain similar flotation results, the amine collecting power was ranked from the strongest to the weakest as

Ether > Quaternary > Primary ≈ Tertiary ≈ Condensate > Secondary (?)

The tolerance for slimes as indicated by the amine consumption during locked cycle tests to achieve quasi-steady state was also evaluated in Part I. The rank of tolerance for slimes from the most tolerant to the least one was

Primary > Tertiary ≈ Condensate > Ether > Quaternary

This paper summarizes the results obtained and ranks the effect of particle size on the performance of four types of fatty acid amines, (primary, secondary, tertiary, and quaternary) an ether amine, and a condensate amine. The evaluation of the effect of particle size on the performance of these amines was based on tests that showed the best flotation results for each amine tested in the first part of this investigation. Therefore, the type, origin, and characteristics of the amine flotation feed sample used influenced the flotation results obtained. Consequently, absolute values of grades and recoveries changed for different amine feeds, but the trends were similar. Moreover, the flotation results used in this evaluation did not correspond to optimum operating conditions with respect to any variable. In addition, the amines selected for this investigation were not custom designed for each amine flotation feed used as is the case in industrial practice.

Consequently, the results in this part of the paper were evaluated on a relative basis using several criteria and indices developed, such as R, D, $P_{Insol\ 50}$, and Insol rejection ratio.

MATERIALS AND METHODS

Flotation Feeds

The amine flotation feeds used to study the effect of particle size on the flotation performance of six different types of amines representing typical Florida phosphate products. Table 1 presents the characterization data of two types of amine flotation feeds: Plant I, Unsized Amine Feed and Plant II, Coarse and Fine Sized Amine Feeds.

To evaluate the effect of particle size on the flotation performance of the different types of amines used in this investigation, batch flotation tests were conducted using a Denver Laboratory Model 3:1 flotation cell with a 2.5-liter tank at 1,200 rpm. From the best batch flotation tests for each amine tested, screen assays of the concentrates and tailings were performed. The following data were obtained: (i) weight frequency of Insol and P_2O_5 as a function of particle size, (ii) cumulative distribution of Insol and P_2O_5 as a function of particle size, (iii) amine consumption required by each type of amine tested to obtain 2% of Insol in the concentrates for Plant I (related to the flotation of coarse silica, 20×28 mesh), and (iv) flotation results for both coarse and fine amine flotation feeds of Plant II.

With this data, the R and D indices were developed as well as the $P_{insol\ 50}$ parameter and the Insol rejection ratio criterion. The index R corresponded to the ratio between the Insol frequency distribution and the P_2O_5 frequency distribution in the tailings for a given particle size. The objective of this index was to access the specificity of the amine for silica at a stated particle size. D Index established the difference between the Insol weight frequency and the plain weigh frequency in the tailings for a given particle size.

$P_{Insol\ 50}$ parameter was defined as the particle size at which the Insol has 50% probability to report in the tailings or the concentrate. This

Table 1 Amine flotation feed characterization data

Characterization Data	Plant I, Unsized	Plant II	
		Sized Coarse	Sized Fine
P_2O_5, %	28.18	30.62	26.88
Insol, %	16.48	9.78	20.17
d_{10}, µm	152	171	113
d_{50}, µm	331	400	200
d_{90}, µm	600	631	427

Table 2 Selected amines

Brand Name	Description	Type	State
Armeen HT	Hydrogenated Tallow Alkylamine	Primary	Solid
Armeen 2HT	Dihydrogenated Tallow Alkylamine	Secondary	Solid
Armeen DMHTD	Dimethylhydrogenated Tallow Alkylamine Distilled	Tertiary	Liquid
Arquad 2HT-75	Dihydrogenated Tallow Alkyldimethyl Ammonium Chloride	Quaternary	Solid
Adogen 185	Ether Amine Acetate	Either	Liquid
Arrmaz	Condensate Amine	Condensate	Liquid

parameter was obtained using the Insol and P_2O_5 distributions for each type of amine tested.

Finally, the Insol rejection ratio corresponded to the ratio between the coarse and fine Insol rejection for each amine tested using Plant II sized amine flotation. This ration was developed to establish the capability of each amine to reject Insol in the coarse amine feed with respect to that in the fine amine feed.

Reagents

The amine selected were 75% neutralized with acetic acid and mixed with 25% of 2-ethylhexanol frother. These amines represented the three major groups of amines used as collectors: fatty acid amines, ether amines, and amine condensates. Even though a sequence based on the substitution of methyl groups was desired, samples of secondary amine with one methyl group were not available. Table 2 presents the amine used in this investigation.

RESULTS AND DISCUSSION

To evaluate the effect of particle size on the performance of the six types of amines tested, the best flotation results for Plant I, Unsized Amine Flotation Feed were used. Table 3 presents the metallurgical balance of these tests, which is the basis for this evaluation. Screen assays of the concentrates and tailings of these flotation tests were carried out, and the data were presented as weight frequency of Insol and P_2O_5 and Insol and P_2O_5 distribution as a function of particle size. In addition, flotation tests for Plant II, Coarse Amine Feed and Plant II, Fine Amine Feed were performed based on the amine additions used in the best tests for each amine for the Plant I, Unsized Amine Feed. Since the dosages used for the flotation of the Plant II samples were not adequate for these types of new feeds, the results were analyzed only on a relative basis.

Table 3 Summary of best flotation results for tests on Plant I, unsized amine flotation feed

Amine		Concentrate			Distribution	
Type	Dosage,lb/ton	%wt.	% P_2O_5	%Insol	% P_2O_5	%Insol
Armeen HT	0.395	83.70	32.88	2.05	97.56	10.85
Armeen 2HT	2.846	90.61	30.81	8.50	99.04	47.65*
Armeen DMHTD	0.440	84.32	32.73	3.93	97.59	20.40
Arquad 2HT-75	0.322	85.53	32.66	3.46	98.72	18.63
Adogen 185	0.163	84.34	33.39	2.54	97.96	13.52
Arrmaz	0.405	83.96	33.30	2.54	97.98	13.17

* Secondary amine was difficult to disperse in water.

The data are presented in Figures 1–12 and Tables 4 to 8. For this purpose, the two types of indices, R and D, were developed and are shown in Table 4. R corresponds to the ratio between the Insol distribution at a given particle size to the P_2O_5 distribution at the same particle size for the tailings (see Figures 1–6). Therefore, R could be related to the specificity of the amine for the Insol at the stated particle size. If R is less or equal to one, it is indicating that loses in P_2O_5 are greater than ($R<1$) or equal ($R=1$) to the rejection of Insol; thus, no specificity for the Insol. For an $R>1$, the amine specifically adsorbs onto the silica surfaces, the silica particles floating preferentially.

The index D was calculated by comparing the plain weight frequency with the Insol weight frequency (see Figures 1–6) in the tailings. The purpose of this index was to determine if the reagent is able to alter the weight distribution by selectively rejecting Insol at a given size fraction. A positive number indicates that the weight frequency has being altered by floating more Insol at the stated size fraction. A zero or negative value indicates that no or less Insol was floated from that size fraction.

Table 4 presents the specificity for Insol of each amine. Primary amine is more efficient in the 65 × 150-mesh particle size range, secondary amine in the 100 × 150-mesh range, tertiary amine in the 35 × 200-mesh particle size range, quaternary amine in the 65 × 200-mesh range, ether amine acetate in the 65 × 200-mesh particle size range, and condensate amine in the 100 × 200-mesh range. Furthermore, it was observed that the broaden the size distribution in the tailings with the Insol frequency distribution lower than that of the P_2O_5 corresponded to a less specificity of the amine at that given particle size fraction. As an example, the weight frequency distribution of Insol and P_2O_5 for tailings of Plant I, Unsized Amine Feed for Armeen DMHTD and Adogen amines were presented in Figures 3 and 5, respectively. These figures showed that for Adogen 185 the Insol frequency distribution was broaden and the weight frequency lower than those of P_2O_5, whereas those for Armeen DMHTD were higher.

Figures 7–12 shows the distribution of both Insol and P_2O_5 in the concentrates and tailings as a function of particle size for the primary, seconary, tertiary, quaternary, ether, and condensate amines, respectively. These figures clearly show the particle size at which the Insol has 50% of probability to report in the tailings or the concentrate ($P_{\text{Insol 50}}$). A coarser $P_{\text{Insol 50}}$ indicates that the amine was able to float coarser particles. Thus, $P_{\text{Insol 50}}$ was used in Table 7 as a parameter to evaluate the effect of particle size on the different types of amines. For example, Figure 7 showed the distribution of Insol and P_2O_5 for concentrate and tailings of Plant I, Unsized Amine Feed using Armeen HT primary amine. It was shown that the cumulative distributions for Insol and P_2O_5 intersected at $P_{\text{Insol 50}}$ particle size of 410 mm. In the case where a tertiary amine (Armeen DMHTD) and a quaternary amine (Arquad 2HT-75), the distributions of

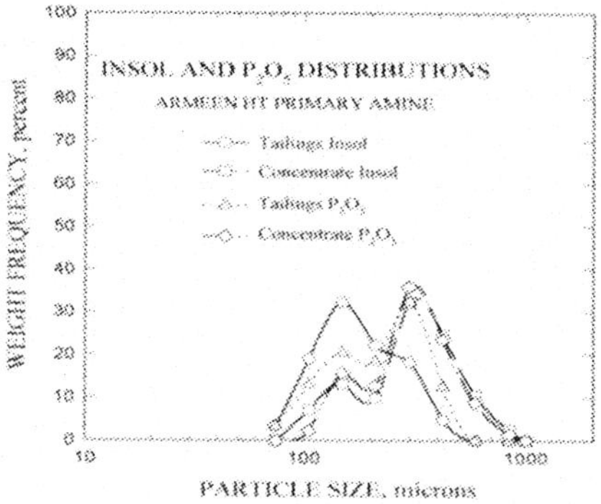

Figure 1 Weight frequency distributions for Insol and P_2O_5 of tailings and concentrates for Plant I, unsized amine feed as a function of particle size using Armeen HT primary amine

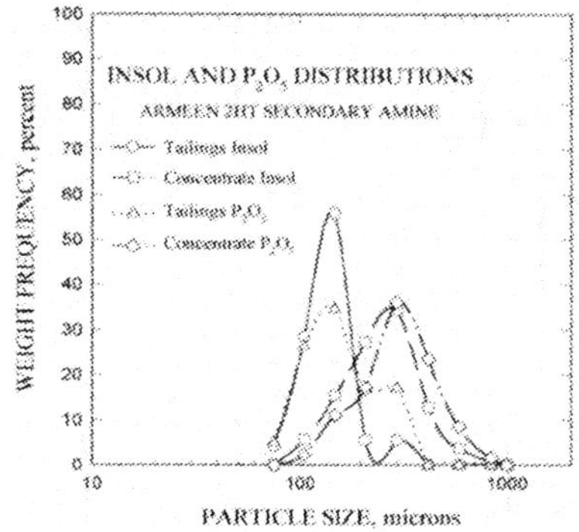

Figure 2 Weight frequency distributions for Insol and P_2O_5 of tailings and concentrates for Plant I, unsized amine feed as a function of particle size using Armeen 2HT secondary amine

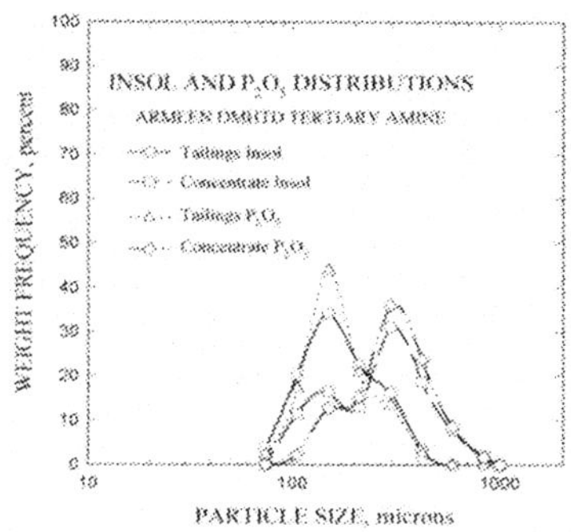

Figure 3 Weight frequency distributions for Insol and P_2O_5 of tailings and concentrates for Plant I, unsized amine feed as a function of particle size using Armeen DMHTD tertiary amine

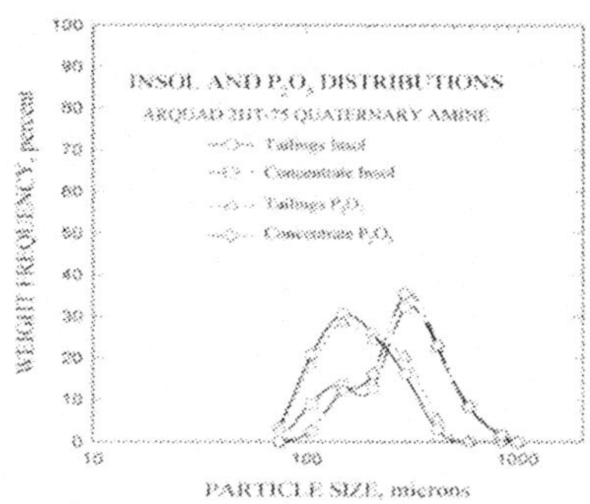

Figure 4 Weight frequency distributions for Insol and P_2O_5 of tailings and concentrates for Plant I, unsized amine feed as a function of particle size using Arquad 2HT-75 quaternary amine

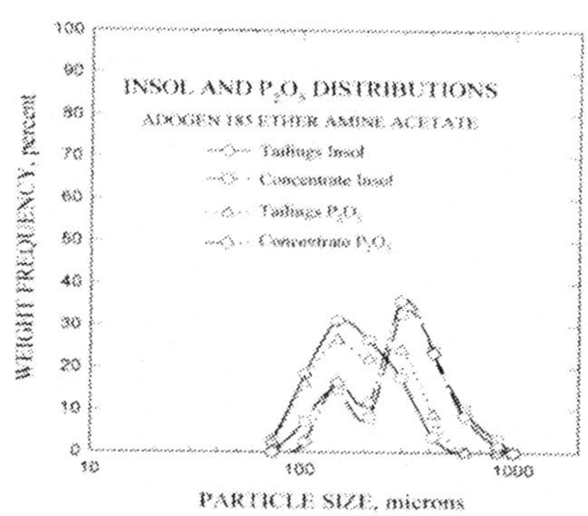

Figure 5 Weight frequency distributions for Insol and P_2O_5 of tailings and concentrates for Plant I, unsized amine feed as a function of particle size using Adogen 185 ether amine acetate

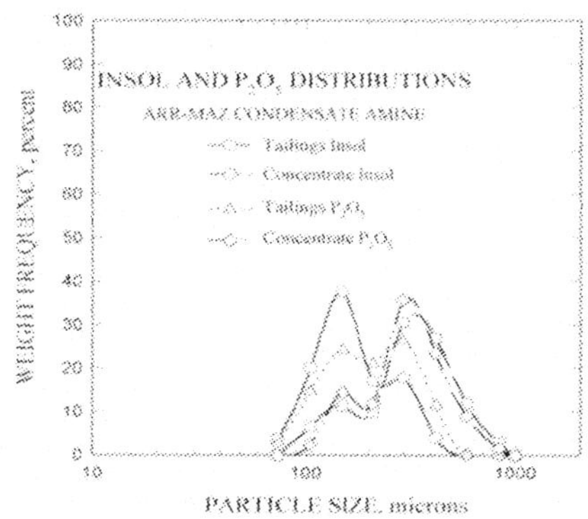

Figure 6 Weight frequency distributions for Insol and P_2O_5 of tailings and concentrates for Plant I, unsized amine feed as a function of particle size using ARR-MAZ condensate amine

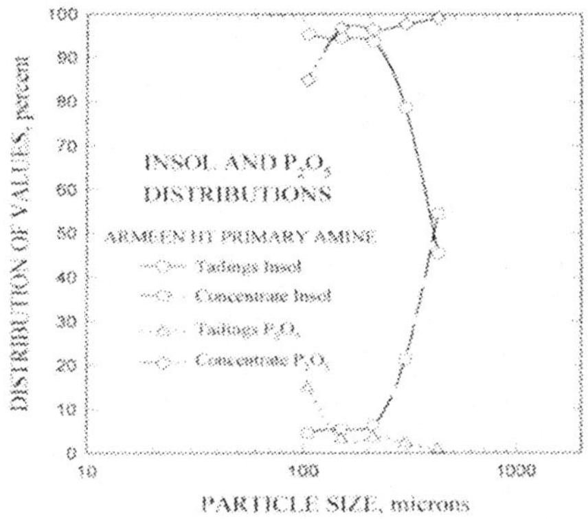

Figure 7 Distributions of Insol and P_2O_5 in both tailings and concentrates for Plant I, unsized amine feed as a function of particle size using Armeen HT primary amine

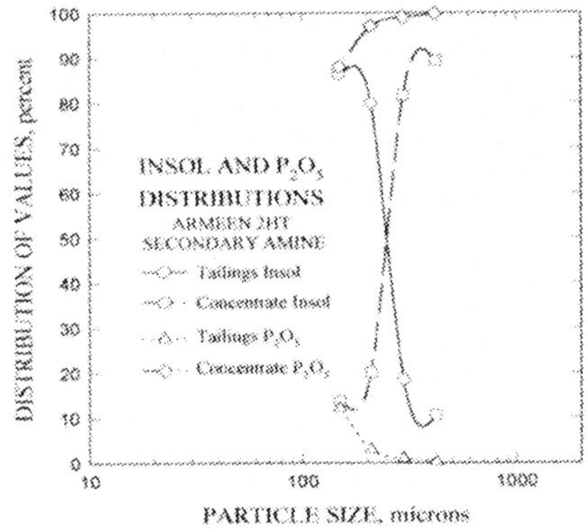

Figure 8 Distributions of Insol and P_2O_5 in both tailings and concentrates for Plant I, unsized amine feed as a function of particle size using Armeen 2HT secondary amine

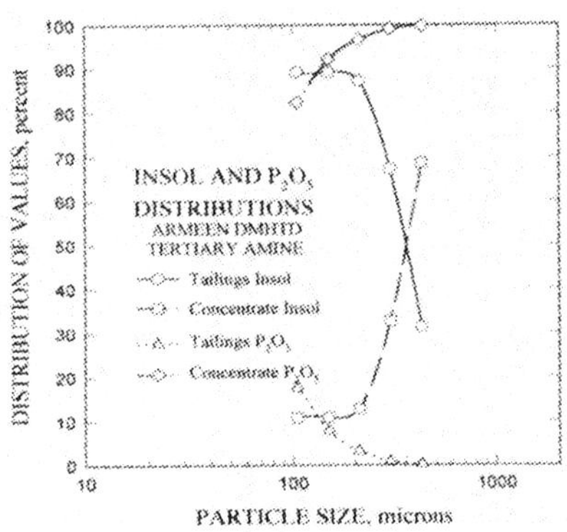

Figure 9 Distributions of Insol and P_2O_5 in both tailings and concentrates for Plant I, unsized amine feed as a function of particle size using Armeen DMHTD tertiary amine

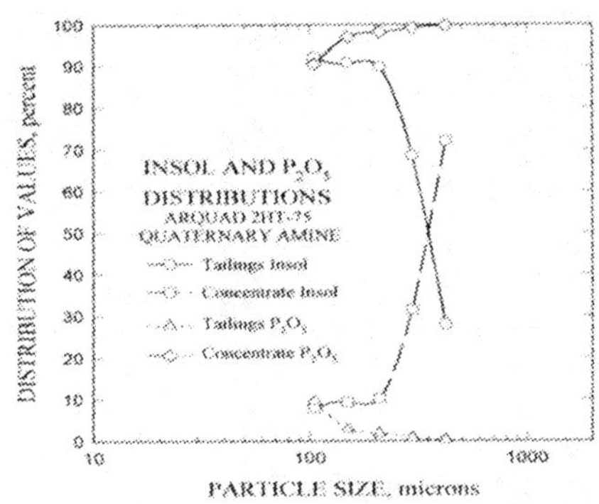

Figure 10 Distributions of Insol and P_2O_5 in both tailings and concentrates for Plant I, unsized amine feed as a function of particle size using Arquad 2HT-75 quaternary amine

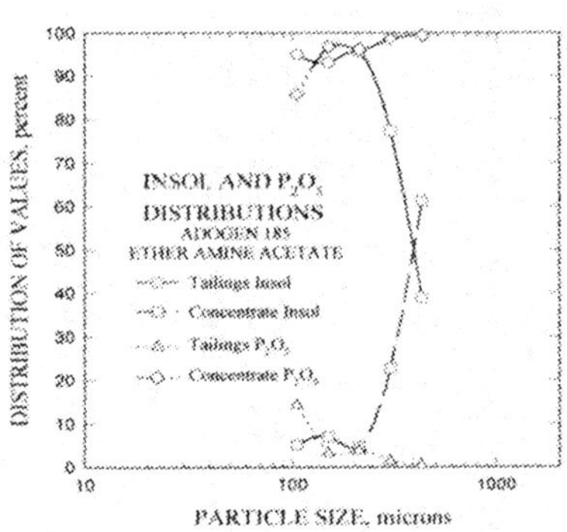

Figure 11 Distributions of Insol and P_2O_5 in both tailings and concentrates for Plant I, unsized amine feed as a function of particle size using Adogen 185 ether amine acetate

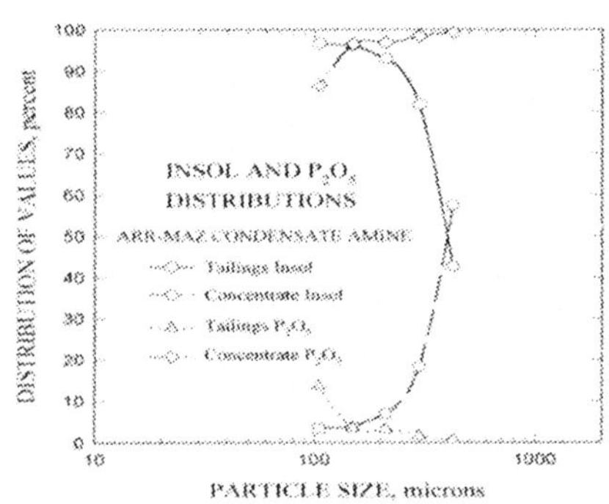

Figure 12 Distributions of Insol and P_2O_5 in both tailings and concentrates for Plant I, unsized amine feed as a function of particle size using ARR-MAZ condensate amine

Table 4 Summary of screen assays for tailings using the relative effect on distributions: R and D

Amine type	Index	Size Mesh					
		35	48	65	100	150	200
Primary	R	0.39	0.56	1.21	1.57	1.46	1.05
	D	-1.02	-1.97	0.56	1.61	0.83	-0.02
Secondary	R	--	0.33	0.33	1.61	1.09	0.85
	D	--	-1.07	1.02	1.99	0.21	-0.11
Tertiary	R	1.25	1.22	1.08	0.77	1.19	1.65
	D	0.02	0.21	0.23	-1.13	0.50	0.17
Quaternary	R	0.53	0.79	1.07	1.08	1.13	1.16
	D	-0.23	-0.46	0.17	0.27	0.24	0.01
Either	R	0.41	0.73	1.21	1.17	1.14	1.12
	D	-0.60	-0.88	0.54	0.62	0.30	0.02
Condensate	R	0.35	0.67	0.81	1.56	1.38	1.33
	D	-0.83	-1.11	-0.41	1.61	0.66	0.08

Table 5 Summary of flotation results for Plant II, coarse amine flotation feed

Amine		Concentrate			Distribution	
Type	Dosage,lb/ton	%wt.	% P_2O_5	%Insol	% P_2O_5	%Insol
Primary	0.395	73.97	33.04	2.41	80.67	18.08
Tertiary	0.440	93.86	32.42	3.82	99.64	38.26
Quaternary	0.322	93.87	32.42	4.47	99.56	42.40
Either	0.163	94.11	32.56	4.15	99.67	41.29
Condensate	0.405	92.51	32.90	3.17	99.21	30.29

Table 6 Summary of flotation results for Plant II, fine amine flotation feed

Amine		Concentrate			Distribution	
Type	Dosage,lb/ton	%wt.	% P_2O_5	%Insol	% P_2O_5	%Insol
Primary	0.395	62.79	33.27	1.60	76.87	5.13
Tertiary	0.440	86.22	30.92	9.00	98.64	38.09
Quaternary	0.322	85.43	31.09	8.99	98.57	36.50
Either	0.163	86.42	30.91	9.03	98.78	38.43
Condensate	0.405	82.14	32.61	4.60	98.06	18.95

Table 7 Summary of semi-quantitative evaluation of the effect of particle size

Amine	$P_{Insol\,50}$, μm	Insol Reject		Specificity Size Range4	Insol rejection Ratio, C/F	Consump. @2%Insol
		Coarse	Fine			
Primary	410	81.92	94.87	65x150 M	0.86	0.4
Secondary	230	---	---	100x150 M	---	---
Tertiary	333	61.74	61.91	32x200 M	1.00	1.1
Quaternary	333	57.60	63.50	65x200 M	0.91	0.4
Either	370	58.71	61.57	65x200 M	0.95	0.2
Condensate	400	69.61	81.05	100x200 M	0.86	0.5

Table 8 Summary of ranking by data and parameter evaluated

Amine	$P_{Insol\ 50}$, µm	Insol Reject Coarse	Specificity Size Range4	Insol rejection Ratio, C/F	Consump. @2%Insol	Overall Rank
Primary	1	1	2	4	2	1
Secondary	5	---	4	---	---	--
Tertiary	4	3	1	1	4	3
Quaternary	4	5	2	3	2	4
Either	3	4	2	2	1	2
Condensate	2	2	3	4	3	3

Insol and P_2O_5 were presented in Figures 9 and 10. These figures showed that the cumulative distributions of Insol and P_2O_5 intersected at $P_{Insol\ 50}$ particle size of 333 mm. The ether amine acetate showed a $P_{Insol\ 50}$ particle size of 370 mm (Figure 11), whereas the condensate amine showed a $P_{Insol\ 50}$ particle size of 400 mm (Figure 12).

In Part I of this paper (Zhang et al., 2002) it was shown that the amine consumption to obtain 2% Insol in the concentrate for Plant I, Unsized Amine Feed changed for each amine used in this investigation. The data showed that it was 0.4 lb/ton for the primary and quaternary amines, 1.1 lb/ton for the tertiary amine, 0.2 lb/ton for the ether amine acetate, and 0.5 lb/ton for the condensate amine. This information was incorporated in Table 7 as a parameter for comparison of the different amines evaluated.

The next step in trying to better understand the effect of particle size on different types of amines involved the use of a sized amine feed source. Since sizing was already performed in the plant before the flotation of the rougher concentrate, the amine flotation results for both the coarse and fine amine flotation feeds indicated the capability of a given amine tested to float large particle size. For this purpose, flotation of the Plant II, Coarse Amine Feed and Plant II, Fine Amine Feeds were carried out. However, the dosages of the different amines used were based on those used for the best tests of Plant I, Unsized Amine Feed. Therefore, the dosages were not adequate for these types of new feeds. As a consequence, the data was analyzed only on a relative basis using the Insol Rejection Ratio between the flotation results for the coarse and fine amine feeds. Table 5 presents the summary of the flotation results for Plant II, Coarse Amine Flotation Feed, and Table 6 presents those for Plant II, Fine Amine Feed.

Apparently, the flotation rank from coarse to fine Insol follows (see Table 7):

Primary > Condensate > Tertiary > Ether > Quaternary

for Plant II, Coarse and Fine Amine Feeds.

The ratio of Insol rejection in the tailings for the flotation tests of Plant II, Coarse and Fine Feed was 0.86 for the primary and condensate amines, 1.00 for the tertiary amine, 0.91 for the quaternary amine, and 0.95 for the ether amine.

Table 7 presented a summary of the parameters and data obtained from the different evaluations, criteria, and indices developed. This table was used to rank the different amines, taking into consideration different aspects related to the effect of particle size on the performance of the amines. Table 8 showed the ranking for different data and parameters, and the overall rank from the flotation of the coarser Insol to the finer Insol. This ranking was

Primary > Ether > Condensate ≈ Tertiary > Quaternary

SUMMARY AND CONCLUSIONS

The evaluation of the effect of particle size on the performance of six types of amines was carried out using two different amine flotation feeds: Plant I, Unsized Amine Feed and Plant II, Coarse and Fine Sized Amine Feed.

Since amines' performance depends on the type, origin, and characteristics of the amine flotation feed, evaluation of the effect of particle size was done using flotation results, screen assays, indirect measurements (amine consumption to obtain 2% Insol in the concentrates Insol Rejection ratio, C/F), and distribution parameters ($P_{Insol\ 50}$, R, and D).

Based on the data and evaluation parameters, the ranking of the amines from flotation of coarser Insol to that of finer Insol was

Primary > Ether > Condensate ≈ Tertiary Quaternary

ACKNOWLEDGMENTS

The authors would like to acknowledge the financial support of the Florida Institute of Phosphate Research, FIPR, for funding the experimental work. We would also like to express our gratitude to Dr. Seng Yap of ArrMaz Custom Chemicals for her valuable suggestions and assistance in preparing the amines.

REFERENCES

Jirong "Patrick" Zhang et al. 2002. Updating the Knowledge of Amine Flotation in Phosphate Processing. I. Selectivity. *Beneficiation of Phosphate: Fundamentals and Technology*. Zhang, P., El Shall, H., Somasundaran, P., and Stana, R., editors. SME, Littleton, Colorado, Chapter 8, pp. 79–91.

CHAPTER 11

Challenging the Crago "Double Float" Process IV: Selectivity Enhancement in All-Anionic Flotation of Florida Phosphates

Robert Snow* and Patrick Zhang*

ABSTRACT

In its continued efforts to develop a simpler, more efficient and environmentally friendlier flowsheet than the conventional "double float" process for phosphate flotation, the Florida Institute of Phosphate Research (FIPR) has invented several all-anionic flowsheets. These flowsheets are based on fatty acid rougher-cleaner flotation, with sizing of flotation feed, rougher concentrate, or rougher tails to reduce coarse phosphate loss. The key to success using these flowsheets is to achieve relatively low insol products while maintaining high P_2O_5 recovery. Numerous flotation modifiers were evaluated for improving selectivity. Tested modifiers included sodium silicate, lignosulfonates, calcium complexing agents, and ground phosphate rock. Proper sizing also enhanced selectivity.

INTRODUCTION

The rational for developing an alternative to the Crago "double float" process has been elaborated in some of our previous publications and presentations (Snow and Zhang, 2002; Zhang et al., 2004). The Crago process involves anionic (fatty acid) rougher flotation, acid scrubbing/washing (de-oiling) of the rougher concentrate, and cationic (amine) flotation. Our current research effort focuses on developing flotation processes with anionic reagents only, thus eliminating the de-oiling and amine flotation steps.

One major problem with most of the previously developed anionic flotation processes for phosphates is the conflict between recovery and concentrate grade. Although this conflict exists in all flotation processes, it is more remarkable in anionic flotation of phosphate. For example, to reduce the Insol in the concentrate from 10% to 6% would generally sacrifice recovery by up to 10%. Coarse phosphate particles are the main reason for the big gap between recovery and grade. Over reagentizing in rougher flotation could ensure better recovery of coarse phosphate particles, but would make it impossible to achieve low Insol product in cleaner flotation. On the other hand, if reagent starvation were practiced in rougher flotation, coarse phosphate particles would not float readily in cleaner flotation. Another problem with some of the previous processes is the high cost for purer and more expensive collectors.

Recognizing the above problems, FIPR initiated a research program a few years ago with

*Florida Institute of Phosphate Research, Bartow, Florida.

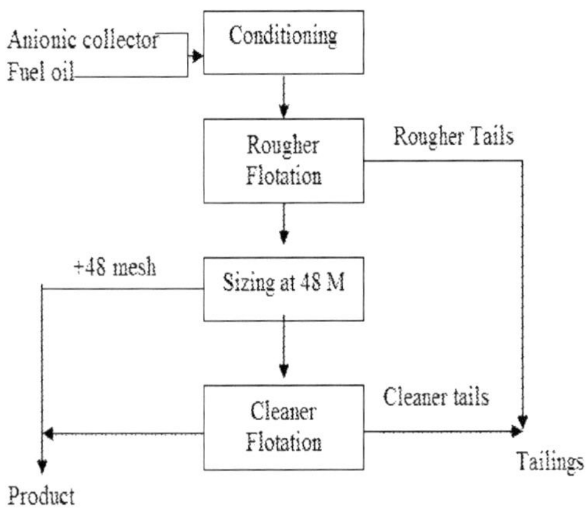

Figure 1 Basic flowsheet #1 of the FIPR process

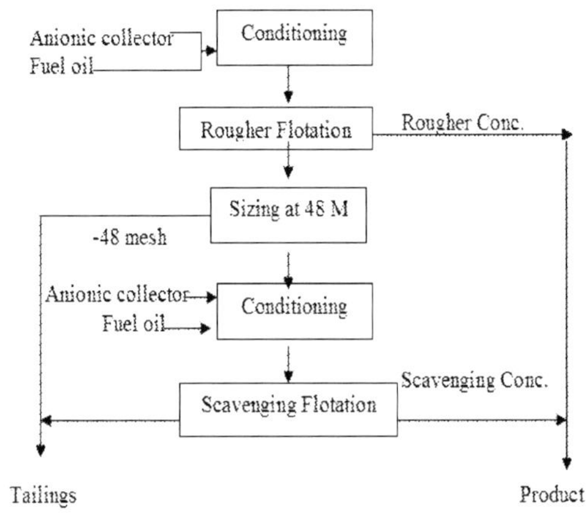

Figure 2 Flowsheet #2, single-collector flotation with scavenging coarse rougher tails

the objective of developing an energy saving and more efficient beneficiation process for phosphate. As a result, four all-anionic flowsheets were developed. In flowsheet #1, shown in Figure 1, a relatively pure, selective yet inexpensive ($0.35/lb) collector is mixed with a typical fatty acid collector used by the phosphate industry. The mixture boosts selectivity and the recovery of coarse phosphate at a reasonable reagent cost.

Sizing reagentized material (the rougher concentrate) could be challenging. In order to overcome this problem, flowsheet #2 was developed. In this flowsheet, Figure 2, rougher flotation is conducted under "reagent starvation" condition to achieve a high-grade rougher concentrate. The coarse phosphate (+48 mesh) reported to the rougher tails is recovered by scavenging flotation.

While flowsheet #1 is somewhat handicapped by sizing the rougher concentrate, flowsheet #2 requires delicate control of the rougher flotation under "reagent starvation" conditions. Since the major challenge of all anionic flotation is to reduce coarse phosphate loss while maintaining a decent Insol level (making sure the fine sands drop out in cleaner flotation without losing coarse phosphate), it should make sense to size the feed into coarse and fine fractions in the

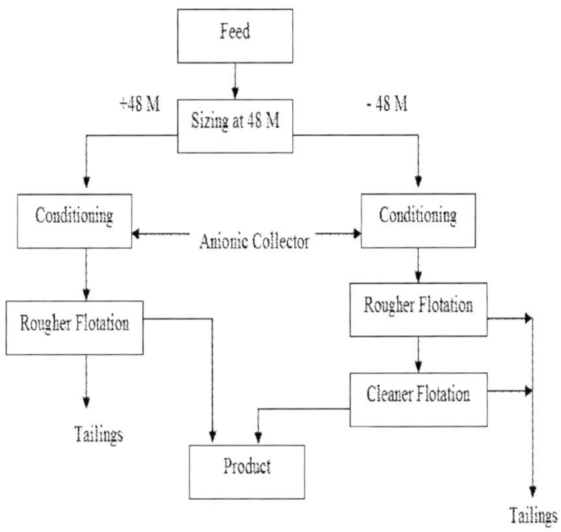

Figure 3 Single-collector flotation flowsheet #3

first place. Flowsheet #3 was thereforc proposed (Figure 3).

It should be pointed out that the optimal cut size for flowsheet #3 may not be 48 mesh, but perhaps somewhere between 35 and 48 mesh. Most recently, FIPR has developed yet another flowsheet. As is shown in Figure 4, this is nearly a true one-step process. "Rinsing" is simply removal of some liquid from the top of the

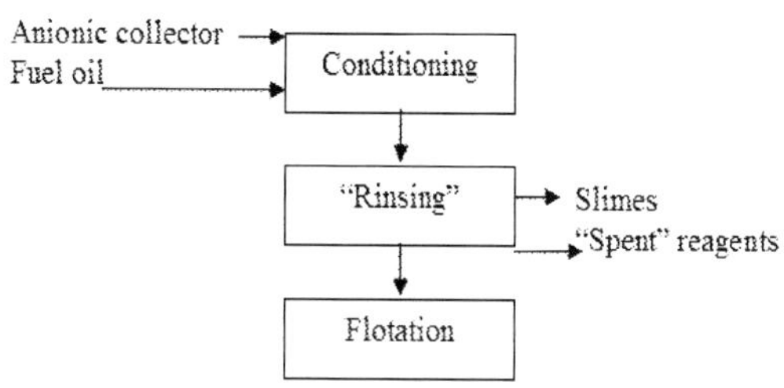

Figure 4 Single-collector lotation flowsheet #4

conditioner prior to flotation. This "rinsing" step improves selectivity significantly by removing both "slime" and the extra fatty acid collector.

Preliminary bench scale tests on numerous flotation feeds have demonstrated the potential of these new processes (Table 1).

These results are extremely encouraging. However, an Insol level of around 10% is not the current industry norm, even though it is regularly accepted by some fertilizer plants. To further reduce Insol while maintaining recovery of 90% or higher, one has to improve selectivity. The following selectivity enhancement methods were tested in this study:

1. Addition of collector-absorbing powders, such as ground phosphate rock and $CaHPO_4$ to take up the extra fatty acids
2. Removal of detrimental Ca^{++} ions using complexing agent oxalic acid
3. Discharge ("rinse") of conditioning liquid to remove secondary slime and residual fatty acid
4. Comparison of flotation modifiers focusing on silica depressants sodium silicate and lignosulfonates
5. Proper sizing of flotation feed, rougher concentrate, and tailings

EXPERIMENTAL

Materials and Reagents

Flotation Collectors. Unless noted, flotation tests were performed using 0.6 lb./TF of N-silicate as usual, a blend of the 1:1 Century MO-5 to Liqro GA fatty acid collector mixed at a 1.0:0.6 ratio with No. 5 fuel oil.

Flotation Feed Samples. Three flotation feed samples were tested, with two collected from CF Industries' Hardee plant and one from Mosaic's South Fort Meade mine. Tables 2–4 show size distribution and assay of the three samples.

Flotation Modifiers. Eight flotation modifiers were selected for initial screening of their effectiveness in improving selectivity in rougher (fatty acid) flotation of phosphate (Table 5).

Collector Adsorbing Powers. The first ground phosphate additive was prepared by wet ball milling and wet screening a sample of laboratory CF flotation concentrate. The product contained 100% -48 mesh particles and had the following size/assay profile:

Product	% Wt.	% P_2O_5	% Insol	% Dist. P_2O_5
+200M	34.3	30.59	9.68	42.5
-200M	65.7	30.95	6.83	57.5
Total	100.0	30.82	7.81	100.0

The second phosphate concentrate additive was prepared by further wet grinding and screening at 325 mesh.

A laboratory grade $CaHPO_4$ and an activated carbon were purchased from Fisher Scientific.

Test Procedures

Rougher Flotation. Standard laboratory 500g rougher flotation tests were performed in a Denver cell using different collector quantities with two minutes conditioning at approximately 73% solids. No. 5 fuel oil at 0.6 FO/FA ratio and soda ash (pH regulator) were used for all tests. Fuel oil was pre-mixed with the fatty acid before adding to the conditioner. Conditioning pH was usually 9.0–9.2 and flotation time was two minutes. Tap water was used for all tests.

Sizing and Cleaner Flotation. Sizing of the rougher concentrate at either 48 or 65 mesh was conducted manually using lab size Tyler screens. The fine fractions from two rougher flotation "floats" were combined for cleaner

Table 1 Some laboratory testing results using various flowsheets

Flowsheet	Product Grade		% P_2O_5 Recovery
	% P_2O_5	% Insol	
#1, Unsized feed	29.60	9.82	91.4
#1, Fine feed	30.28	8.00	81.3
#1, Coarse feed	30.71	9.55	94.7
#2, Coarse feed	32.61	7.10	94.7
#2, Fine feed	31.59	10.10	91.1
#3, Coarse & fine combined	30.20	9.98	90.7
#3, Unsized feed	29.66	9.41	84.9
S3, fine feed	30.06	9.88	92.7

Table 2 Size distribution and analysis of the CF feed #1

Size, Tyler mesh	% wt.	Cumulative %wt.	% P_2O_5	% Insol
+20	2.2	2.2	22.64	24.56
20 by 28	2.9	5.1	19.48	37.52
28 by 35	8.0	13.1	13.57	57.75
35 by 48	21.4	34.5	9.18	71.94
48 by 65	34.3	68.8	6.46	80.45
65 by 100	22.8	91.6	5.66	83.34
100 by 150	6.8	98.4	4.83	85.12
-150	1.6	100.0	3.96	84.17
Total	100.0	---	7.98	75.41

Table 3 Size distribution and analysis of the CF feed #2

Size, Tyler mesh	% wt.	Cumulative %wt.	% P_2O_5	% Insol
+20	0.8	0.8	22.61	22.61
20 by 28	1.6	2.4	19.42	37.12
28 by 35	5.3	7.7	13.41	56.92
35 by 48	18.2	25.9	8.60	71.72
48 by 65	39.3	65.2	5.84	80.33
65 by 100	25.3	90.5	5.00	83.77
100 by 150	8.3	98.8	4.56	85.22
-150	1.2	100.0	4.00	85.19
Total	100.0	---	6.76	77.69

Table 4 Size distribution and analysis of the Mosaic feed

Size, Tyler mesh	% wt.	Cumulative %wt.	% P_2O_5	% Insol
+20	0.4	0.4	---	---
20 by 28	0.5	0.9	23.53	23.80
28 by 35	2.2	3.1	14.44	53.16
35 by 48	11.4	14.5	9.81	69.01
48 by 65	27.8	42.3	8.78	70.21
65 by 100	35.5	77.8	10.66	67.40
100 by 150	17.8	95.6	9.51	70.44
-150	4.4	100.0	7.72	72.01
Total	100.0	---	9.91	68.41

Table 5 Flotation modifiers tested

Modifier	Solution Concentration	pH
Ca lignosulfonate (Lignotech D-712)	5%	4.22
Na lignosulfonate (Lignotech D-750)	5%	8.11
Na thiosulfate (LabChem 25150-1)(cont's. Na_2CO_3	2.0 N	9.76
Na silicate (N-Brand 3.22:1)	36° Be	11.70
Na hypochlorite (Clorox)	6.15%	9.65
Sulfonated oleic acid (Na salt – Tenn. Corp. OA-5)	5%	3.93
Oreprep 606	Pure	---
Tergitol NP-10	5%	6.64

flotation with a flotation time up to a minute without adding new reagents.

Addition of Sodium Silicate. N-silicate was used in many of the tests and was added during the final 20 seconds of conditioning, usually at 0.6 lb per ton of feed.

Addition of Modifiers. These various "modifiers" were added to the feed before fatty acid conditioning.

Flotation Procedure with Solid Powder Additives. Flotation conditioning (approx. 73% solids) with soda ash and 1:1 Century MO-S/Liqro GA mix plus No. 5 fuel oil at a 0.6:1.0 ratio with the fatty acid blend was performed for 1.5 minutes. Next, the ground phosphate rock "additive" slurry (at 65% solids) was added and conditioning continued for 0.5 minutes. In the case of $CaHPO_4$, it was added as a dry power. Finally, 0.6 lbs/TF of N-silicate was added and conditioning continued for 0.5 minutes. Total conditioning time was 2.5 minutes. Subsequent rougher flotation was performed for 2 minutes. The froth product, after drying, was dry screened (Ro-Tap) for 10 minutes using Tyler 48 mesh as the coarse and fine concentrate split size. Total rougher concentrate yield, grades and P_2O_5 recovery were calculated from the coarse and fine concentrate data.

RESULTS AND DISCUSSION

Effect of Lignosulfonates

Flotation tests were performed using varying levels of collector plus 0.6 lb/TF of Ca lignosulfonate or Na lignosulfonate added with stirring for 30 seconds before standard conditioning. The primary objective of these tests was to hopefully block active clay particle sites and/or "tie up" cations such as Ca^{+2} that could consume collector. The results for these tests are presented in Figures 5 to 7.

Figure 5 indicates significant recovery improvement by adding lignosulfonates. The benefit of lignosulfonate is more remarkable at lower collector dosages. For example, at a collector dosage of one pound per ton of feed, flotation recovery was improved by over 20% with D-750 (sodium lignosulfonate) and about 10% with D712 (calcium lignosulfonate).

Not only did lignosulfonates increase phosphate recovery, but they also enhanced selectivity, as is shown obviously in Figure 6. For example, at a low collector dosage of 0.9 pounds per ton, the rougher concentrate analyzed nearly 31% P_2O_5 with an Insol of less than 8%, when the sodium lignosulfonate was used. It may also be seen that Insol content in the rougher concentrate increases faster with collector dosage without lignosulfonates.

Sodium silicate is widely used in minerals flotation to improve selectivity as a depressant for silica. Silicate is also used at some phosphate beneficiation plants in Florida for its effectiveness in depressing sands and low price.

In Figure 7, the test points with lignosulfonates are all above the selectivity curve for sodium silicate in the high recovery range from 92–96%. Therefore, lignosulfonates are more selective silica depressant than silicate.

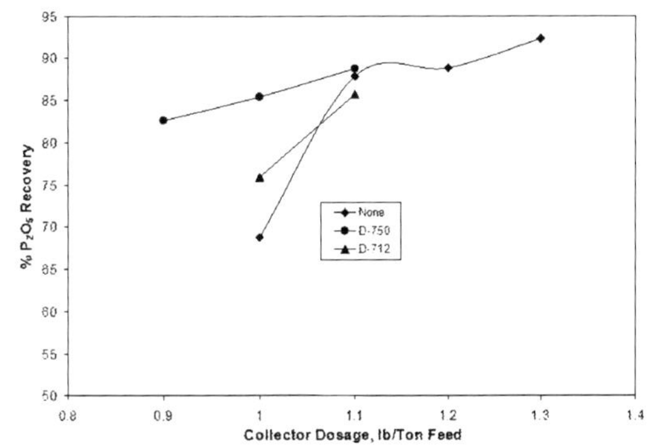

Figure 5 Effect of lignosulfonates on flotation recovery

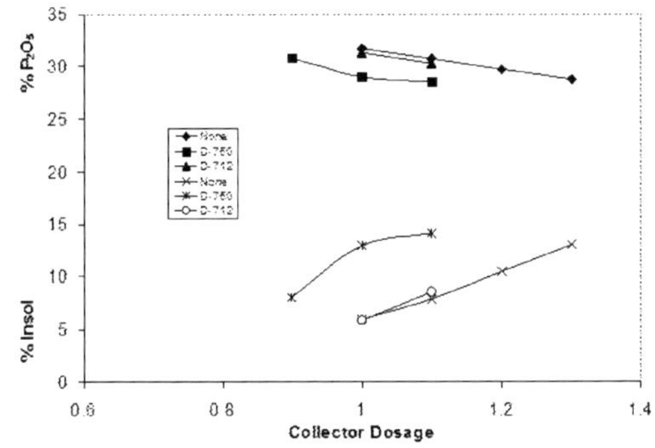

Figure 6 Effect of lignosulfonates on concentrate grade

Effect of Frothers

Two exploratory float/size tests were performed, using the standard 1.0 lb/TF of collector, wherein a small quantity of frother, Oreprep M-600 was added at the start of rougher flotation or Tergitol NP-10 was added during the final 15 seconds of rougher flotation. Final concentrates analyzed 31.57% P_2O_5/6.34% Insol at 64.8% P_2O_5 recovery, and 31.98% P_2O_5/6.10% Insol at 73.3% P_2O_5 recovery, respectively, for these two tests. No truly significant improvement resulted in either final concentrate grade or P_2O_5 recovery compared to the standard test in which no frother was used. The P_2O_5 recovery differences observed were probably the result of experimental error encountered using minimal collector.

Two preliminary float/size tests were also performed in which 0.5 lb/TF of additional fuel oil, either No. 5 or the aromatic Philflo was added to the conditioning stage, using 1.0 lb/TF of collector at a 1.0:0.6 ratio with No. 5 fuel oil as usual, to determine if increased P_2O_5 recovery was possible with no serious decrease in final concentrate grade. The use of additional No. 5 fuel oil resulted in a final concentrate analyzing 30.89% P_2O_5/7.49% Insol at 79.4% P_2O_5 recovery. These results represent an apparent possible increase of about 10% in overall P_2O_5 recovery over the comparison standard, with very little drop in final concentrate grade, and are believed to be significant. When Philflo oil was used, the

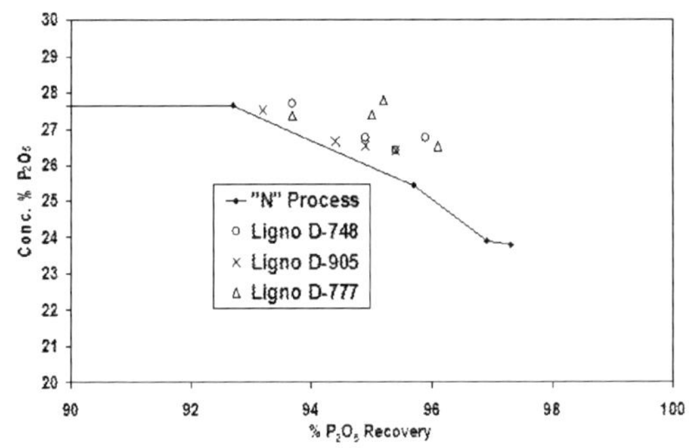

Figure 7 Selectivity comparison of silicate with lignosulfonates

final concentrate analyzed 30.98% P_2O_5/8.14% Insol at 74.3% P_2O_5 recovery.

Effectiveness of Various Solid Powers

Twenty-two laboratory rougher flotation tests were performed to compare the concentrate grades and P_2O_5 recoveries obtained with and without additions of ground phosphate concentrate (65% -200 mesh) or finer ground (100% -325 mesh), and/or monocalcium phosphate powder during high-solids conditioning. Rougher concentrates produced analyzed >30% P_2O_5 <10% Insol for eight of the tests at P_2O_5 recoveries ranging from about 61–84%. The remaining 17 tests produced rougher concentrates analyzing 27–30% P_2O_5 and 10–19% Insol at 81–95+% P_2O_5 recovery. Some of the best test results are shown in Table 6.

Table 6 Effectiveness of phosphate powers as flotation modifiers in rougher flotation of phosphate at 1.1 lb/ton collector

Test condition	Concentrate Grade		% P_2O_5 Recovery
	% P_2O_5	% Insol	
Blank	30.10	10.70	83.2
5lb/ton -200 M rock	30.40	9.00	84.0
5lb/ton -325 M rock	31.02	7.07	69.8
2lb/ton $CaHPO_4$	30.90	8.20	69.4
5lb/ton activated carbon	30.80	8.10	82.5

The +48 mesh size fraction of 24 of the 25 rougher concentrates produced using various levels of "additive" analyzed >29.8% P_2O_5 <10% Insol. Additive levels used were 2, 5, 10, and 20 pounds of ground phosphate concentrate (65% -200 mesh); 5, 10, and 20 pounds of the extra ground phosphate concentrate (-325 mesh); and 2 and 5 pounds of $CaHPO_4$ per ton of flotation feed.

All tests performed utilized 0.6 lbs/TF of N-Brand sodium silicate as the primary selectivity enhancer. Fatty acid collector used ranged from 1.0–1.3 lbs/TF at a 1.0:0.6 ratio with No. 5 fuel oil.

"The Rinse Process"

Flotation tests were performed using the conditioner discharge "rinse" procedure. N-Brand sodium silicate was also used during these tests. The feed sample for these later tests was obtained from CF Industries. These tests were performed to obtain preliminary data illustrating the effect of "rinsing" the conditioner discharge before rougher flotation.

These tests were conducted using the general flotation test procedure. Rougher flotation conditioning with soda ash and Century MO-5/ Liqro GA (1:1 mix) plus No. 5 fuel oil at a 0.6:1.0 ratio with the fatty acid blend was employed for all tests. When oxalic acid was used to complex Ca and Mg ions, this modifier was added before the pH regulator and collector mix. Conditioning time with the oxalic acid was 15 seconds. Additional conditioning time with the standard collecting reagents was 2 minutes. N-silicate was added during the final 30 seconds of conditioning for all tests.

Table 7 "Rinse" process performance

Test No.	Concentrate Grade		% P_2O_5 recovery
	% P_2O_5	% Insol	
4C2-1	30.06	9.88	92.7
4C2-3	31.19	6.37	79.6
4C2-4	30.91	7.25	84.7

Each rougher concentrate was dried and ro-tap screened at 48 mesh. The +48 and -48 mesh size fractions were analyzed separately. Total rougher concentrate yield, grades, and P_2O_5 recovery were calculated from the sized concentrate data. Table 7 shows some typical results on a fine (from the Four Corners) flotation feed analyzing 5.43% P_2O_5 and 83.36% Insol.

Rougher Flotation Using Oxalic Acid Modifier

Oxalic acid solution (10%) was added at the beginning of high-solids conditioning, and N-Brand sodium silicate was added during the final 30 seconds of conditioning. Conditioning pH measurements were performed after the 15-second oxalic acid conditioning stage was completed and at the end of the total conditioning stage immediately before rougher flotation.

Laboratory rougher flotation tests were performed to determine the effect of oxalic acid addition at the start of conditioning on subsequent rougher concentrate grade and P_2O_5 recovery at different levels of collector addition. Collector addition varied from 1.0 to 1.3 lb/TF, and oxalic acid levels ranged from 0 to 0.8 lb/TF.

The flotation material balances obtained using CF feed #1 are presented graphically in Figures 8A (% P_2O_5 recovery), 8B (Concentrate % P_2O_5) and 8C (Concentrate % Insol). The best overall test produced a rougher concentrate analyzing 30.05% P_2O_5/9.0% Insol at 88.0% P_2O_5 recovery.

The flotation material balances obtained using the CF feed #2 are presented graphically

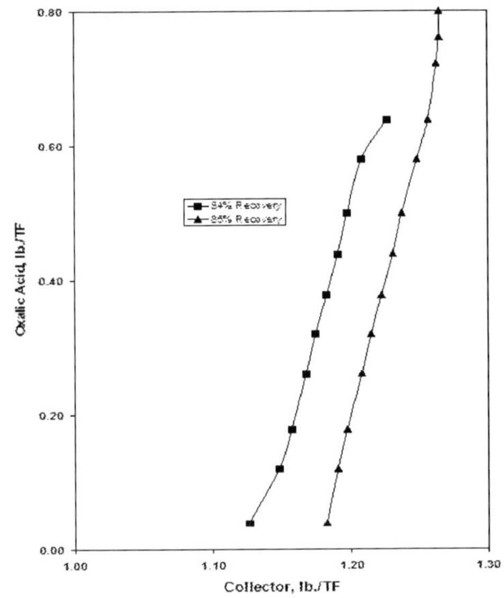

Figure 8A Concentrate % recovery P_2O_5 for CF feed #1 with oxalic acid

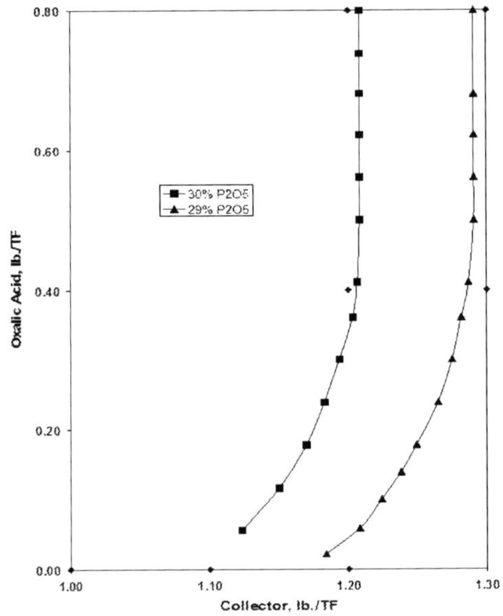

Figure 8B Concentrate % P_2O_5 for CF feed #1 with oxalic acid

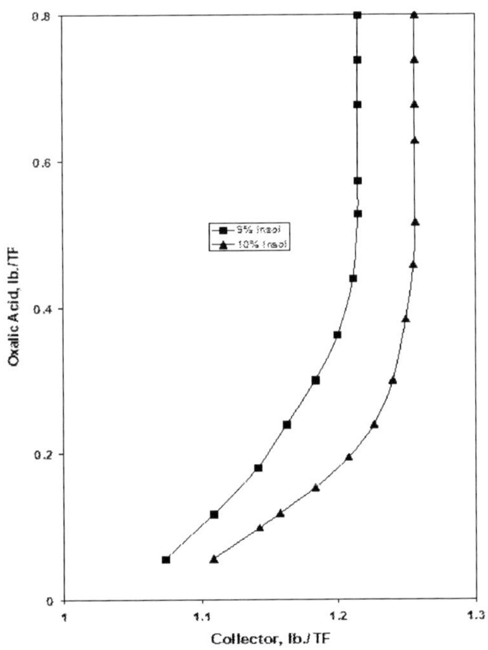

Figure 8C Concentrate % Insol for CF feed #1 with oxalic acid

in Figures 9A (% P_2O_5 recovery), 9B (Concentrate % P_2O_5) and 9C (Concentrate % Insol). The best tests produced a rougher concentrate analyzing about 29.30% P_2O_5/9.90% Insol at 85.9% P_2O_5 recovery.

Using the Cargill feed, a total of 3 rougher flotation tests, using oxalic acid as a potential selectivity modifier, were performed (CA-1, 2, and 3) along with three comparison standard tests using no oxalic acid (CA-4, 5, and 6). Oxalic acid addition was 0.8 lb/TF when this reagent was used, and collector level once again ranged from 1.0–1.3 lb/TF, and N-silicate use was 0.6 lb/TF. The flotation results for the Cargill feed are presented in Figure 10.

These figures show that consistent high recovery with acceptable concentrate grade is achievable by adding oxalic acid as a flotation modifier. It is postulated that this selectivity enhancement is due to complexing of detrimental ions, such as calcium and magnesium, with oxalic acid.

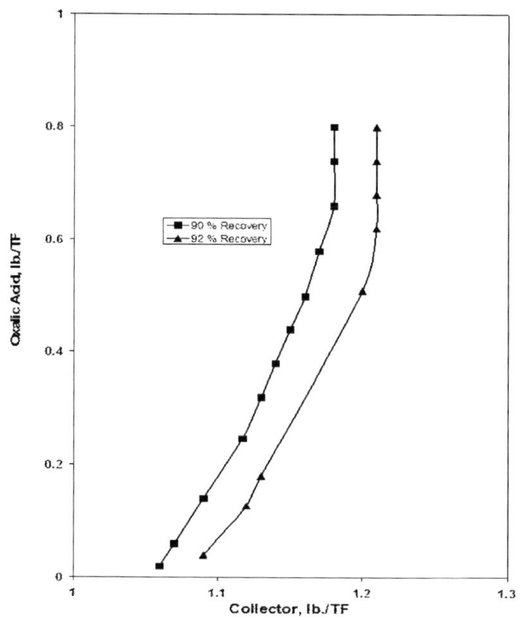

Figure 9A Concentrate % recovery P_2O_5 for CF feed #2 with oxalic acid

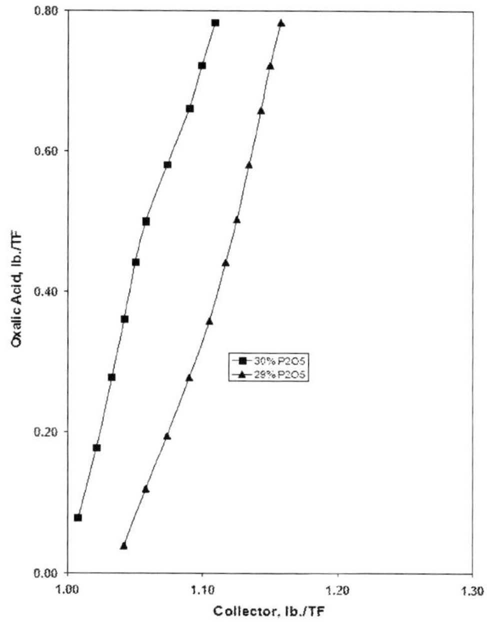

Figure 9B Concentrate % P_2O_5 for CF feed #2 with oxalic acid

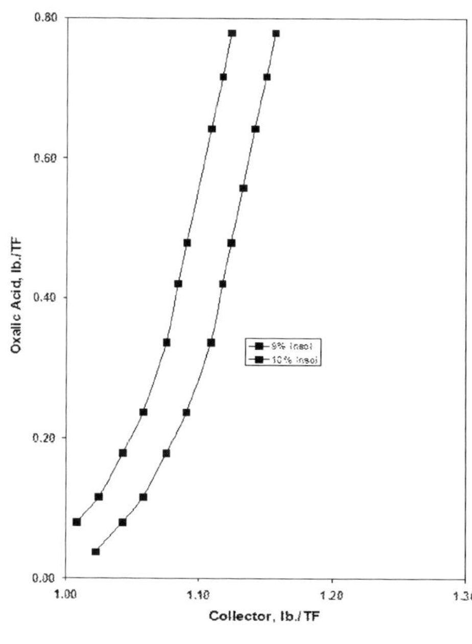

Figure 9C Concentrate % Insol for CF Feed #2 with oxalic acid

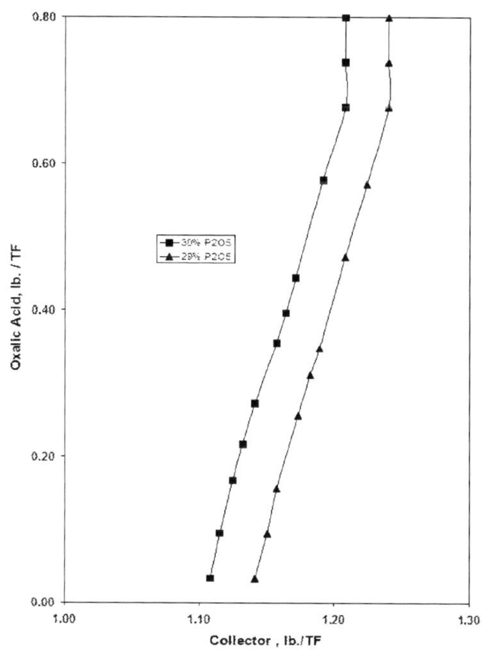

Figure 10 Concentrate % P_2O_5 Cargill fine feed (JPP) with oxalic acid

Sizing as a Means to Improve Selectivity and Grade

One major problem with most of the previously developed anionic flotation processes for phosphate is the conflict between recovery and concentrate grade. Although this conflict exists in all flotation processes, it is more remarkable in anionic flotation of phosphate. For example, to reduce the Insol in the concentrate from 10% to 6% would generally sacrifice recovery by up to 10%. Coarse phosphate particles are the main reason for the big gap between recovery and grade. Over reagentizing in rougher flotation could ensure better recovery of coarse phosphate particles, but would make it impossible to achieve low Insol product in cleaner flotation. On the other hand, if reagent starvation were practiced in rougher flotation, coarse phosphate particles would not float readily in cleaner flotation.

This problem could be significantly reduced by manipulating sizing of feed, tails, or rougher concentrate. For example, in the flowsheet #1, sizing of the rougher concentrate prior to cleaner flotation not only reduces coarse phosphate loss in cleaner flotation, but also lowers sand entrainment thus improving grade. Table 8 shows the benefit of sizing prior to cleaner flotation.

CONCLUSIONS

Lignosulfonates offer the most potential for improving selectivity in anionic flotation of phosphate by acting as both "slime blinder" and silica depressant. They could also reduce fatty acid consumption to some extent.

Ground phosphate rock is some what helpful in enhancing selectivity by taking up extra fatty acid.

Oxalic acid is an effective complexing agent for calcium in flotation water, but its cost and potential toxicity may prohibit its use.

Discharging conditioning water is very effective, but its practicality and economic feasibility require further investigation.

Table 8 Effect of cleaner cut size on concentrate grade and recovery

Feed and Process	Concentrate Grade		% P_2O_5
	% P_2O_5	% Insol	Recovery
Fine Feed A			
Rougher	27.31	20.61	96.8
Cleaner at 65 M	28.87	16.40	96.0
Cleaner at 48 M	31.18	9.96	96.2
Fine Feed B			
Rougher	28.88	11.30	86.1
Cleaner at 65 M	10.28	8.0	81.3
Unsized Feed			
Rougher	28.98	11.23	92.1
Cleaner at 65 M	30.37	7.28	85.7
Cleaner at 48 M	29.60	9.82	91.4

REFERENCES

Zhang, P. and Snow, R., 2002, Challenging the Crago Double Float Process III. Beneficiation of Siliceous Phosphate Using the FIPR/SAPR Process. *Beneficiation of Phosphates: Fundamentals and Technology*, SME, Littleton, Colorado. pp. 175–185.

Zhang, P., Snow, R., Miller, J., and Kohmuench, J., 2004, Improving Phosphate Flotation with New Chemistry, Smart Flowsheet and Novel Equipment. Presented at the SME Annual Meeting, Feb. 23–25, Denver, Colorado.

CHAPTER 12

Study of Pure Fatty Acid Compounds in Phosphate Flotation

Robert Snow,* Patrick Zhang,* and Michael Bogan*

ABSTRACT

This program was designed to compare results obtained using various C12–C22 fatty acids as Florida phosphate collectors. The unsaturated C20 fatty acid, eicosenoic acid, was of particular interest since its use as a flotation reagent has not been found in the literature.

Also, preliminary laboratory flotation evaluation of an "extract" of fatty acids obtained by acidulation and washing of the comparatively cheap cottonseed soapstock sample was considered to be a worthwhile endeavor. Flotation tests were performed using the "extract" alone and also blended in various amounts with Liqro GA tall oil that was previously shown to be a very good phosphate collector for Florida phosphate.

INTRODUCTION

A literature review of C8–C22 fatty acids properties and sources was performed in order to prepare reference tables for selection of collectors for preliminary flotation comparison testwork. Several technical literature references were found pertaining to Florida phosphate flotation using C12–C18 fatty acids, iso-stearic acid, a C18 halogenated fatty acid, a C36 dimer acid and a selected petroleum oxidate. Laboratory fatty acid comparison flotation tests results were described in 1986 by Finch and Riggs. In addition, extensive laboratory and pilot plant test results comparing the four collector types cited with a selected tall oil were presented in 1982 articles by Hsieh and Lehr. Of particular interest are the economic reagent cost comparisons presented in the pilot plant testing report. None of the reagents compared favorably with tall oil from a cost/usage standpoint. A brief list of these articles is shown in the References section. Some properties of fatty acid compounds are listed in Tables 1 and 2.

Laboratory rougher flotation tests were performed on a phosphate flotation feed sample using selected C12–C22 fatty acids with No. 5 fuel oil as the phosphate collector. The saturated fatty acids were found to be very poor collectors when using the standard conditioning procedure; whereas, the unsaturated C16–C22 fatty acids showed fair to good phosphate collecting ability. The C18 fatty acids oleic, impure oleic, linoleic and the C20 eicosenoic acid were shown to be the best phosphate collectors with and without N-silicate addition.

Flotation concentrates analyzing approximately 15–25% P_2O_5 at 78–87% P_2O_5 recovery were produced using these four fatty acid

*Florida Institute of Phosphate Research, Bartow, Florida.

Table 1 Physical values for various fatty acids

No.C	Name	Formular	M. Wt.	Form	Saturation	M.P., °C
8	Caprylic	$C_7H_{15}COOH$	144.2	Liq.	Sat.	16.3-16.7
10	Capric	$C_9H_{19}COOH$	172.3	So.	Sat.	31.4
12	Lauric	$C_{11}H_{23}COOH$	200.3	So.	Sat.	44.1
14	Myristic	$C_{13}H_{27}COOH$	228.4	So.	Sat.	54.2
16	Palmitic	$C_{15}H_{31}COOH$	256.4	So.	Sat.	62.8
16	Palmitoleic	$C_{15}H_{29}COOH$	254.4	Liq.	Unsat.	–0.5 - +0.5
16	Palmitolic	$C_{15}H_{27}COOH$	252.4	Sol.	Unsat.	42.0
18	Stearic	$C_{17}H_{35}COOH$	284.5	Sol.	Sat.	69.6
18	Oleic	$C_{17}H_{33}COOH$	282.5	Liq.	Unsat.	14.0
18	Elaidic	$C_{17}H_{33}COOH$	282.5	Sol.	Unsat.	51.0-57.0
18	Linoleic	$C_{17}H_{31}COOH$	280.5	Liq.	Unsat.	-5.0
18	Linolenic	$C_{17}H_{29}COOH$	278.4	Liq.	Unsat.	-11.0
18	Eleaostearic	$C_{17}H_{31}COOH$	280.4	So.	Unsat.	48.0-49.0
20	Arachidic	$C_{19}H_{39}COOH$	312.5	So.	Sat.	75.0-77.0
20	Eicosenoic	$C_{19}H_{37}COOH$	312.5	So.	Unsat.	23.0-24.0
22	Behenic	$C_{21}H_{43}COOH$	340.6	So.	Sat.	80.0
22	Erucic	$C_{21}H_{41}COOH$	338.5	So.	Unsat.	34.0
22	Brassidic	$C_{21}H_{41}COOH$	338.4	So.	Sat.	48.0-49.0

Table 2 Specific gravity (S.G.) and source for various fatty acids

No. of C	Common Name	S.G.	Typical Natural Sources
8	Caprylic	0.910	Palmoils, milk fat
10	Capric	0.895	Palm oils, whale head oils
12	Lauric	0.883	Palm oils, laural family oils, milk fats
14	Myristic	0.858	Milk fats, nutmeg butter
16	Palmitic	0.853	Soybean, corn, peanut, cottonseed oils
16	Palmitoleic	0.864	Marine animal oils, amphibia&reptile fats
16	Palmitolic	---	(Synthetic)
18	Stearic	0.847	Animal and milk fats, marine oils
18	Oleic	0.895	Almost all plant & animal fats
18	Elaidic	0.851	(Oleic acid stereoisomer)
18	Linoleic	0.903	Almost all plant & animal fats
18	Linolenic	0.914	Soybean, safflower, walnut & other seed fats
18	Eleaostearic	0.903	Tung oil, China wood oil
20	Arachidic	0.824	Peanut oil & rel. plants
20	Eicosenoic	0.883	Marine oils & jojoba wax
22	Behenic	0.822	Rapeseed & peanut oils
22	Erucic	0.860	Cruciferae species
22	Brassidic	0.859	(Erucic acid stereoisomer)

collectors. Inferior flotation results were obtained using the unsaturated C16 palmitoleic acid and the unsaturated C22 erucic acid. Two mixtures containing 15% and 25% commercial grade palmitic acid added to commercial grade oleic acid were observed to perform somewhat similar to the oleic acid as phosphate collectors. Palmitic acid was selected as the saturated fatty acid for addition to oleic acid because of it's presence in most natural fatty acid mixtures produced from vegetable oils.

A fatty acid extract prepared by sulfuric acid addition to a heated, rapidly stirred cottonseed soapstock solution, followed by multiple settling and washing stages, was observed to be a very poor, unselective collector for phosphate during flotation. The reason for this poor performance is presently unknown. Mixtures of the soapstock "extract" with Liqro GA tall oil performed poorly as a phosphate collector. Both concentrate grade and P_2O_5 recovery decreased markedly as the "extract" percentage was increased from 10% to 33% by weight. Size/assay analyses were performed on selected laboratory flotation tailings produced during tests using the C16–C22 fatty acid collectors.

EXPERIMENTAL

Phosphate Collectors Evaluated

The C16–C22 unsaturated fatty acids and the C12–C22 saturated fatty acids were selected for brief laboratory flotation testing in order to compare their collecting strength and selectivity with and without the use of N-brand sodium silicate depressant. The various fatty acids used for the comparison flotation tests are shown in Table 3.

The unsaturated fatty acids (except for the C22 erucic acid) were liquid at room temperature. These reagents were mixed by stirring at a 1.0:0.6 ratio with No. 5 fuel oil for medicine dropper calibration in drops per gram. The solid saturated fatty acids were warmed, mixed by stirring at a 1.0:0.6 ratio with No. 5 fuel oil, and each mixture further warmed to obtain a fluid mix for dropper calibration in drops per gram. The commercial grade oleic acid 401 was reported to contain approximately 71% oleic acid, 6% palmitic acid, 10% linoleic acid and 2% linolenic acid.

Table 3 Fatty acids used for comparison flotation tests

Fatty Acid	Producer/Supplier
Unsaturated:	
C16 Palmitoleic (98%)	Sigma Aldrich, Milwaukee, WI
C18 Oleic (99%)	Aldrich Chem. Co., Milwaukee, WI
C18 Oleic 401 (71%)	Welch, Holme & Clark Co.,Newark, NJ
C18 Linoleic (99%)	Acros Organics, Fisher Sci. Co., NJ
C20 Eicosenoic (98%)	Aldrich Chem. Co., Milwaukee, WI
C22 Erucic (90%) (Prifac 2990)	Unichema International, N.A.
Saturated:	
C12 Lauric (Prifac 2922)	Unichema International, N.A.
C14 Myristic (Prifac 2940)	" " "
C16 Palmitic (Prifac 2960)	" " "
C18 Stearic (Prifac 2981)	" " "
C22 Behenic (Prifac 2987)	" " "

Other phosphate collectors used for comparison flotation testwork included two oleic acid/palmitic acid mixtures, an "extract" of fatty acids obtained by dilute, hot sulfuric acid acidulation (followed by settling and washing) of a 10% by weight of as-received cottonseed soapstock solution, and three different mixtures of the fatty acid "extract" with Liqro GA tall oil. The as-received cottonseed soapstock was previ ously shown to be a very unselective phosphate collector.

Flotation Feed Sample

A fine flotation feed from IMC's Four Corners mine was used for this research. Table 4 shows size and chemical analysis of this sample.

Flotation Procedure

Standard laboratory 500g rougher flotation tests were performed in a Denver cell using different collector quantities with two-minute conditioning at approximately 73% solids. No. 5 fuel oil at 0.6 FO/FA ratio and soda ash (pH regulator) were used for all tests. Fuel oil was pre-mixed with the fatty acid before adding to the conditioner. Conditioning pH was usually 9.0–9.2 and flotation time was two minutes. Tap water was used for all tests.

Table 4 Size and chemical analysis of test sample

Size, Mesh	%wt.	$\%P_2O_5$	%Insol
+20	0.6	27.00	19.23
20x28	1.4	18.50	44.45
28x35	8.9	8.79	73.71
35x48	30.9	5.29	84.03
48x65	31.3	4.77	85.73
65x100	16.7	3.94	88.15
100x150	7.8	2.72	91.61
-150	2.3	4.74	89.01

RESULTS AND DISCUSSION

Phosphate Flotation Tests Using Various "Pure" Fatty Acids

A total of 11 fatty acids were briefly tested as phosphate collectors for the IMC Four Corners plant feed. Flotation tests were performed using 1.6 lbs. of fatty acid and 0.96 lbs. of fuel oil at a 1.0:0.6 ratio mixture as the phosphate collector, with and without the addition of N-Brand sodium silicate. Soda ash consumption for conditioning pH regulation ranged from about 1.4–1.6 lbs. per ton of feed, and the resulting conditioning pH was 8.9–9.2 for these tests. Flotation test results using the unsaturated fatty acids without N-silicate and with N-silicate are presented in Tables 5 and 6, respectively. These results are also shown in bar graph form in Figures 1–4.

Figures 1 and 2 show that, when no N-silicate was used, the highest grade concentrates (15–21% P_2O_5) and highest P_2O_5 recoveries (82–87%) were obtained using the two oleic acids and the linoleic acid. Eicosenoic acid yielded results very similar to linoleic acid. Figures 3 and 4 show that, when N-silicate was used to improve selectivity, the highest grade concentrates (20–25% P_2O_5) and highest P_2O_5 recoveries (78–82%) were obtained using the linoleic acid and the eicosenoic acid. The oleic acid 401 (71%) yielded results very similar to eicosenoic acid. Flat, non-persistent froths were observed during most flotation tests using N-silicate for improving selectivity.

The C12–C22 saturated fatty acids were granular, white, waxy solids at room temperature. When warmed and mixed at a 1.0:0.6 ratio with No. 5 fuel oil, the mixtures were also solid upon cooling to room temperature and required additional heating to produce fluids for dropper calibration. None of these mixtures performed as phosphate collectors when used at the standard 1.6 lbs. of fatty acid plus 0.96 lbs. of fuel oil per ton of feed and a conditioning pH = 9.0–9.2. The warmed, fluid C14–C22 fatty acid/fuel oil mixes did not disperse when added to the flotation conditioner. These reagent mixes separated from the flotation slurry during flotation and were observed to form small, black, sticky balls that floated at the air/pulp interface. Contact of the warm collector/fuel oil mix with the colder slurry in the conditioner apparently caused reagent agglomeration resulting in poor dispersion and probably poor saponification during conditioning. The C12 lauric acid did not produce the same visual reagent separation, but did not perform effectively as a phosphate mineral collector.

Additional laboratory flotation tests were performed using 1.6 lbs. per ton of two mixed unsaturated/saturated fatty acid blends containing 85% oleic acid 401 plus 15% palmitic acid, and 75% oleic acid 401 plus 25% palmitic acid, with No. 5 fuel oil and with and without N-silicate addition. Fatty acid:fuel oil ratio remained at 1.0:0.6. The test results are compared to tests using oleic acid only as the fatty acid with No. 5 fuel oil in Table 7. Very flat froths were observed particularly when N-silicate was used to improve flotation selectivity. The Table 7 results were considered to be somewhat erratic; however, P_2O_5 recoveries of 80% or higher were obtained for most tests. The 15% palmitic acid and the 25% palmitic acid mixtures with oleic acid were waxy solids that liquefied when heated to about 29–31 degrees C. and 34–36 degrees C., respectively. The 1.0:0.6 ratio mixtures with fuel oil were fluids at 23 degrees C. and 25 degrees C., respectively.

Table 5 Phosphate flotation material balance for tests using various unsaturated C16–C22 fatty acid collectors at 1.6 lb collector and 0.96 lb fuel oil per ton of feed

Collector	Flotation Product	%wt.	$\%P_2O_5$	%Insol	$\% P_2O_5$ Distribution
Palmitoleic (98%)	Conc.	31.3	12.75	61.13	77.0
	Tail.	68.7	1.73	94.37	23.0
	Feed	100.0	5.18	83.96	100.0
Oleic (99%)	Conc.	20.8	20.81	38.17	85.9
	Tail.	79.2	0.90	96.88	14.1
	Feed	100.0	5.04	84.67	100.0
Oleic (71%)	Conc.	26.8	16.71	49.88	87.3
	Tail.	73.2	0.89	96.82	12.7
	Feed	100.0	5.13	84.24	100.0
Linoleic (99%)	Conc.	27.9	14.88	55.95	82.3
	Tail.	72.1	1.24	95.97	12.7
	Feed	100.0	5.04	84.80	100.0
Eicosenoic(98%)	Conc.	30.9	12.91	61.38	82.6
	Tail.	69.1	1.21	95.99	17.4
	Feed	100.0	4.83	85.40	100.0
Erucic	Conc.	24.8	8.52	74.44	42.0
	Tail.	75.2	3.87	88.17	58.0
	Feed	100.0	5.02	84.76	100.0

Table 6 Phosphate flotation material balance for tests using various unsaturated C16–C22 fatty acid collectors and N-silicate (0.5 lb/t) at 1.6 lb collector and 0.96 lb fuel oil per ton of feed

Collector	Flotation Product	%wt.	$\%P_2O_5$	%Insol	$\% P_2O_5$ Distribution
Palmitoleic (98%)	Conc.	19.9	16.39	50.85	66.1
	Tail.	80.1	2.08	93.50	33.9
	Feed	100.0	4.93	85.01	100.0
Oleic (99%)	Conc.	16.4	21.58	60.40	70.9
	Tail.	83.6	1.74	94.50	29.1
	Feed	100.0	4.99	84.91	100.0
Oleic (71%)	Conc.	19.4	20.31	39.53	78.2
	Tail.	80.6	1.36	95.63	21.8
	Feed	100.0	5.04	84.75	100.0
Linoleic (99%)	Conc.	16.9	25.43	24.75	84.0
	Tail.	83.1	0.99	96.77	16.0
	Feed	100.0	5.12	84.60	100.0
Eicosenoic(98%)	Conc.	18.7	21.55	39.03	78.6
	Tail.	81.3	1.35	95.55	21.4
	Feed	100.0	5.13	84.98	100.0
Erucic	Conc.	16.4	12.85	61.58	41.5
	Tail.	83.6	3.55	89.13	58.5
	Feed	100.0	5.08	85.10	100.0

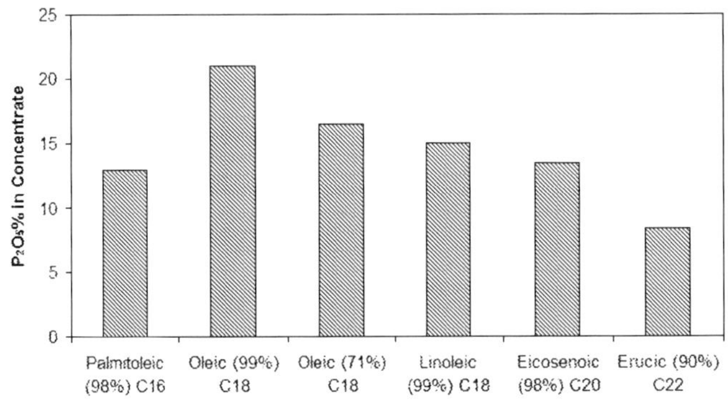

Figure 1 Concentrate grades using various unsaturated C16–C22 fatty acids at collector dosage of 1.6 lb/ton feed with 0.96 lb fuel oil

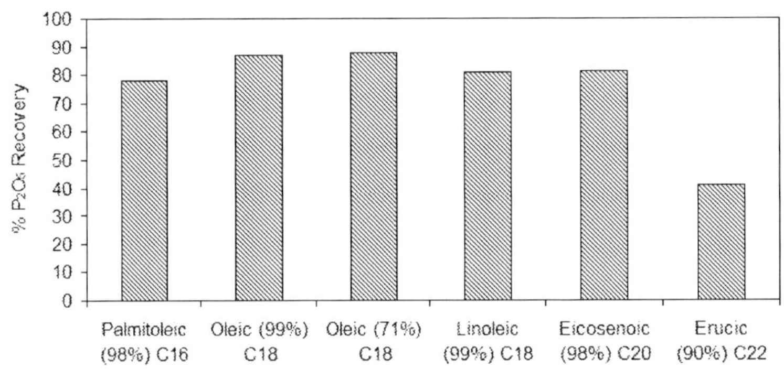

Figure 2 Flotation recovery using various unsaturated C16–C22 fatty acids at collector dosage of 1.6 lb/ton feed with 0.96 lb fuel oil

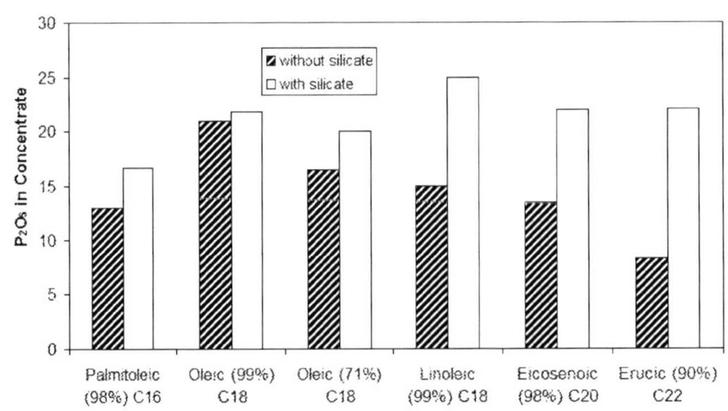

Figure 3 Effect of silicate on concentrate grade using unsaturated C16–C22 fatty acids at 1.6 lb/ton feed with 0.96 lb fuel oil and 0.5 lb silicate

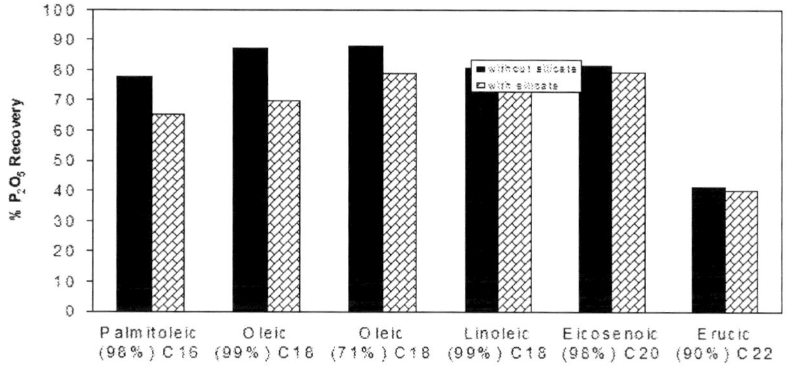

Figure 4 Effect of silicate on flotation recovery using unsaturated C16–C22 fatty acids at 1.6 lb/ton feed with 0.96 lb fuel oil and 0.5 lb silicate

Table 7 Material balance using various oleic (OA)/palmitic (PA) acid mixes

Collector	Flotation Product	%wt.	%P_2O_5	%Insol	% P_2O_5 Distribution
Oleic only	Conc.	26.8	16.71	49.88	87.3
	Tail.	73.2	0.89	96.82	12.7
	Feed	100.0	5.13	84.24	100.0
OA:PA=85:15	Conc.	22.3	17.95	47.77	79.5
	Tail.	77.7	1.33	95.67	20.5
	Feed	100.0	5.03	84.99	100.0
OA:PA=75:25	Conc.	28.8	14.78	55.98	85.0
	Tail.	71.2	1.05	96.47	15.0
	Feed	100.0	5.01	84.81	100.0
Oleic only With silicate	Conc.	19.4	20.31	39.53	78.2
	Tail.	80.6	1.36	95.63	21.8
	Feed	100.0	5.04	84.75	100.0
OA:PA=85:15 With silicate	Conc.	21.5	19.32	42.54	82.3
	Tail.	78.5	1.14	96.21	17.7
	Feed	100.0	5.04	84.67	100.0
OA:PA=75:25 With silicate	Conc.	20.3	20.51	39.14	82.5
	Tail.	79.7	1.11	96.29	17.5
	Feed	100.0	5.04	84.69	100.0

Phosphate Flotation Using Cottonseed Soapstock Fatty Acid Extract

The cottonseed soapstock fatty acid extract previously described was tested as a 1.0:0.6 ratio mixture with No. 5 fuel oil as the phosphate collector. The collector mixture was fluid at a room temperature of 25 degrees C. Usage levels were 1.2, 1.4 and 1.6 lbs. of fatty acid extract per ton of feed. Conditioning pH was 9.0–9.1. This reagent was a very poor and unselective phosphate collector. Concentrates analyzed less than 4% P_2O_5, and P_2O_5 recoveries never reached as high as 50%. Flotation concentrate silica sand appeared to be stained a light brown color when this collector was used.

Table 8 Size distribution of tails and P_2O_5 recover from different size fractions for tests using various unsaturated C16–C22 fatty acid collectors and N-silicate (0.5 lb/t) at 1.6 lb collector and 0.96 lb fuel oil per ton of feed

Collector	Tails Size Distribution (%wt.)			%P_2O_5 Recovery of Different Sizes		
	+28 Mesh	28x35 M	-35 M	+28 Mesh	28x35 M	-35 M
Palmitoleic (98%)	2.1	8.2	89.7	24.5	62.1	70.8
Oleic (99%)	1.8	8.0	90.2	36.8	70.1	74.0
Oleic (71%)	1.7	7.4	90.9	48.2	82.9	81.5
Linoleic (99%)	1.7	7.6	90.7	47.7	82.6	88.0
Eicosenoic(98%)	1.8	6.9	90.3	42.0	75.1	82.9
Erucic	2.3	9.0	88.7	3.6	34.1	49.0

Table 9 Losses of P_2O_5 in various flotation tailings fractions

Collector	%P_2O_5 Distribution in Tailing		
	+28 Mesh	28x35 M	-35 M
Palmitoleic (98%)	6.0	5.6	22.3
Oleic (99%)	5.0	4.4	19.7
Oleic (71%)	4.3	2.7	14.8
Linoleic (99%)	4.3	2.5	9.2
Eicosenoic(98%)	4.6	3.7	13.1
Erucic	8.0	10.2	40.3

Phosphate Flotation Using Liqro GA Tall Oil/Cottonseed Fatty Acid Extract Mixtures

Three mixtures of the cottonseed soapstock fatty acid extract with Liqro GA tall oil were compared with 100% Liqro GA as the phosphate collector in this test series. The mixtures tested were 10%, 20% and 33% "extract" with Liqro GA tall oil. The mixtures were blended with No. 5 fuel oil at 1.0:0.6 ratio and were used at the 1.4 lbs. of tall oil/"extract" mix plus 0.84 lbs. of fuel oil per ton of feed level. Each collector mix with fuel oil was fluid at room temperature. Conditioning pH was 9.0–9.2, and N-silicate was used for all tests. The data illustrates that the addition of "extract" to Liqro GA tall oil resulted in a decrease in concentrate grade and a loss of P_2O_5 recovery as the percentage of "extract" was increased. Concentrate grade dropped from 24% P_2O_5 to as low as 10% P_2O_5 while P_2O_5 recovery decreased from 89% to about 67% when the "extract" comprised 33% of the collector mixture.

Size/Assay Analyses of Selected Flotation Tails

Abbreviated dry screen analyses were performed on the six selected flotation tailing samples produced using the various unsaturated fatty acids with fuel oil and N-silicate. Sizing the tailings at 28 and 35 Tyler mesh, using a 15 minute Ro-tap time, was performed. The tailings size/assay results are shown in Table 8. The Table 8 results support the general trend that highest % P_2O_5 tailing values are present in the coarsest (+28 mesh) size fractions and the lowest % P_2O_5 tailing values are present in the finer (-35 mesh) size fractions for each fatty acid collector tested. Accordingly, the concentrate % P_2O_5 recoveries in the -35 mesh size fractions are higher than for the coarser size fractions. Linoleic acid, followed by oleic acid 401 and eicosenoic acid, were shown to be the most effective phosphate collectors for each size fraction. Table 9 presents % P_2O_5 distribution within the flotation tailing size fractions for each of the tests listed in Table 8.

None of the individual fatty acids tested recovered 90% or higher of the phosphate present in the -35 mesh feed size fraction, or 85% or higher of the phosphate present in the 28/35 mesh size fraction. These recovery levels were previously obtained and exceeded using Liqro GA tall oil only. The better frothing character of the Liqro GA containing some rosin acids is believed to account in part for its superior performance.

CONCLUSIONS

Unsaturated fatty acids were found to be more suitable for phosphate flotation.

C18 unsaturated fatty acids, such as oleic and linoleic acids, perform better than other compounds.

The relatively pure fatty acids are not as powerful as the tall oil type collectors commonly used by the Florida phosphate industry.

REFERENCES

Kirk Othmer, Encyclopedia of Chemical Technology, Vol. 5, 4th edition, John Wiley & Sons, 1993, pp. 149–150.

Lange's Handbook of Chemistry, J.A. Dean, 10th edition, McGraw-Hill Book Co., 1967.

Tall Oil & Its Uses, L.G. Zachary et al., F.W. Dodge Co., McGraw-Hill Book Co., 1965, pp. 25–26.

Fats & Oils, H.G. Kirschenbauer, Rheinhold Publishing Corp., 1960, pp. 6–17; 169–171.

Industrial Fatty Acids, E.S. Pattison, Rheinhold Publishing Corp., 1959, pp. 1–10; 29–33.

An Introduction to the Chemistry of Fats & Fatty Acids, F.D. Gunstone, Chapman & Hall Ltd., 1958, pp. 1–21, 57–67, 76–82.

Fatty Acids & Their Derivatives, A.W. Ralston, John Wiley & Sons, 1948, pp. 35, 96–150, 261–269.

Kirk Othmer, Encyclopedia of Chemical Technology, Vol. 6, Interscience Publishers, Inc., 1947, pp. 20–37.

Fatty Acids, K.S. Markley, Interscience Publishers, Inc., 1947, pp 20–37.

Merck Index, 5th edition, Merck & Co., 1940, pp. 92, 170, 217, 319.

Cottonseed Soapstock, J.J. Ganucheau, Thesis, Tulane University, 1931.

The Chemistry of Cottonseed Oil, G.S. Jamieson and W.F. Baughman, *J. Am. Chem. Society*, Vol. 42, 1920, pp. 1197–1204.

Finch, E. and Riggs, W.F., *Fatty Acid Collectors—A Selection Guide, Chemical Reagents in the Mineral Processing Industry*, SME, Littleton, CO, 1986, pp.95–98.

Zhang, P., Albarelli, G.R. and Stewart, K.J., Phosphate Beneficiation Bibliography, FIPR Pub. No. 96-02-114, 1999, pp. 158 and 159.

Hsieh, S.S., Flotation Studies on Carboxylic Acid Components of Tall Oils, SME, *Transactions AIME*, Vol. 272, 1982, pp. 2013–2017.

Hsieh, S.S. and Lehr, J.R., *Pilot-Plant Evaluation of Alternative Flotation Reagents*, Division of Chemical Development, National Fertilizer Development Center, TVA, Muscle Shoals, AL, 1982, 17 pgs. and 7 tables.

SECTION 4

Technology Innovations: Processing

CHAPTER 13

Column Flotation: A Key to the Production of Apatite Concentrates from Slimes

Rogerio C. Guimarães,* Antonio E.C. Peres,† and Armando C. de Araujo†

ABSTRACT

The relevance of the production of apatite concentrates from slimes is discussed, taking into consideration the Brazilian scenario. Apatite recovered from slimes represents 12% of an overall production of 5.1 Mt/year. Bunge Araxá was the first concentrator to produce apatite concentrates from slimes, the contribution to their total production being at present 34%. They were also the pioneers regarding the introduction of flotation columns in the Brazilian phosphate industry. Their current circuit consists of six columns, the only circulating load being the tailings of the coarse apatite columns, which require re-grinding prior to being floated again. In other concentrators column flotation is utilized in the ultrafines circuit either in all stages or, at least, in the rougher and cleaner stages.

INTRODUCTION

Brazilian phosphate ores reserves are predominantly from igneous origin, representing 80% of the total reserves. Complex mineralogy and high degree of weathering, with consequent production of fines, are characteristics of all deposits. Despite some attempts of quantifying slimes as the minus 10 mm fraction, regarding industrial practice the expression slimes refers to the fraction of the feed of a concentrator that impairs significantly the flotation performance. This fraction must be rejected in a desliming operation, most commonly performed in hydrocyclones. Desliming at -325# or 44 μm was the usual Brazilian phosphate industry plant practice in the early eighties. The first industrial experiences of producing apatite concentrates from the slimes fraction started at Arafertil, currently Bunge Fertilizantes SA, in Araxá, still in the eighties. In 1985, the production of concentrate from slimes reached 197,000 t/year, corresponding to 24% of the overall concentrates production. The performance of slimes flotation and the process control, at Bunge's concentrator, were improved after the introduction of column flotation technology, in the first half of the nineties. This production corresponded in November 2004 to 34% of the total production of concentrates in the concentrator (830,000 t/year) in a circuit consisting of desliming in 40mm microcyclones and flotation in columns, corn starch and fatty acids being utilized as reagents. The column flotation technology for slimes recovery was expanded to other Brazilian concentrators still in

*Bunge Fertilizantes SA and Centro Universitário do Planalto de Araxá Uniaraxá.
†Federal University of Minas Gerais UFMG.

the nineties: Fosfertil (Catalão – GO), Bunge (Cajati – SP), Fosfertil (Tapira – MG) and Copebrás (Catalão – GO). At present the Brazilian production of phosphate concentrates is approximately 5.1 Mt/year, with a strong trend for a market increase. Brazilian technology is utilized for the processing of slimes, the concentrates representing approximately 12% of the overall production.

ARAXÁ'S PHOSPHATE MINERALIZATION

Arafértil SA, at present Bunge, was constituted in 1971, following economic feasibility studies of the apatite deposit carried out in the late 1960s. Preliminary results from a pilot plant encouraged the commissioning of a semi-industrial plant with a phosphate concentrate production capacity of 10,000 t/year. Although the plant commenced operation in September 1977, the official start up was not until August 1978.

The shape of Barreiro carbonatite complex is approximately circular (4.5 km diameter). The host rocks consist mainly of carbonatites and glimerites. Bunge's concession covers 20% of the complex total area, located in its northwest portion. Weathering on glimerite, that presents an essentially micaceous composition, caused a partial alteration to vermiculite; other silicates were transformed into chlorite group minerals; iron bearing minerals were altered; and part of the apatite was transformed into amorphous hydrated phosphates. The carbonatite suffered, beyond the above mentioned phenomena, leaching of carbonates and precipitation of iron, barium, phosphate, and silicon ions. The ore bearing zone consists mainly of apatite, clay type minerals, titanium minerals (ilmenite and anatase), carbonates (calcite and dolomite), among others. The mineralogical composition of the different "ore" types is shown in Table 1.

Currently mining is restricted to the oxidized ore, with small participation in the feed blend of the silicate ore. The size and chemical distribution (see Table 2) of the ore feeding the concentrator shows a predominance of phosphorus bearing minerals in the coarse fraction, while iron minerals, quartz, and silicates prevail in the fine fractions. Nevertheless, the ore is friable and altered, exhibiting a significant percentage of apatite in the fraction -37mm (-400# Tyler).

PHOSPHATE CONCENTRATES PRODUCTION FROM SLIMES AT BUNGE

At the beginning of Bunge's concentrator operation, only the coarse fraction of the ore (+44 μm or +325# Tyler) was utilized, causing significant phosphate losses in the slimes fraction (natural slimes, produced in the rod mill and slimes from grinding in the ball mill). These slimes were stored in tailings ponds from 1977 to 1984. Figure 1 shows the simplified flowsheet of the concentrator at this stage, including the changes from the original circuit.

From 1984 the concentration of slimes was performed after desliming in 40 mm microcyclones. Natural slimes (after grinding in a rod mill) and fines from grinding (produced in a ball mill) were floated in mechanical cells, as illustrated in Figure 2. A significant increase of 10% phosphorus recovery was achieved at a phosphate concentrates production level of 150,000 t/year.

From 1992 on, the concentrator suffered major changes with the introduction of flotation columns (CESL, at present CPTI), causing significantly enhanced recovery, reduction in reagents consumption, resulting in a much simpler flowsheet (Guimarães and Peres, 2000), illustrated in Figure 3. A supervisory control system was introduced in 1996. The 40 mm microcyclones were replaced by Mosley 2 in. cyclones in 1999.

The importance of producing phosphate concentrate from slimes is stressed in Figure 4. At present, 35% of the total production comes from slimes. Due to market problems, the slimes were not processed in 1989.

The success of the pioneering experience at Bunge's concentrator in Araxá encouraged people from other major phosphate plants in Brazil to process their slimes. Table 3 illustrates the main characteristics of these operations. Bunge Cajati operates a flotation circuit consisting of columns only for the ultrafines and mechanical cells for the coarser sizes, both in a rougher, cleaner, scavenger configuration. The old

Table 1 Approximate mineralogical composition of the different ore types

Minerals	Ore types		
	Oxidized	Silicate	Silico-carbonate
Apatite	28	24	25
Barite	7	1	2
Carbonates	3	4	19
Iron oxides	28	22	15
Quartz	10	7	7
Magnesium silicate minerals	14	32	26
Other minerals	10	10	6
Total	100	100	100

Table 2 Typical cumulative size and chemical distribution of the feed ore to the concentrator

Opening	Weight %	P_2O_5 %	Fe_2O_3 %	Al_2O_3 %	MgO %	SiO_2 %
1"	6.41	10.28	3.43	3.22	3.45	4.02
½"	17.14	25.73	9.22	9.79	11.37	12.89
3/8"	20.94	31.71	10.92	12.34	14.00	17.08
10#	35.73	50.71	22.78	20.66	21.96	31.03
48#	51.24	64.62	41.24	33.88	33.86	47.72
65#	56.31	69.38	47.07	39.33	39.12	53.39
100#	59.79	72.62	51.01	43.31	43.10	57.05
200#	70.49	82.70	62.93	56.51	55.69	67.88
400#	78.65	89.53	71.22	69.21	69.69	76.61
-400#	100.00	100.00	100.00	100.00	100.00	100.00

(before 1999) and the current (since 1999) flowsheets are illustrated in Figures 5 and 6.

Fosfertil's concentrator in Catalão operates three circuits, each one floating a different size range: coarse, fines, and ultrafines. The coarse apatite circuit consists of four stages, all of them operating with mechanical cells: barite preflotation, rougher, cleaner, and scavenger. The fines circuit consists of barite pre-flotation in mechanical cells and rougher apatite column flotation. The rougher concentrate feeds the coarse apatite rougher mechanical cell. The ultrafines circuit consists of rougher and cleaner column flotation stages, and barite pre-flotation and scavenger stages in mechanical cells.

Fosfertil's concentrator in Tapira consists of two major circuits, concentrating two differenore types: grainy and friable (Barros and Ferreira, 2004). The friable circuit is divided in coarse, fine, and ultrafine stages. The grainy

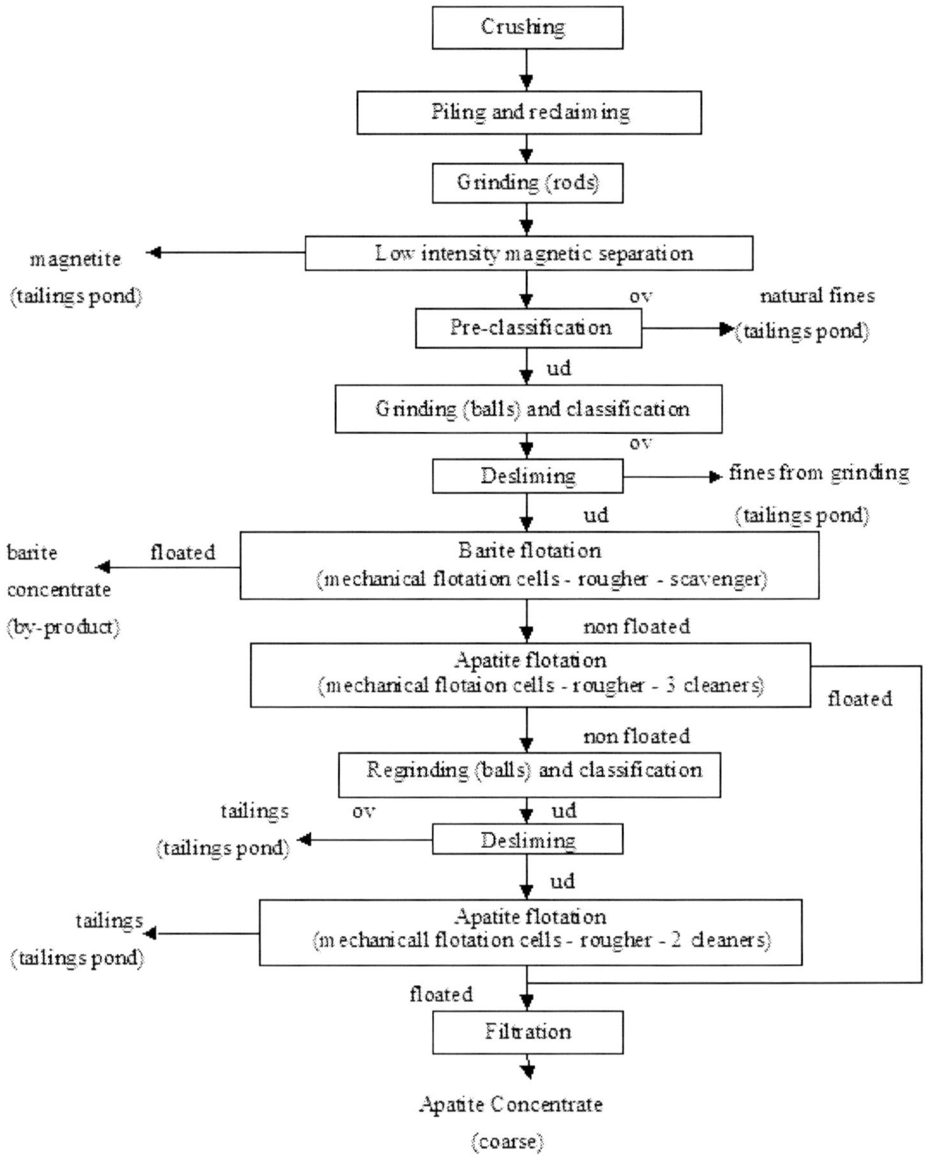

Figure 1 Bunge–Araxá concentrator flowsheet (1977–1984)

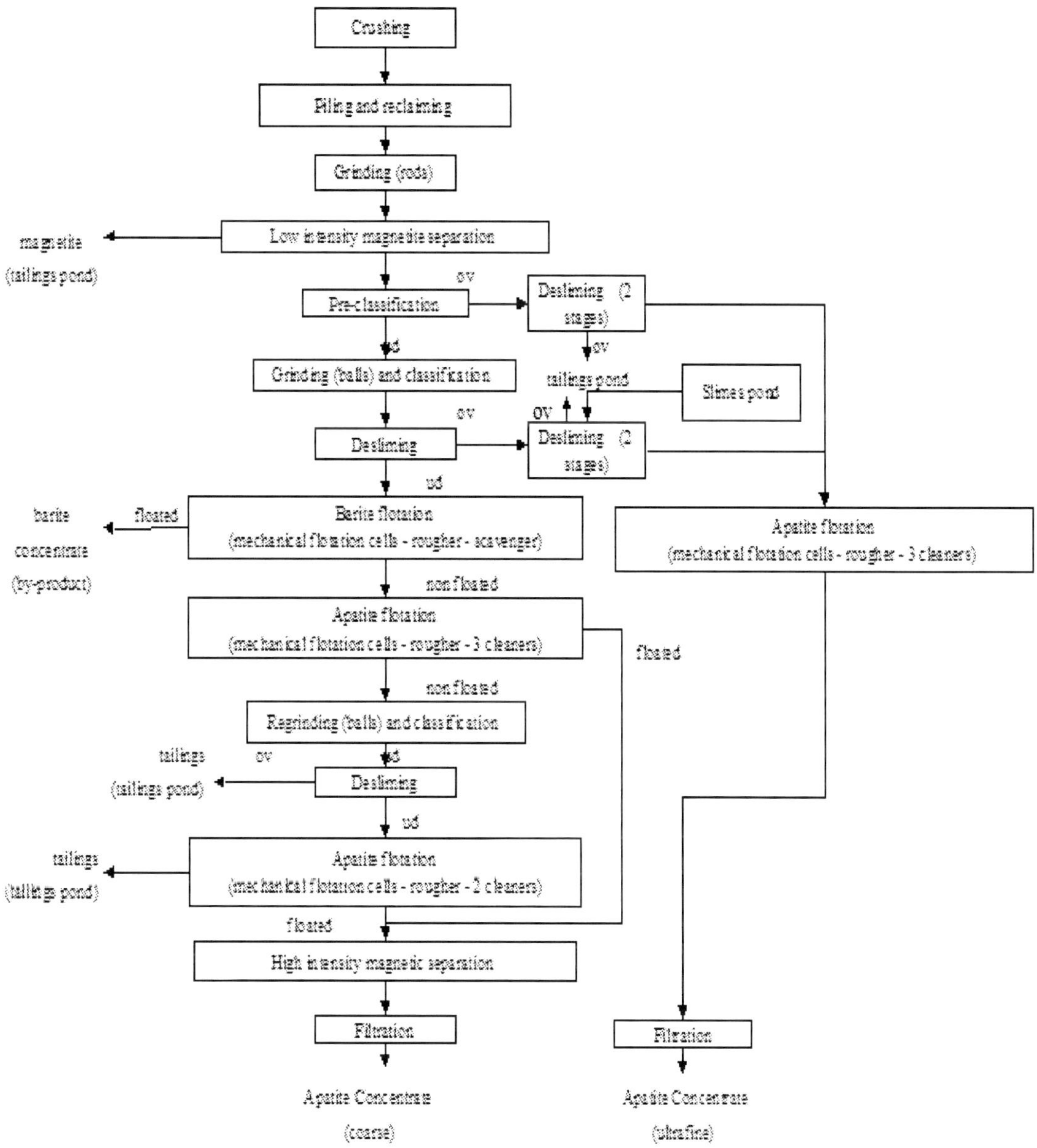

Figure 2 Bunge–Araxá concentrator flowsheet (1984–1991)

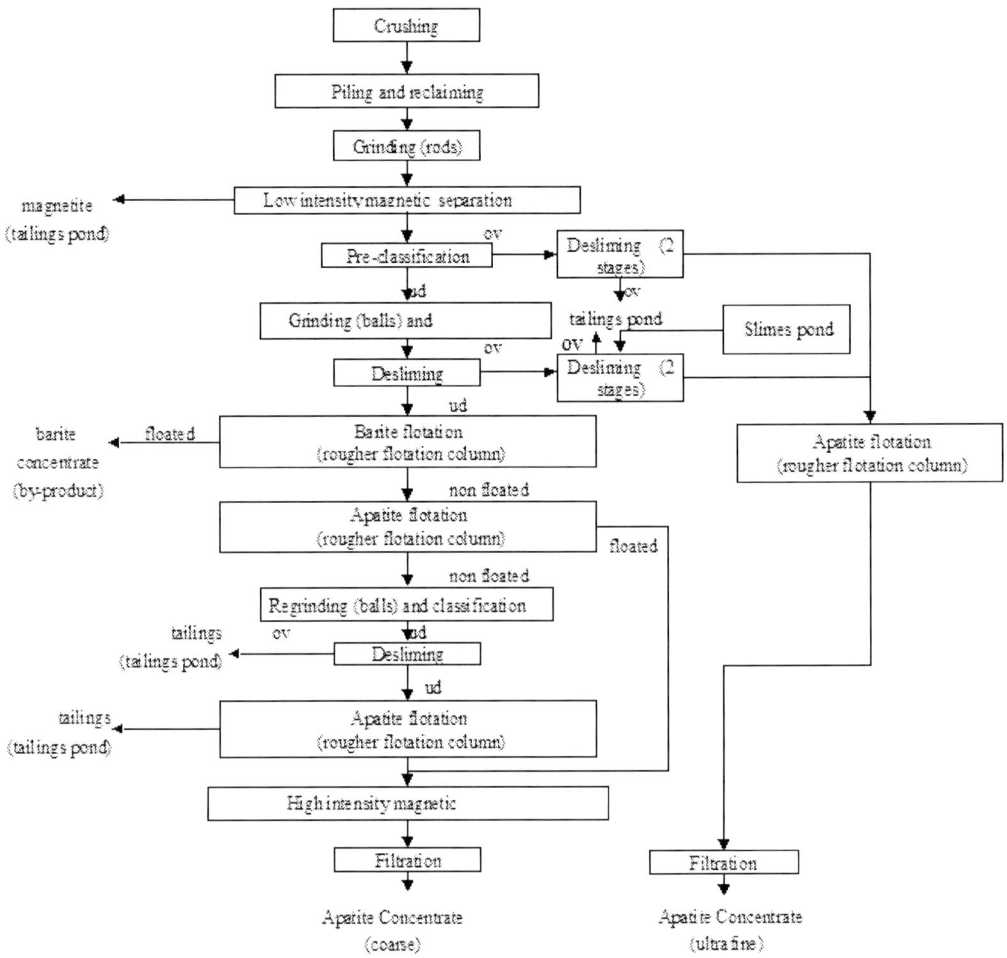

Figure 3 Bunge–Araxá concentrator flowsheet (1992–2004)

ore, coarse and fine friable ores circuits consist of mechanical cells in rougher, cleaner, recleaner, and scavenger configurations. The ultrafine friable circuit consists of rougher and cleaner flotation columns and mechanical cells scavenger stage. Flotation columns were utilized in the past in the grainy ore circuit and in the friable circuit treating fractions other than ultrafines. The operation of these columns is predicted to resume after the 2005 expansion of the concentrator.

CONCLUSIONS

Column flotation plays a prominent role in the Brazilian scenario of apatite concentrates production from slimes, representing 12% of the overall production of 5.1 Mt/years.

Different circuit configurations are utilized, ranging from columns in all stages, to a columns only ultrafines circuit, and also a combination of columns and mechanical cells in the ultrafines circuit, columns being employed in the rougher and cleaner stages.

Table 3 Main characteristics of processes producing phosphate concentrates from slimes in Brazil

Item	Bunge		Fosfertil Tapira	Fosfertil Catalão
	Araxá	Cajati**		
• Head grade - P_2O_5 (%)	12 – 14	4.5 – 5.0	7.5 – 7.8	10.2 – 11.0
• Nominal throughput (t/year)				
coarse fine	540,000	500,000	1,516,000	880,000
ultrafine	290,000	30,000	84,000	120,000
total	830,000	530,000	1,600,000	1,000,000
• Microcyclones				
supplier	Mozley	Mozley	Krebbs	Mozley – Multotec
diameter	2"	2"	2"	40mm
stage	2 (natural slimes) 2 (slimes from grinding)	2	3	3
desliming size (µm)	5 (natural slimes) 10 (slimes from grinding)	5 – 10	5 - 10	5 – 8
• Reagents system				
collector	rice bran oil soap	sarcosinate	rice bran oil soap + sulphosuccinate	rice bran oil soap
depressant	local corn starch "fubá"	pure corn starch	pure corn starch	pure corn starch
additive	-	-	-	ethoxylated nonyl phenol
• Flotation				
machine type	columns	columns	columns + mechanical	columns + mechanical
columns supplier	CESL (CPTI)	CESL(CPTI)	CESL (CPTI)	CESL (CPTI)
stages	rg*	rg – cl - sv	rg – cl - sv	rg – cl – sv
• P_2O_5 recovery in the ultrafine fraction	48	40	45	45
• Concentrate quality				
P_2O_5	33.0 – 34.0	33.0	33.5 – 34.0	33.5 – 35.0
F_2O_3	5.0 – 6.5	1.0 – 2.0	1.2 – 1.6	2.8 – 3.5
Al_2O_3	0.4 – 0.7	-	0.2 – 0.4	0.4 – 0.8
MgO	0. – 0.6	2.0	0.6 – 1.6	1.2 – 1.6
SiO_2	1.5– 2.0	-	2.5 – 4.5	4.0 – 5.5
application	fertilizer	H_3PO_4 – di-calcium phosphate	H_3PO_4 – fertilizer	fertilizer
• Start of operation	1984	1999	2002	1995

* rg = rougher; cl = cleaner; sc = scavenger

** At Bunge-Cajati concentrator the coarse and ultrafine concentrates constitute one single product.

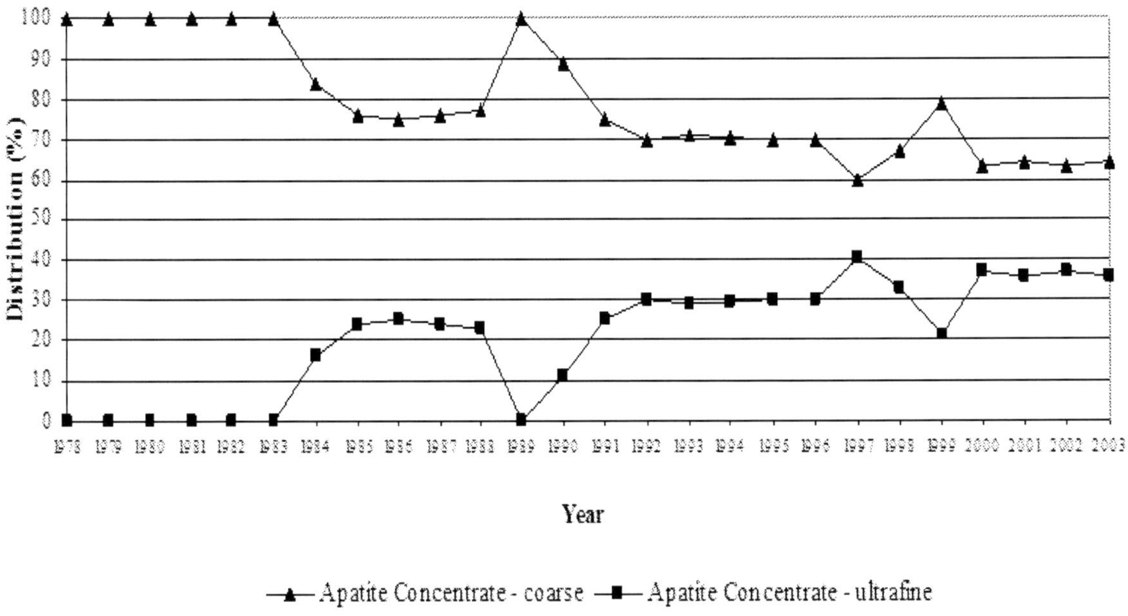

Figure 4 Profile of phosphate concentrate production at Bunge–Araxá (1978–2003)

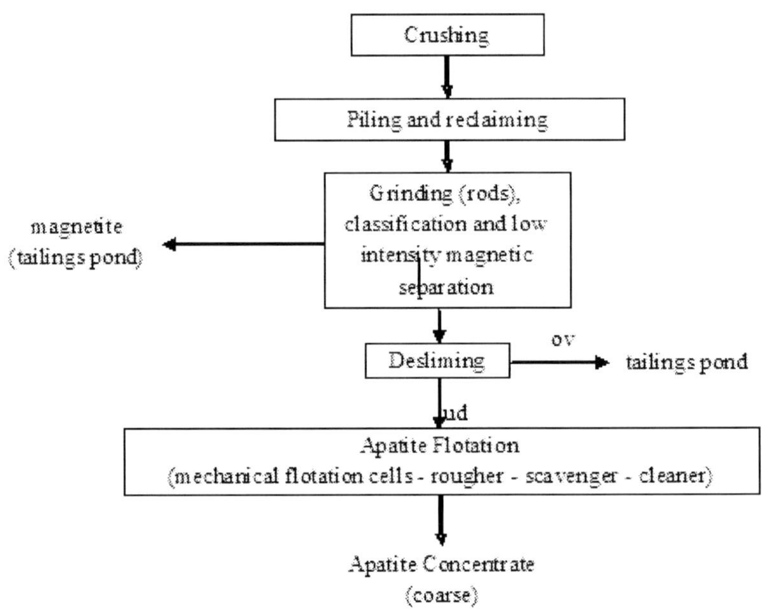

Figure 5 Bunge–Cajati concentrator flowsheet (before 1999)

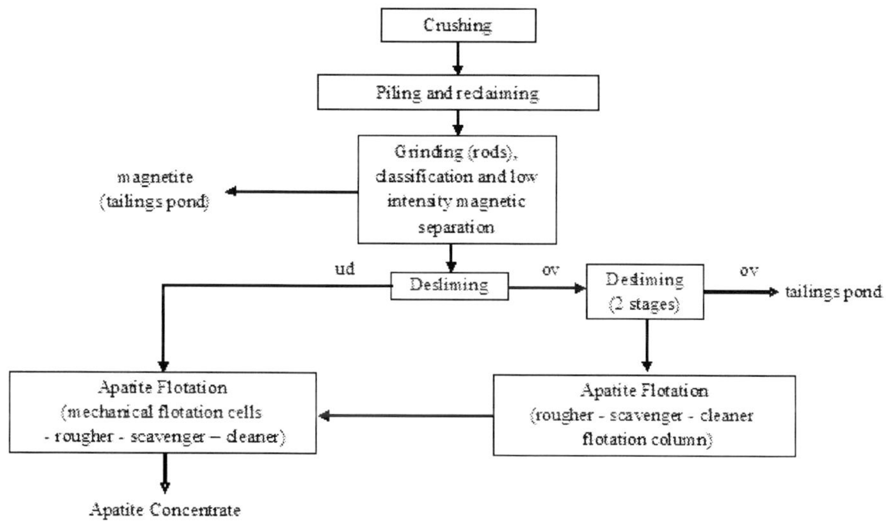

Figure 6 Bunge–Cajati concentrator flowsheet (1999–2004)

REFERENCES

Barros, L.A.F. and Ferreira, E.E., 2004. Personal communication.

Guimarães, R.C. and Peres, A.E.C., 2000. Industrial practice of phosphate ore flotation at Serrana–Araxá, Brazil. XXI International Mineral Processing Congress. Rome, Italy. pp B9-17–B9-23.

CHAPTER 14

Structural Features of Burenkhaan Phosphate Rocks and Their Mechanochemical Treatment

Kh. Maam*

ABSTRACT

The structural features of Burenkhaan phosphate rocks (Mongolia) and studies on the preparation of phosphorus fertilizers through mechanochemical activation of natural phosphate rocks are reported. Natural and activated phosphate rocks were characterized with inductively coupled plasma spectrometer, X-ray diffractometer, Fourier transform infrared spectroscopy and BET-N2 techniques. It was found that the Burenkhaan phosphate rock belonged to quartz-apatite type of ore consisting of flourapatite and hydroxyapatite with a hexagonal symmetry and a space group of P63/m and SiO_2 (quartz) having a trigonal symmetry and a space group $P3_22_1$. As a result of mechanically induced structural changes, the solubility of milled powders in citric acid solution was increased 1.7–2.9 times compared to that of initial phosphate rocks, demonstrating their potential as fertilizers for agronomic purposes.

INTRODUCTION

As main source of inorganic phosphorus in nature, phosphate rocks in terms of calcium apatite have attracted much attention not only for their technological importance in the production of phosphoric acid, phosphor-organic compounds and fertilizers but also for their potential in biomedical usage such as bioceramics: artificial bones, dental care, etc.[1].

Mongolia is one of the countries rich in the phosphate ores. As a reference, Burenkhaan phosphate ore was considered to be one of the largest phosphorite-bearing basins in the world, with predicted reserves of more than 4.5 billion tons of high and middle quality ores[2]. However, the deposit is situated in the specially protected area of the unique lake of Khubsgul, which is connected with the lake Baikal by the river Eg. Thus, exploration, mining and processing of this deposit requires special attention from an environmental point of view.

In this regard, much work has dealt with high energy ball-milling of minerals, particularly phosphate rock, as a non-equilibrium solid state process, which is not only important for practical applications but gives fundamental insight in mechanochemistry, in general[3]. Phosphorus fertilizers prepared in this way from phosphate rocks of low reactivity can be modified to provide a high rate of dissolution in soil giving readily available phosphorus for initial plant growth, in addition to providing vital micronutrients

*Institute of Chemistry and Chemical Technology, Mongolian Academy of Sciences, Ulaanbaatar, Mongolia.

such as Zn, Mo, Co, Mn, and Cu as well as a long-term residual effect.

M.B. Chaikina found that the reactivity of the phosphate rock varied depending on the operating conditions of the mill and its ball size as well as duration of treatment[4]. B.B. Boldyrev et al. demonstrated that the reactivity variation of the phosphate rock associated with mechanically induced distortion of the crystal structure in CaO_6F complexes as well as PO_4^{3-} tetrahedrons of the apatite structure[5]. It is well known that the presence of a second component such as quartz has an influence with an increase in the lattice distortion of phosphate minerals[6]. On the basis of field and growth chamber trials for wheat, M.B. Chaikina et al. reported that the mechanically activated phosphate rocks performed as effectively as triple super phosphate in various kinds of soils low in available phosphorus[7].

With the intention of processing Burenkhaan phosphate rocks in environmentally friendly ways, the present research is focused on the preparation of phosphorus fertilizers through mechanochemical treatment of natural phosphate rock and their structure/properties studies. The ability of natural and activated phosphate rock for release of nutrients are studied by means of inductively coupled plasma spectrometer (ICP), X-ray diffractometer (XRD), Fourier transform infrared spectroscopy (FTIR) and BET-N2 techniques.

EXPERIMENTAL

Coarse grained phosphate rocks of Burenkhaan deposit were used in this study. For mechanical activation, the sample was ground in air for different periods of time, using either a vibration mill, VCM-3 or planetary mill, ϑH-2×150. In both cases, the milling media were steel balls with diameter of 8 mm, the mass ratio of balls to powder being 100:1.

Chemical compositions of the phosphate rock were determined with wet chemical methods[8]. Contents of the microelements were determined by an ICP spectrometer. Reactivity of the samples was examined by the contents of P_2O_5 of which soluble in citrate solution (2 wt%) as well as neutral ammonium citrate solution. Since the associated free carbonates in phosphate rock may affect the solubility of apatite in citrate solution, the following double extraction was used. A 1 g sample was extracted with 100 ml of neutral ammonium citrate solution at 65°C for 1 h with constant agitation. The mixture then was filtered, and the residues were extracted again with 100 ml of fresh neutral ammonium citrate solution. Phosphorus concentrations in the two filtrates were determined by the molybdovanadate method[9].

The XRD pattern was recorded on a Siemens D500 powder diffractometer operating in reflection geometry with Cu Ka radiation (a=1.54439Å). The data were collected over the 2θ range from 10° to 80° with a step size of 0.02° and scanning rate of 30 seconds per step. FTIR spectra in a region of 400–4000 cm^{-1} were obtained with a Shimadzu FTIR-8300 spectrophotometer (KBr, pellet). Specific surface area was determined by BET method on a DeSorb 2300A particle analyzer (Micromeritics).

RESULTS AND DISCUSSION

Table 1 shows chemical analysis of the Burenkhaan phosphate rock. It contained 35.5% P_2O_5 of which 5.35% is citrate soluble and 13.4% SiO_2, belonging to quartz-apatite type of ore. Flourine content of the sample was found to be 2.41, value smaller than that of corresponded to the stoichiometric composition (3.12%). Probably this difference was caused by the fact that a chemical substitution occurs naturally within the crystal structure of the apatite where some fluorine ions have replaced carbonate, hydroxyl ions and, for calcium ions occur partial isomorphic substitution with strontium, barium etc. Contents of the accompanying microelements such as Zn, Mn, Co, Cu, and Mo appeared to be an efficient source of the vital nutrients required by plants.

Figure 1 shows FTIR profiles. The native phosphate rock showed a couple of sharp absorption peaks at 570, 605 cm^{-1} corresponding to the P-O deformational oscillations and at around 1,000 cm^{-1} arising from P-O valence

Table 1 Chemical analysis of the Burenkhaan phosphate rock

Component	Main component, %									Microelements, ppm					
	P_2O_5	CaO	CO_2	SiO_2	MgO	Al_2O_3	Fe_2O_3	Na_2O	F	Mo	Zn	Cu	Nb	Zr	Ba
Content	35.0	45.8	4.4	8.36	0.84	0.76	0.97	0.5	2.41	11	55	75	3	70	11

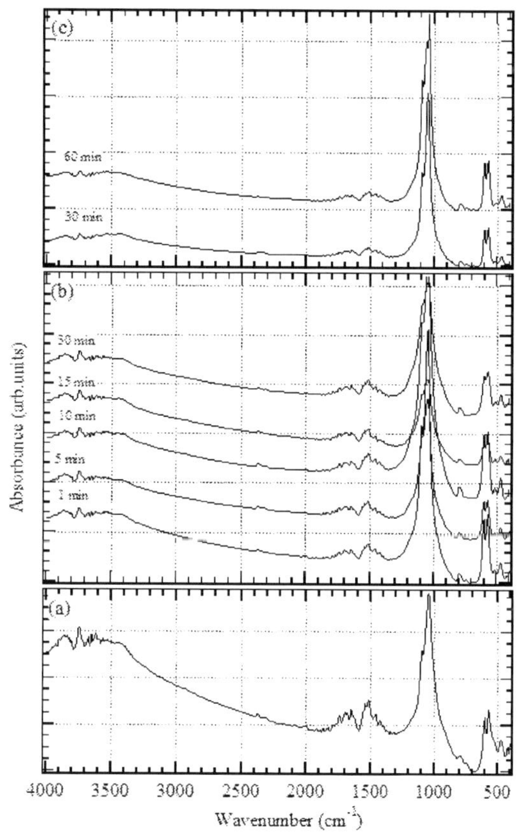

Figure 1 FTIR spectrums of native phosphate rock (a) and phosphate rock powders grinded in either planetary (b) or vibration (c) mills for different periods of time

oscillations of the PO_4^{3-} moieties. The absorption bands resulting from silica and carbonate appeared at 474, 690 cm^{-1} due to Si-O stretching of silica and at 1,419, 1,630 cm^{-1} resulting from C-O valence oscillations of CO_3^{2-} moieties, respectively. It is believed that absorption bands of fluorine ions are overlapped at 400–500 cm^{-1}. It can be seen from the Figure 1(b,c) that the frequency of any absorption band remains practically unaltered for the samples which were mechanically activated at either vibration and planetary mills for different periods of time.

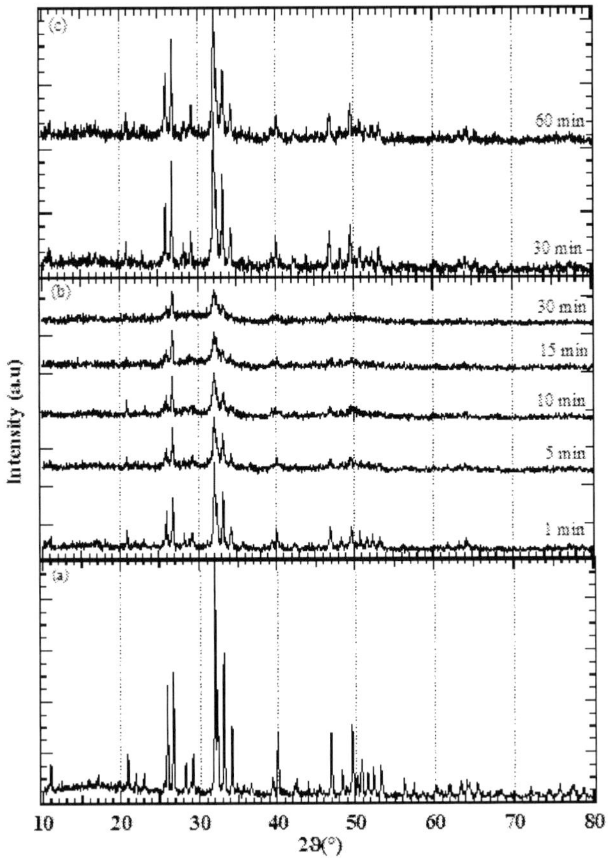

Figure 2 XRD patterns of native phosphate rock (a) and phosphate rock powders grinded in either planetary (b) or vibration (c) mills for different periods of time

Table 2 Characteristics of powders

		Surface area, m^2/g	Solubility in citric acid (2%), %	Solubility in ammonium citrate, %
Native rock phosphate		3.24	6.1	0
Activated at planetary mill for	1 min		11.8	4.7
	5 min	6.84	14.9	9.9
	10 min		16.6	10.6
	15 min	7.5	17.0	11.6
	30 min		17.9	12.5
Activated at vibration mill for	30 min		9.6	6.25
	60 min		10.6	7.5

However, the variations in intensities of the absorption bands corresponding to PO_4^{3-} moieties relative to the other spectral features are most probably due to order–disorder evolution with the particles.

XRD patterns of the native and as milled phosphate rocks are presented in Figure 2. Phosphate rocks of the Burenkhaan consists of two phases. The first phase corresponds to flourapatite and hydroxyapatite with a hexagonal symmetry and a space group of $P6_3/m$. By using Rietveld method, unit cell parameters of the crystalline structures were determined as follows: a=b=9.3777Å, c=6.8868Å and γ= 120.0176°. The second phase was identified as SiO_2 (quartz) having a trigonal symmetry and a space group $P3_22_1$.

As shown in Figure 2(b,c), the samples activated at vibration mill for 30 and 60 min exhibited broad diffraction peaks compared to that of the native rock phosphate reference. For the case of planetary mill, more intensive broadening of the diffraction peaks were observed with increase of milling time from 1 min to 30 min. These results indicate that nanocrystallization and/or amorphization occurred as milled samples. It should be noted that mechanical activation by planetary mill reduces average particle size in relatively short periods of time.

As already mentioned, the purpose of this study was a preparation of phosphorus fertilizers from the natural phosphate rock through mechanical activation. The efficiency of such fertilizers is mainly characterized by an environmental reactivity, i.e., their solubility in citric acid (2 wt%) and ammonium citrate solution. Since natural apatite has a lower solubility in neutral and alkaline mediums and it is hardly accessible for plants if used directly as fertilizers. However, the solubility of the phosphate rock can be increased as a result of mechanically induced changes in surface structure during grinding.

The most striking behaviour displayed by the mechanically activated powders was seen in reactivity in terms of solubility in citric acid and ammonium citrate solution given in Table 2. The main feature in Table 2 is that the citric acid solubility increases as surface area of the as milled powders increases. In comparison with native phosphate rocks, the solubility of the samples activated at planetary mill is increased 2.9 times whereas 1.7 times for samples grinded in vibration mill. This could be explained by the increased specific surface area of the as milled powders as well as the mechanically induced structural changes of phosphate rocks. Due to pressure and shear, a large number of new surfaces having serious inhomogenities, like lattice distortion, reduced symmetry and broken exchange bonds for surface atoms, were formed during mechanical activation. Such surface disorder is the major driving force for increased reactivity of the as milled powder since symmetry breaking lead a destabilization of electronic structure for surface atoms. This view is supported by the fact that dissolution rate of as milled phosphate rock increased as specific surface area increased as well as X-ray intensities broadened. Note that X-ray reflections are more sensitive for smaller crystals.

A preliminary agronomic trial was also conducted at a growth chamber for tomatoes. It was demonstrated that no significant difference was found in the growth and yield of the tomato, regardless of applying mechanically activated phosphate rock or triple super phosphate.

CONCLUSION

We could show that it is possible to prepare phosphorous fertilizer via mechanical activation of low soluble phosphate rock. As a result of high energy ball milling, solubility of phosphate rock in citric acid and neutral ammonium citrate solution increased 1.7–2.9 times compared to that of initial phosphate rock. A preliminary agronomic trial demonstrated that the as milled phosphate rock can be used as phosphorous fertilizer.

ACKNOWLEDGMENTS

The author gratefully acknowledges the contributions of G. Batdemberel, School of Materials Technology, Mongolian University of Science and Technology, Ulaanbaatar, Mongolia; and J. Amgalan, Institute of Chemistry and Chemical Technology, Mongolian Academy of Sciences, Ulaanbaatar.

REFERENCES

1. Senamaud, N., et al., 1997, *Solid State Ionics*, 101–103, p.1357.
2. Kaljuvee, T., Kuusik, R., and Veiderma, M., 1995, *International Journal of Mineral Processing*, 43, p. 113.
3. Raynaud, S., et al., 2002, *Biomaterials*, 23, p. 1065.
4. Dorozhkin, S., 1997, *J. Cryst. Growth*, 182, pp. 125–133.
5. Amgalan, J., et al., 2003, Crystal structure of phosphorite of Mongolian Burenkhan deposit and its mechanochemistry. *Comm. of Mongolian Academy of Sciences*. N2, p. 3–15. 2003 (in Mongolian).
6. Amgalan, J., et al., 2005, Investigation of crystal structures of natural and mechanochemically activated phosphorite minerals from some Mongolian deposits. International School on Contemporary Physics (ISCP-III), Ulaanbaatar, Mongolia. August 8–15, *ISCP-III Proceedings*, p. 120.
7. Amgalan, J., 1999, Agrichemistry microelements (Mn, B, Cu, Zn, Mo, Co, Ni) of Mongolian soil and mechanochemical technology of phosphoric fertilizers production. Mongolia, Ulaanbaatar, 1999, p. 144–246 (in Russian).
8. Amgalan, J., Chaikina, M.V., and Bilegbaatar, A., 2003, Mechanochemical technology of natural phosphorites in Mongolia. INCOME-2003, Fourth International Conference on Mechanochemistry and Mechanical Alloying, Braunschweig, Germany.
9. Rietveld, A., 1998, Profile Matching and Integrated Intensiteis Refinement of X-ray and/or Neutron Data (powder and/or single-crystal), Version 3.5d Oct98-LLB-JRC, Juan Rodriguez-Carvajal, Laboratories Leon Brillouin (CEA-CNRS).

CHAPTER 15

Flotation Alternatives for Florida Phosphate

Glenn A. Gruber*

ABSTRACT

Over the last 50 years the Crago process has been used almost exclusively to maximize the extraction of high-grade phosphate rock from Florida phosphate ores. The success of this reverse flotation process is dependant upon direct rougher flotation with anionic reagents, de-oiling of the rougher concentrate with sulfuric acid, and inverse cleaner flotation using cationic reagents.

A pilot plant program to investigate alternative flotation processes that use only anionic reagents was funded by the Florida Institute of Phosphate Research and performed by Jacobs. Four flotation processes were tested using feed from three mines. Some tests were performed with plant process water.

The results indicate recovery improvements and reductions in reagent costs are possible with an anionic flotation process. The direct anionic flotation schemes produced a concentrate with higher acid insoluble content than the Crago process.

INTRODUCTION

Phosphate mining has been a vital industry in Central Florida for more than a century. The phosphate mineral occurs in all particle sizes of the ore; however, only the enriched pebble fraction (>1 mm material) can be recovered as a commercial product by simple washing. The advent of froth flotation made beneficiation of the feed fraction (1/0.1 mm material) a commercial reality. Many flotation processes were examined, but the patented Crago process (Crago, 1942) became the standard method of recovering phosphate from the feed fraction of Florida phosphate ores. The Crago process allowed significant quantities of phosphate resources to be reclassified as ore, and a significant period of industry growth was initiated.

It should be noted that the Crago process evolved in an industry where the economics of easily recovered pebble and the requirement for high-grade concentrate dominated beneficiation. Although the Crago process is used exclusively, a re-evaluation is warranted because of the following evolutionary changes in the phosphate industry.

Demand for high-grade concentrate has been significantly reduced because:

- Phosphate is now exported in the form of high analysis fertilizers rather than high-grade rock.

*Jacobs Engineering Group, Lakeland, Florida.

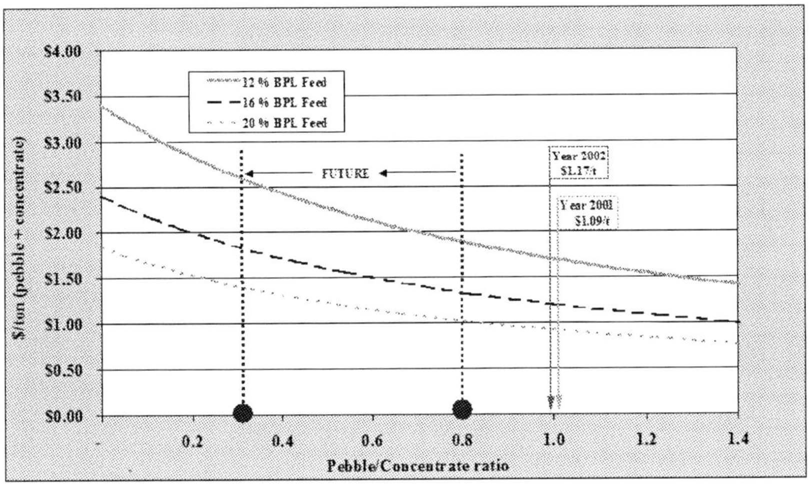

Figure 1 Impacts of ore characteristics on reagent costs

- DAP and MAP fertilizers have displaced TSP in the market place. The former do not require high-grade rock.
- Florida phosphate deposits having a high yield of low-cost pebble are being depleted. The remaining ore reserves contain relatively more phosphate rock as concentrate than as pebble. Consequently the costs of applying the reagent suite required by the Crago process will continue to increase.

The Crago process utilizes three sets of reagents. Anionic flotation reagents (fatty acids, fuel oils, and alkaline pH modifiers) are used to recover phosphate by rougher flotation; sulfuric acid is used to remove anionic reagents from the rougher concentrate; and cationic flotation reagents (amines and possibly diesel oil) are used to remove quartz by cleaner flotation. The costs of sulfuric acid and the cationic reagents typically range from 30–50% of the total reagent cost. Only anionic flotation reagents were used for the alternative processes described in this paper. Figure 1 illustrates the impacts of ore richness on reagent cost, expressed as $/ton of total product. Reagent cost will increase as future mining extracts leaner and leaner ore unless compensating changes are made in the flotation process.

METHODOLOGY

Development Testing

Prior to this project, six variations of anionic flotation were identified for flotation of Florida phosphate. The six process options are illustrated on Figure 2. Jacobs performed comparative testing of options 1 through 4, while the Florida Institute of Phosphate Research performed comparative testing of options 1, 5, and 6. This development testing led to the process options selected for testing by this program.

Preliminary Laboratory Tests

More than 100 laboratory flotation tests were performed on samples of phosphate flotation feed collected from four plants operated by Cargill Crop Nutrition, CF Industries, IMC Phosphates, and PCS Phosphate. The laboratory equipment used for this testing included a Labmaster LIU08 mixer operated at 350 rpm for anionic conditioning and a DECO Model D-12 laboratory flotation machine operated at 1,500 rpm for anionic flotation. The tests compared the collectors used by the four plants at that time with various promising anionic collectors. The test data were evaluated to select effective collectors for the pilot-scale evaluation

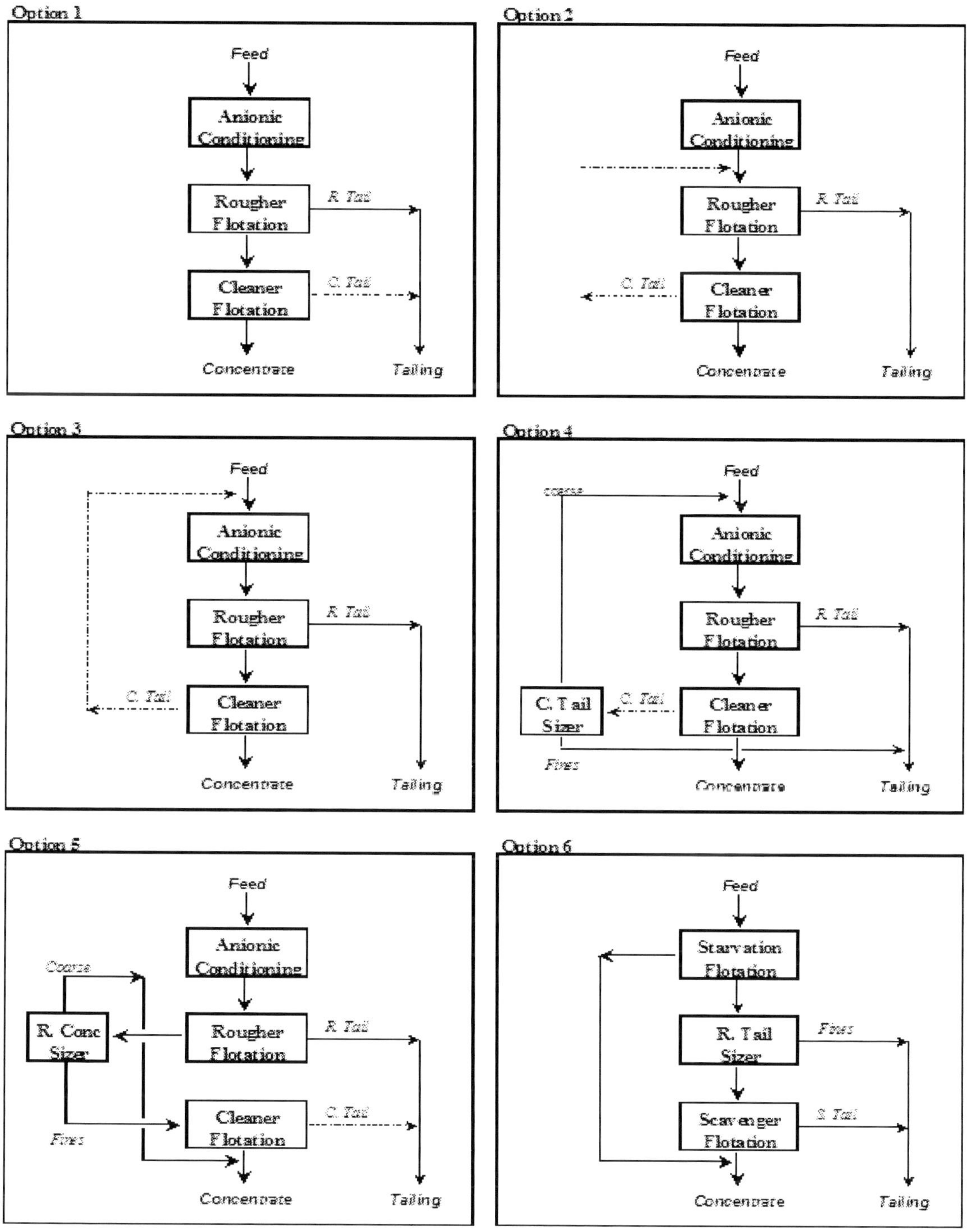

Figure 2 Process options for anionic flotation

Table 1 Major stream data

	Head Sample	Tyler Mesh Fractions >28	28/35	35/48	48/65	65/100	100/150	<150
Slurry kg/hr	yes	no	no	no	no	no	no	no
Solids kg/hr	yes	no	no	no	no	no	no	no
% weight	no	yes	yes	yes	yes	yes	yes	yes
% BPL	yes	yes	yes	yes	yes	yes	yes	yes
% Insol	yes	yes	yes	yes	yes	yes	yes	yes

of anionic flotation process alternatives to the Crago process.

Test Materials and Quantities

Feed samples collected from three phosphate flotation plants operating in Florida and transported by a 15-ton haul truck to Jacobs' pilot plant in Lakeland, Florida. The quantity of sample from each plant follows.

- Plant 1: 2 loads
- Plant 3: 1 loads
- Plant 4: 2 loads

The feed was dumped onto a clean paved area near the pilot plant. A Bobcat was used for blending and reclaiming the feed.

Plant reagents were used for Crago process flotation tests. Anionic collectors, selected on the basis of laboratory test results, were obtained for rougher-cleaner flotation testing. Soda ash was used as the pH modifier for the three feed samples. Sodium silicate was used for some tests as a depressant.

Some testing was performed with plant process water, which was collected and transported to the pilot plant by tanker truck.

Table 1 lists the major parameters analyzed on processing streams.

DISCUSSION OF TEST RESULTS

Development

Referring back to Figure 2, Jacobs bench scale development testing concluded the following:

- Option 1 produces higher-grade concentrate and lower recovery than Option 2
- Option 3 gives a relatively higher recovery of concentrate at an unacceptable grade
- Option 4 gave the best performance of the rougher-cleaner configurations and a very stable recycle of middlings
- The high-grade operating mode of the Crago process had higher reagent costs and lower recovery than the anionic options
- The high-recovery operating mode of the Crago process had a comparable recovery but higher reagent costs than Option 4

The Florida Institute of Phosphate Research lab tests demonstrated that options 5 and 6 had potential to improve recovery relative to normal rougher-cleaner flotation and provide acceptable concentrate grade. Consequently, options 4, 5, and 6 were selected for pilot testing. Not shown on Figure 2 for option 6 are two required anionic conditioning steps, the first prior to starvation flotation, and the second prior to scavenger flotation.

Reagent Tests

The results of testing the four plant feeds with different anionic reagents are shown on Figures 3, 4, 5, and 6. The coefficient of separation was plotted against collector usage to compare reagent selectivity. Collector usages, expressed as pound of collector per ton of new feed to flotation (Lbs/t) are compared over a range of 0.2 to 1.2 Lbs/t.

(Coefficient of Separation = % BPL recovery – % Insol recovery)

The usages shown are for collector only. Results obtained from testing four plant feeds are discussed below.

From Figure 3 it is apparent that Sylfat FA12 was more selective than the plant collector and the other collectors tested on feed from plant 1. The results from testing plant 2 feed are illustrated on Figure 4, showing that CENTURY MO-5 was the most selective of the collectors tested. Two separate samples of plant 3 feed were tested and the results showed that Sylfat FA12, CENTURY MO-5, and the plant collector had similar selectivity. The data for plant 3 are displayed on Figure 5. From Figure 6 it is apparent that CENTURY MO-5 was more selective than the collector used by plant 4 and more selective than the other reagents tested, except for Sylfat FA12 at higher reagent usage.

No single anionic reagent demonstrated superior performance for all feeds tested.

Anionic Processes Pilot Tested

Based on the development work, the three most promising anionic flotation options were identified and pilot scale tested. The pilot plant flow diagrams for these processes are illustrated on Figures 7, 8, and 9. Options 4, 5, and 6 each use size classification to enhance the flotation performance. Option 4 sizes the cleaner tailing and recycles the oversize to conditioning and rougher flotation. Option 5 sizes the rougher concentrate and sends only the fine concentrate to cleaner flotation. Option 6 sizes the rougher tailing and sends the coarse tailing to conditioning and scavenger flotation.

The equipment numbers below are also shown on the flow diagrams corresponding to option 4 (Figure 7), option 5 (Figure 8), and option 6 (Figure 9).

Item 1: Variable speed screw feeder (fabricated item, 500–1,500 pounds/hr)

Item 2: Feed transfer pump (H-Q model P-003)

Item 3: Feed dewatering screw (fabricated item)

Item 4: Anionic conditioner (2 tanks at 10-inch diameter × 10 inches high)

Item 5: Rougher flotation machine (Denver No. 8, two cells)

Item 6: Cleaner flotation machine (Wemco model 18, 4 cells)

Item 7: Cleaner tails transfer pump (HR 1-inch sand pump)

Item 8: Cleaner tails sizer (Derrick model J24-36ms^{-1}, with DF66 cloth)

Item 9: Coarse cleaner tails recycle pump (H-Q model P-003)

Item 10: Rougher concentrate transfer pump (HR 1-inch sand pump)

Item 11: Rougher concentrate sizer (11a: Derrick model J24-36ms^{-1}, with DF66 cloth) (11b: CFS Density Separator model 8x8)

Item 12: Rougher tails transfer pump (H-Q model P-003)

Item 13: Rougher tails sizer (Derrick model J24-36ms^{-1}, with DF66 cloth)

Item 14: Scavenger feed transfer pump (H-Q model P-003)

Item 15: Scavenger feed dewatering screw (fabricated item)

Item 16: Anionic conditioner (2 tanks at 8-inch diameter × 8 inches high)

Item 17: Scavenger flotation machine (Wemco model 18, 4 cells)

The sampling points for the major streams of each process are depicted with a diamond on the flow diagrams.

Pilot Plant Data

Flotation feeds from three plants were tested. Anionic options 4, 5, and 6 were tested and comparative data for the Crago process were obtained. The results from each feed are discussed below and the data are summarized on Table 2. Generally the use of plant process for flotation instead of city water required more reagents but improved recovery and reduced concentrate grade. The Crago process recovers less fine phosphate and less fine quartz relative to the anionic processes, which accounts for the lower grade and recovery of concentrate from the latter.

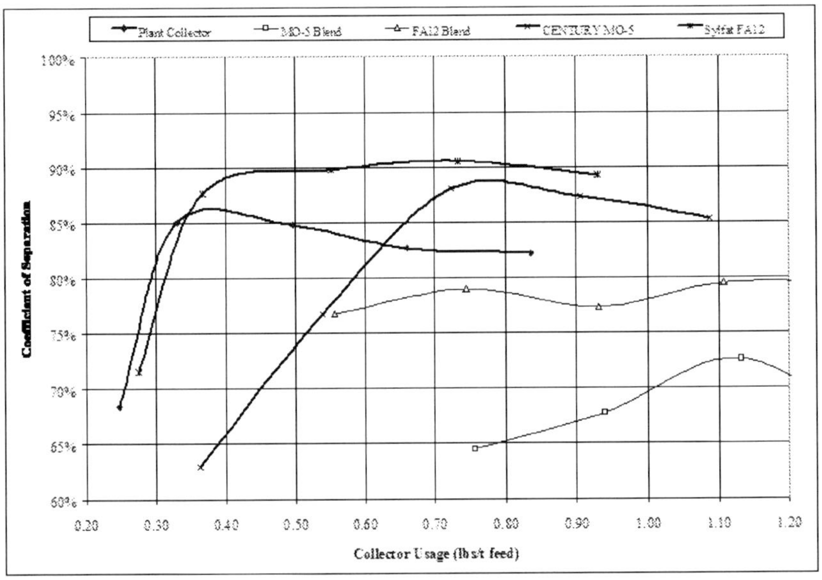

Figure 3 Collector selectivity—Plant 1 feed

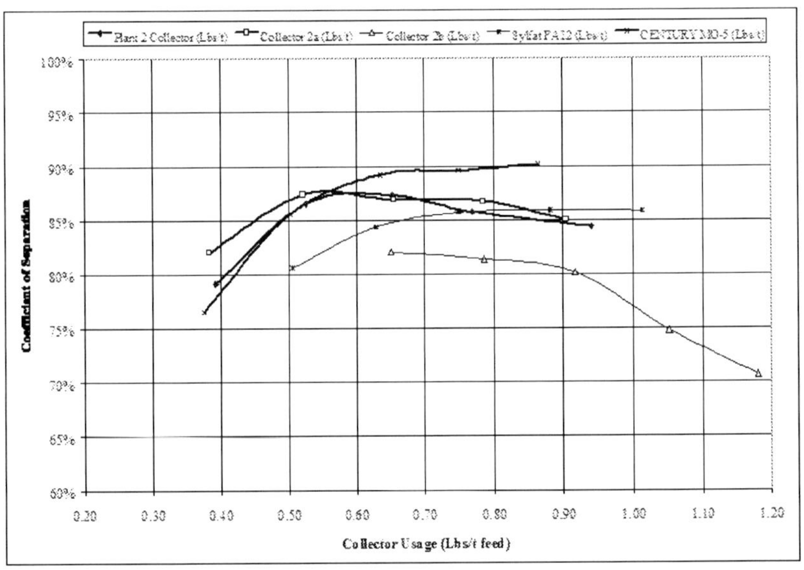

Figure 4 Collector selectivity—Plant 2 feed

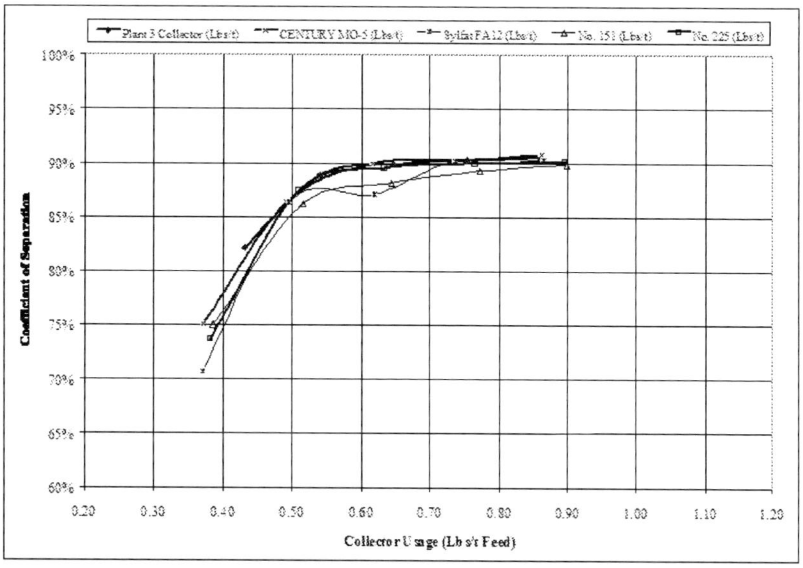

Figure 5 Collector selectivity—Plant 3 feed

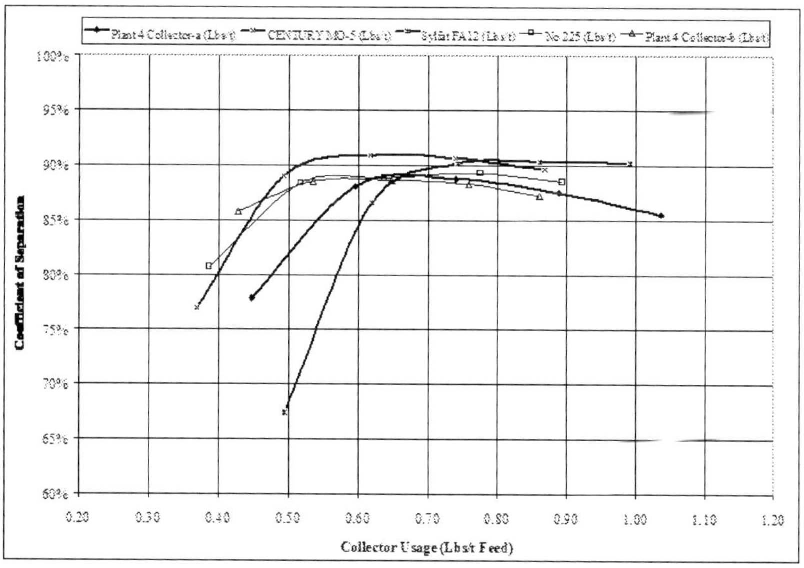

Figure 6 Collector selectivity—Plant 4 feed

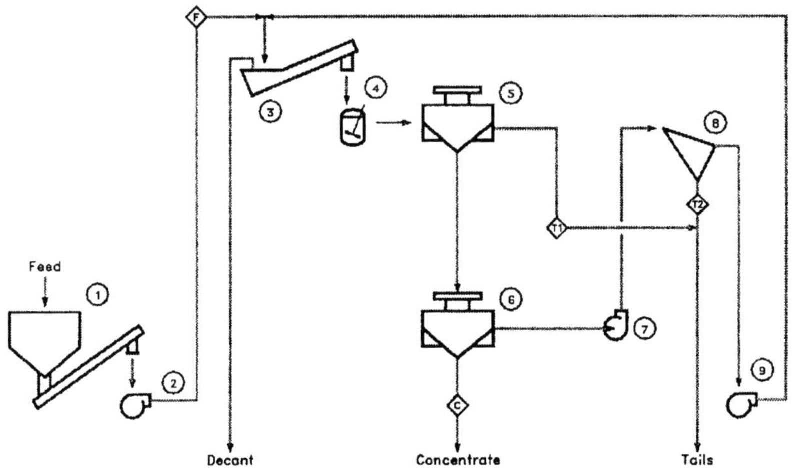

Figure 7 Option 4

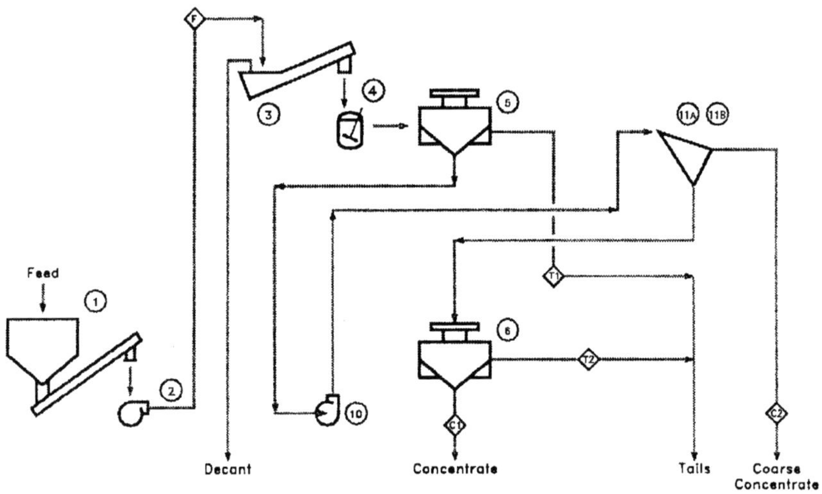

Figure 8 Option 5

Plant 1 Feed

Ten formal runs were performed on this feed. This particular feed sample was difficult to process. Based on informal discussions with plant personnel, our pilot plant Crago results were in line with plant performance data around the time the bulk sample was obtained. The Crago process recovered 72% of the BPL in a concentrate analyzing 70% BPL and 5% Insol.

Anionic option 4 recovered 85% of the BPL in a concentrate containing 59% BPL and 20% Insol. BPL recovery losses to the rougher tail and discarded fraction of the cleaner tail averaged 13% and 2% respectively, for option 4. For material balance for anionic process option 5, 80% of the BPL was recovered in a concentrate analyzing 56% BPL and 21% Insol. The anionic process options were able to recover 8–13% more

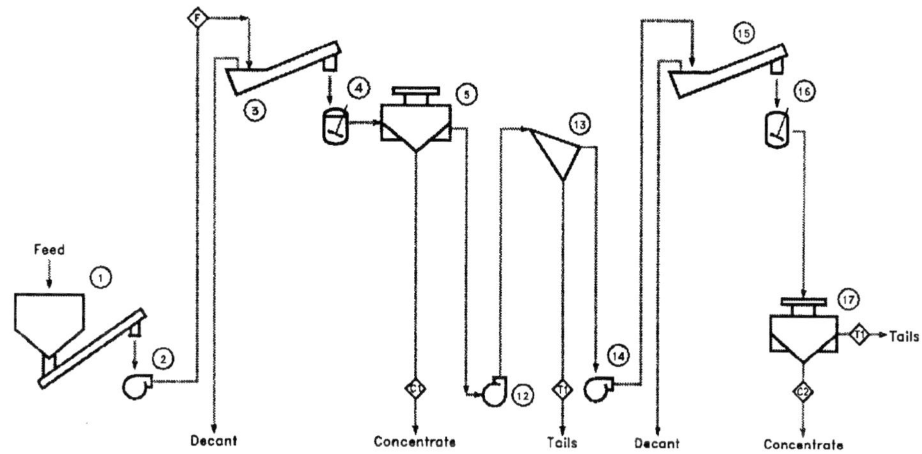

Figure 9 Option 6

Table 2 Summary of pilot plant test results

	Crago	Anionic Processes		
Feed grade	Process(1)	Option 4(1)	Option 5(1)	Option 6(2)
% BPL	16.7	17.5	17.3	17.9
% Insol	76.0	75.2	75.0	74.4
BPL % Recovery				
All Plant Feeds	81.9	88.6	85.6	80.2
Excluding Plant 1 Feed	87.1	90.6	88.3	80.2
Concentrate % BPL				
All Plant Feeds	69.3	62.9	61.0	63.7
Excluding Plant 1 Feed	69.2	65.1	63.4	63.7
Concentrate % Insol				
All Plant Feeds	4.1	12.9	14.1	9.1
Excluding Plant 1 Feed	3.8	9.5	10.9	9.1
Reagent (Lbs/t)				
Anionic collector	1.03	0.98	1.23	0.73
Fuel oil	0.43	0.40	0.37	0.60
Soda ash	0.57	0.37	0.57	0.78
Sodium silicate	0.35	0.11	0.27	0.78
Sulfuric acid	3.69	-	-	-
Cationic collector	0.29	-	-	-

(1) data are averaged for feeds from plants 1, 3, & 4.
(2) data obtained from plant 4 feed only

phosphate from plant 1 feed than the Crago process; however, their concentrates were significantly more diluted by Insol.

Plant 3 Feed

The Crago process recovered 86% of the BPL in a concentrate analyzing 70% BPL and 4% Insol. Anionic process option 4 recovered 90% of the BPL in a concentrate containing 66% BPL and 9% Insol. Anionic process option 5 recovered 87% of the BPL in a concentrate analyzing 63% BPL and 10% Insol. Anionic process options 4 and 5 were able to recover 1–4% more phosphate from plant 3 feed than the Crago process. The concentrates from these two processes contained from 6.5–11.4% Insol. Option 4 gave 3% higher recovery and slightly better concentrate grade than option 5.

Plant 4 Feed

The Crago process recovered 88% of the BPL in a concentrate analyzing 69% BPL and 4% Insol. Anionic option 4 recovered 91% of the BPL in a concentrate containing 64% BPL and 10% Insol. Option 5 recovered 90% of the BPL in a concentrate analyzing 63% BPL and 11% Insol. Anionic process option 6 was tested on this feed. From the viewpoint of the flotation operator, this process was the most difficult to operate. An overall BPL recovery of 80% was attained, on average, with a concentrate containing 64% BPL and 9% Insol.

SUMMARY

The purpose of this program was to demonstrate an alternative flotation process for Florida phosphate that, relative to the Crago process, could achieve certain goals. The goals and corresponding results are listed below.

1. Increase flotation recovery by 2–4%. Anionic process option 4 increased flotation recovery by 3–7%.
2. Reduce the cost of flotation reagents by 33%. Anionic process option 4 reduced reagent costs as follows:
   - 30% per ton of feed
   - 42% per ton of concentrate
3. Eliminate sulfuric acid and cationic reagent usage. Goal will be accomplished if an anionic flotation process is adopted.
4. Eliminate process requirements for good quality water (deep-well water). Goal will be accomplished if an anionic flotation process is adopted.

The program objective was achieved and the results met identified goals. Additional evaluation of anionic flotation by operating companies may be warranted if the production of low-Insol concentrates is not an operating requirement.

ACKNOWLEDGMENTS

The research described in this paper was funded by the Florida Institute of Phosphate Research. The Institute's financial support and the technical contributions were much appreciated. Plant access by the operating companies and the cooperation of the operating personnel and plant metallurgists is gratefully acknowledged.

REFERENCES

Crago, Arthur, inventor; 1942. U.S. Patent 2,293,640.

Gruber, G., Hicks, F., and Kelahan, M. (Zellars-Williams, Inc.). 1989. Anionic flotation of Florida phosphate. Bartow, Florida: Florida Institute of Phosphate Research. Publication nr 02-063-071. 135 p.

Zhang, P,. Snow, R., and Yu, Y. 2002. Beneficiation of siliceous phosphate using the FIPR/SAPR process. In: *Beneficiation of Phosphates: Fundamentals and Technology*. Zhang, P., El-Shall, H., Somasundaran, P., and Stana, R., eds. Littleton, Colorado: SME. p. 175–185.

Luttrel, G.H., and Eisenmann, M.E. 2002. A simple method for evaluating data from phosphate beneficiation plants. In: *Beneficiation of Phosphates: Fundamentals and Technology*. Zhang, P., El-Shall, H., Somasundaran, P., and Stana, R., eds. Littleton, Colorado: SME. p. 135–143.

CHAPTER 16

Electrostatic Beneficiation of Phosphate Ores

C.B. Wu,* B.Y. Jiang,† X.X. Duan,† and H.X. Dai†

ABSTRACT

A novel dry beneficiation process that adopted new equipment called the suspension electrostatic separator has been implemented to handle phosphate ore in Yunnan, China. Compared with flotation, the process is economically competitive and environmentally safe and has great potential for regions lacking water or under strict environmental protection. The dry electrostatic separation process was recommended based on the analysis of the ore properties and extensive experiments of grinding, desliming, separation, and optimization of separation parameters. The results indicated the following optimal conditions for the process: (a) grinding fineness for separation is 92.9% below 100; (b) feed temperature is 110°; and (c) voltages for roughing, cleaning and scavenging are 80, 60 and 90 kilovolts respectively. The test results showed that the concentrate grade reached 30.23% P_2O_5 with 83.26% recovery from an ore containing about 25% P_2O_5. The contents of harmful impurities expressed as SiO_2, the ratio of magnesium and phosphate (MgO/P_2O_5) and R_2O_3 ($Al_2O_3+Fe_2O_3$) were reduced from 27.21%, 7.27% and 3.63% to 11.48%, 2.58% and 2.81%, respectively.

INTRODUCTION

Phosphate ore is an important raw material for manufacturing phosphoric acid and phosphate fertilizer and usually requires beneficiation before chemical processing. Currently, unbeneficiated phosphate rock accounts for about 10% of the total consumption of phosphate rock worldwide. The other mid- to low-grade ores are upgraded using various beneficiation techniques, including desliming, gravity separation, flotation, magnetic separation, , and electrostatic separation (Wu et al. 2000, 2002, 2003). The flotation process is most broadly used and has played a leading role processing phosphate minerals for over half a century. However, flotation has the following disadvantages: (1) use of various chemicals, (2) use of large amount of water, and (3) potential damages to the environment.

Many of the above listed processes are limited to laboratory study at present, with the exception of electrostatic separation (Ciccu, et al. 1993). The dry process could avoid the

*Department of Environmental Engineering, East China Jiaotong University, Nanchang, China.
†Department of Mineral Processing, Kunming University of Science and Technology, Kunming, China.

disadvantages of flotation and is especially suitable for regions lacking water or under strict environmental protection.

In the case of this project, the phosphate deposit is located near two pristine lakes, the Dianchi Lake and Xing Yuan Lake, in Yunnan province, where environmental protection prohibits the use of flotation for processing the phosphate.

EXPERIMENT

Properties of the Test Sample

The sample used in this study was collected from a phosphate deposit in Yunnan, where the estimated reserve is about 200 million tons. At present only the high-grade ore is directly shipped to wet phosphoric acid plant. This high-grade fraction accounts for about a quarter of the total mined ore body. The remaining ores are piled up around the mine field and not utilized. These piles are causing surface runoff and earth bleeding thus damaging the ecological systems.

Analyses conducted on the sample included spectrum analysis, multi-element chemical analysis and mineral analysis using x-ray diffraction. The results are shown in Tables 1–3.

At nearly 25% P_2O_5, the phosphate grade of the sample is not bad at all. However, harmful impurities such as MgO, SiO_2, Al_2O_3, and Fe_2O_3 are on the high side, with a ratio of magnesium to phosphate (MgO/P_2O_5) of 0.072 and MER of 0.22. Obviously, this ore is not suitable for the wet acid process without further treatment.

The sample was also observed under microscope. It was found that phosphorus mineral exists mainly in the shape of colloidal crumb and seldom of oölitic, lithic and phanerocrystalline. Usually the maximum grain size can be up to 1.08mm and the minimum one down to 0.04mm. Silicon mineral exists mainly in quartz as oxide, with secondary presence in feldspar, mica and zircon as silicate. Soluble SiO_2 was found to be 1.32%. The maximum grain size for quartz is 0.3 mm. Interestingly, apatite is not only associated intimately with quartz, but some colloidal crumb apatite is included in fine size quartz, as shown in Figure 1. This makes it difficult to obtain high grade products from this ore even with beneficiation techniques.

Experimental Apparatus

The test equipment called the suspension electrostatic separator, invented by Kunming University of Science and Technology, differs from the conventional drum type separator (Zhang, Z.H. and Zhang, G.F. 2002, 2003). The system is unique in keeping the feed particles suspended in an airflow on their completely fixed orbits. The particles of different minerals are homogeneously dispersed in the airflow and travel along their orbits. This means that different minerals can be separated efficiently. In addition, both particle circulation and separation are carried out in a closed system to avoid dust problems.

This system design takes full advantage of friction as a means of electrifying the particles. When the two minerals are frictionated each other, the mineral with a larger dielectric constant will be positively charged, while the one with a smaller dielectric constant will be negatively charged. Since apatite has a dielectric constant of 3.2 and quartz of 2.5, respectively, when they are frictionated each other, apatite will be positively changed and quartz negatively, thus achieving separation in an electric field. Figure 2 shows the flowsheet used in the experiment, which is also the flowsheet recommended for commercial application.

RESULTS AND DISCUSSION

Grinding Testing

Table 4 indicates that grinding is essential for separating apatite from quartz. In the process of grinding, water content of the feed and the fineness are critical factors. Extensive one-pass (roughing) electrostatic separation tests were conducted to investigate the effect of grinding fineness on separation efficiency. The results, shown in Figure 3, demonstrate that both the grade and recovery of concentrate can be improved by increasing grinding fineness. However, overgrinding could reverse the trend. The optimal grinding fineness was found to be 92.9% passing 100 mesh for the test feed.

Table 1 Spectrum analysis of the sample

Element	Al	Si	P	Mn	Mg	Pb	Fe	Cr	Ti	Ca	Cu	Zn	Ni
Content /%	1	3~10	>10	0.1	0.3	0.003~0.01	1	0.01	0.1	>10	0.03	0.1	0.003

Table 2 Chemical multi-elements analysis of the sample

Component	P_2O_5	F	SiO_2	Fe_2O_3	CaO	MgO	Al_2O_3	CO_2	Acid Insol	LOI	Total
Content /%	24.47	1.55	27.21	1.29	31.82	1.78	2.34	4.44	28.68	4.24	99.14

Table 3 Mineral compositions of the sample

Mineral	Apatite	Quartz	Limonite	Dolomite +Calcite	Mica +Feldspar +Zircon	Total
Content	57.96	24.25	1.44	9.30	6.85	100.00

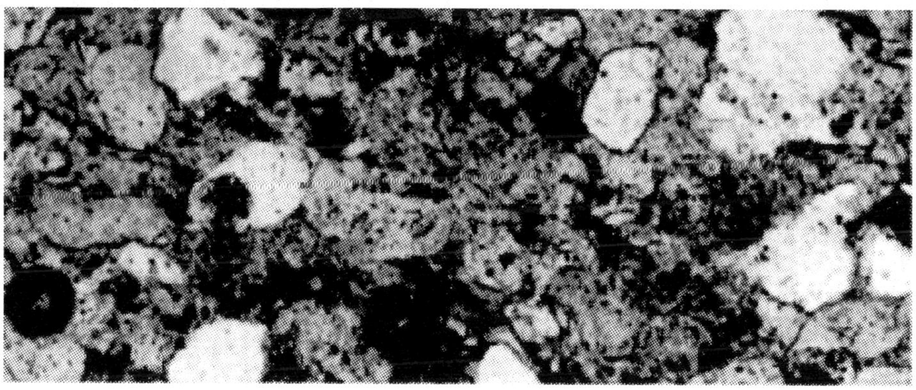

Figure 1 The inclusion phenomenon of apatite and quartz ×90

Coincidentally this optimal grinding fineness is nearly the same as required for wet phosphoric acid production. Therefore, the grinding cost could be charged against the acid plant, thus making the dry process more appealing economically.

The grinding aid, a carbon powder, was found to be effective in reducing aggregation and improving grinding efficiency.

Desliming Test

In order to achieve better separation efficiency for weathered phosphate ore, a desliming test was carried out. Table 5 shows separation test data with and without desliming at 400 mesh. It can be seen that the concentrate grade with desliming has in fact increased over 10%. However, phosphate recovery suffered about 32%

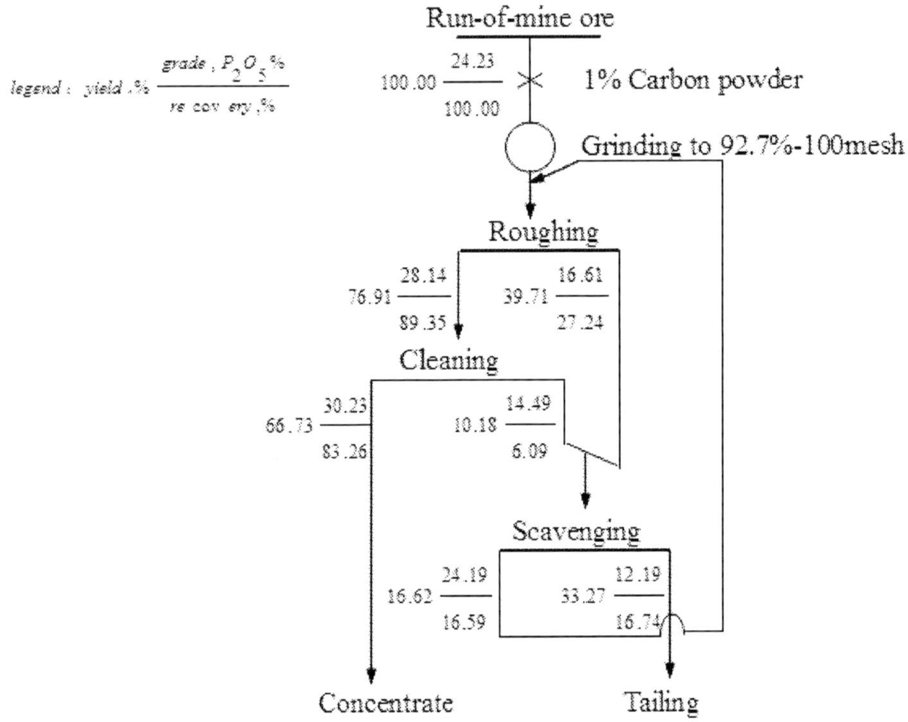

Figure 2 The flowsheet recommended and its test results

Table 4 Size characteristic and phosphorus distribution of samples

Size range /mm	Yield /%	Grade /%	Recovery/%
3-2	8.3	27.84	9.47
2-1	12.6	26.98	13.93
1-0.5	13.9	25.89	14.75
0.5-0.3	10.3	25.43	10.73
0.3-0.15	14.9	25.25	15.47
0.15-0.1	10.5	24.84	10.69
0.1-0.076	8.5	22.77	7.93
-0.076	21.0	19.79	17.03
Total	100.0	24.41	100.00

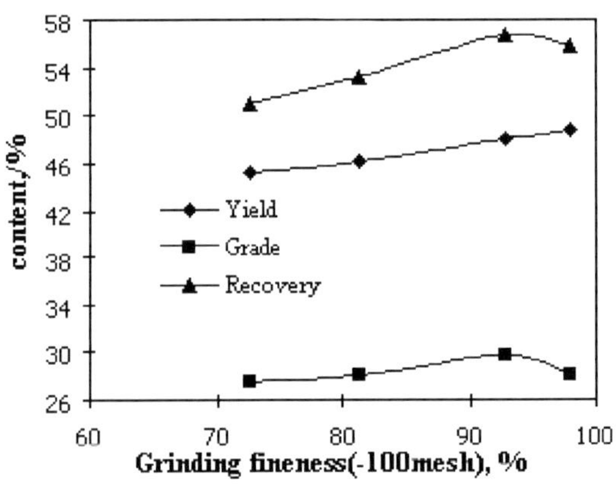

Figure 3 Results of grinding fineness test

Table 5 Results of desliming test

Conditional	Product	Yield /%	Grade /%	Recovery to raw ore
Conventional test	Concentrate	49.29	28.57	58.12
	Tailing	50.71	20.01	41.88
	Feeds	100.00	24.23	100.00
Desliming test (-400mesh)	Concentrate	59.78	29.64	46.49
	Tailing	40.13	20.27	21.30
	Feeds	100.00	25.88	67.79

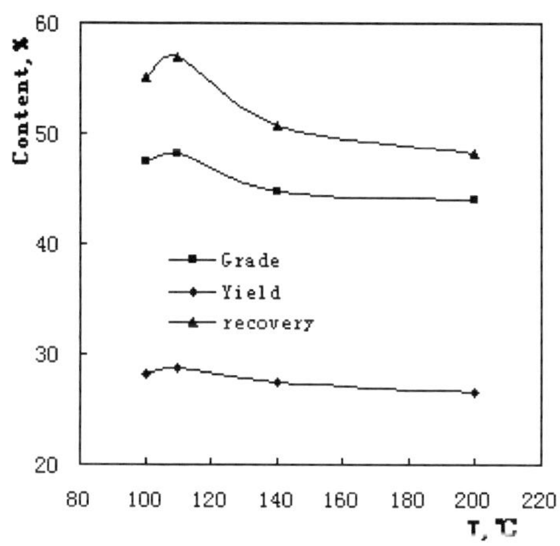

Figure 4 Results of different temperature

due to loss of phosphate in the slime. Therefore, desliming was not investigated further.

Effect of Separation Steps

Separation tests showed that the roughing tailings and cleaning middling were high in phosphate. Thcrefore, a scavenging operation is added to separate the mixed feed consisting of the roughing tailing and the cleaning middling to maximize the total recovery of apatite. Since the grade of the scavenging concentrate is low, it is sent back to the roughing operation to form a closed circuit as shown in Figure 2.

Effect of Ore Temperature

Ore temperature has impact on both water content and electric charge energy. In order to achieve the maximum charge and optimum separation efficiency, some experiments were carried out at different temperatures. The results are shown in Figure 4.

Figure 4 shows that temperature has an effect upon concentrate grade and recovery, and that the optimal temperature is about 110°C for the test feed.

High Voltage Test

Applied voltage also has great influence on the separation efficiency. Roughing experimental results at different voltage levels are shown in Figure 5, which indicates that the optimum voltage for roughing is 80 kilovolts. Similar experiments showed optimum voltages of 60 and 90 kilovolts for cleaning and scavenging, respectively.

Flowsheet Testing

The final testing results using the complete flowsheet are shown in Table 6. The contents of impurities in the final concentrate are also listed in Table 6.

The results in Table 6 show that the contents of all major harmful impurities such as MgO, Al_2O_3, Fe_2O_3, and SiO_2 were reduced to a certain extent, with the total minor element ratio, MER, reduced from 0.22 to 0.12.

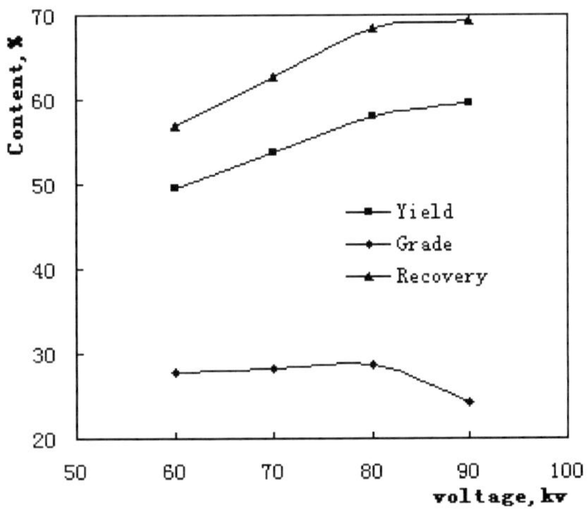

Figure 5 Results of different voltages in roughing

The same flowsheet was also tested on two lower grade samples, achieving a concentrate of 28.78% P_2O_5 at 79.64% recovery from a feed analyzing 19.85% P_2O_5, and a concentrate of 27.60% P_2O_5 at 62.73% recovery from a feed analyzing 14.76% P_2O_5.

CONCLUSIONS

A feasible flowsheet has been developed for processing a medium grade ore contaminated with MgO and Al_2O_3 using the electrostatic separation technology.

The suitable grinding fineness for the test ore was found to be 92.9% minus 100 mesh. However, complete liberation was not possible because some colloidal crumb apatite is included in finely sized quartz.

The optimum ore temperature is 110°C, and the optimum voltages for roughing, cleaning and scavenging are 80, 60 and 90 kilovolts, respectively.

The dry process, the innovative equipment and the recommended flowsheet have great potential for industrial application in regions lacking water or under strict environmental protection.

Table 6 Results of multi-elements analysis of phosphorus concentrate

Component	P_2O_5	SiO_2	Al_2O_3	Fe_2O_3	MgO	CaO	CO_2
Content /%	30.23	11.48	1.59	1.22	0.78	43.78	1.26

ACKNOWLEDGMENTS

The authors gratefully acknowledge the contribution of H.L. Dai, Department of Mineral Processing, Kunming University of Science and Technology, Kunming, China, to this paper.

REFERENCES

Wu, C.B., and Duan, X.X., 2000. Studies on processing phosphate ores in China. *Yunnan Metallurgy*, No. 4, 19–21.

Wu, C.B., et al., 2003. New beneficiation method for mid-low-grade phosphate rock-dry electrostatic separation. *Industrial Minerals & Processing*, No. 9, 7–9.

Wu, C.B., et al., 2002. Experimental study on phosphorus ore by electrostatic separation, proceedings of 1st international symposium on dry coal cleaning, clean coal technology. Xuzhou, China, July, pp. 291–298.

Ciccu, R., Ghiani, M., Peretti, R., et al., 1993. Phosphate beneficiation with electrostatic techniques. *XVIII Int. Miner. Process. Cong*. Vol. II Sydney, Australia, May, pp. 430–438.

Zhang, Z.H., Zhang, G.F., Gao, L.K., 2002. Research on suspension type electrostatic separation in manufacture and application of dry coal separation, Proceedings of 1st international symposium on dry coal cleaning, clean coal technology. Xuzhou, China, July, pp. 323–330.

Zhang, G.F., Zhang, Z.H., Gao, L.K., 2003. Research on test of separation fine particle in suspension type electrostatic separator. *Non-ferrous Metals* (Part: Mineral Processing). No. 2, 25–28.

SECTION 5

Processing of Complex Ores

CHAPTER 17

Beneficiation of Sedimentary Dolomitic Phosphate Ore by Dense Medium Separation and Flotation

Zhengxing Gu,* Shibo Zheng,† Xianjun Cao,† Wenyi Song,† and Felicia Peng*

ABSTRACT

A pilot plant test was conducted to beneficiate a sedimentary high MgO phosphate ore using a combined process of Tri-Flo dense medium separation and flotation. The dolomitic phosphate ore containing 21.7% P_2O_5 and 5.0% MgO was crushed to pass 12 mm and sized at 1 mm. The pilot plant tests were carried out with two flowsheets. In the first flowsheet, the coarse fraction was treated with Tri-Flo dense medium separator to obtain coarse particle concentrate, while the minus 1 mm fraction was further ground and concentrated with flotation to float dolomite. In the second flowsheet, the middlings of Tri-Flo dense medium separator were combined with minus 1 mm portion and subjected to the grinding and flotation process. The test results showed that a composite phosphate concentrate containing 32.9% P_2O_5 and 1.2% MgO could be obtained at the P_2O_5 recovery of 78.4% using the first flowsheet, and 31.8% P_2O_5 and 1.0% MgO at the P_2O_5 recovery of 88.3% using the second flowsheet, respectively.

INTRODUCTION

Dense medium separation is considered the most efficient process of all gravity separation techniques. It is widely used to clean coal and extensively for the concentration of a variety of ores such as those of iron spodumene, complex Pb-Zn ore, nickel laterite, buaxite, barite, petalite, fluorspar, feldspar, cassiterite, etc. Dense medium separation process has many potential advantages (Wills, 1997). Those advantages include sharp separation, easy adjustment of specific gravity to meet changing conditions of ore, continuous operation, broad size range of feed, low operation and maintenance costs, and high operation capacity. The dense medium separation process can be applied directly to produce final products, to generate a concentrate and a middling of different quality, or as a preconcentration step to reject tailings at coarse particle size ranges.

In the phosphate industry, flotation process is most widely used to process dolomite phosphate ores, and research has been conducted worldwide. Although the density difference between apatite and dolomite minerals is very small compared to other minerals, some research works have been carried out to process well liberated dolomite phosphate ores by attempting to obtain phosphate concentrate at coarse particle sizes. Lawver et al. (1983) used dense medium cyclones to process -6.7+1.0 mm (-3+16 mesh) dolomite pebbles. In their study, the underflow product contained 25% P_2O_5 and 2.3% MgO,

*Department of Mining Engineering, West Virginia University, Morgantown, West Virginia.

†China Lianyungang Design and Research Institute, Lianyungang, Jiangsu, China.

and the overflow product contained 8.4% P_2O_5 and 13.0% MgO from a feed of 20.7% P_2O_5 and 5.1% MgO. The underflow product of dense medium cyclones was ground to -0.6 mm (-28 mesh) and deslimed at 100 μm (150 mesh) for flotation to further reject silica and carbonate minerals. A demonstration plant using dense medium cyclones with magnetite as dense medium in the capacity of 200,000 t/y of feed was constructed in China to treat high MgO phosphate ore (Fu and Fan, 1993). The ore was crushed to 15 mm and deslimed at 1 mm and then separated in the dense medium cyclones. The feed ore contained 23.9% P_2O_5 and 3.2% MgO. After one rougher and one cleaner separation, a phosphate concentrate of 31.9% P_2O_5 and 1.48% MgO was produced at 81.4% P_2O_5 recovery. Warren Spring Laboratory (WSL) (Ferrara et al., 1994) conducted a laboratory feasibility study on applying Tri-Flo dense medium separator (TFS250) to enrich a low-grade sedimentary phosphate ore of 22.1% P_2O_5 and 3.0% MgO from China. The sink product of Tri-Flo dense medium separator contained 33.9% P_2O_5 and 1.2% MgO, and the P_2O_5 recovery was up to 90%.

In this study, the combined process of Tri-Flo dense medium separation and froth flotation was applied to process a sedimentary dolomitic phosphate ore. The objective was to produce a portion of acceptable phosphate concentrate and/or pre-concentrate to discard tailings at coarse size range prior to grinding in order to reduce the overall operation cost. The minus 1 mm fraction and middlings were ground for further froth flotation.

EXPERIMENTAL

Materials

A 20-ton sedimentary dolomitic phosphate sample was collected from a phosphate deposit in Hubei, China. The phosphate rock consisted of phosphorite, apatite, dolomite, quartz, and a small amount of iron and aluminum oxides. The phosphate deposit existed in banded structure. The phosphate minerals and main gangues such as dolomite and quartz were in separate bands. The thickness of most of those bands was more than 10 mm. The chemical composition of the phosphate ore sample from the deposit is given in Table 1. It can be seen that it is a highly dolomitic phosphate ore with 5% MgO content. Prior to the beneficiation, the dolomitic phosphate ore was crushed to pass 12 mm. The size distribution is given in Table 2. For the crushed product, the P_2O_5 and MgO grade are rather evenly distributed except -12+10 mm size fraction which has higher MgO and lower P_2O_5 grades compared with other size fractions.

Equipment

Dense medium separator. A Tri-Flo dense medium separator (Tri-Flo DMS), TFS250, from Inpromin, Italy, was used in the pilot plant test. The separator is a multistage dense-medium dynamic device consisting of two cylindrical compartments connecting through an axial orifice as shown in Figure 1 and Figure 2. The separator has a diameter of 242 mm, and the length of 725 mm for the first compartment and 970 mm for the second compartment. Each compartment has a medium inlet and a sink outlet. The feed is fed through a hopper into the medium vortex in the first compartment with 5% of the total medium flow rate for feed flushing. The sink I comes out from the outlet of the first compartment, and the float of the first compartment is directly fed to the second compartment. The sink II and the final float are generated in the second compartment. Because there was only one-density medium preparation unit for this pilot plant facility, both compartments used the dense medium slurry with the same relative density.

Atomized ferrosilicon, Cyclone 40, produced by the Hoechst Co. (Germany), was used as dense medium. Fine magnetite was also added to increase the viscosity of the dense medium slurry for stabilization of dense medium suspension. The ratio of ferrosilicon and magnetite was 4:1. The size distributions of the dense media are given in Figure 3.

Mechanical flotation machine and cell. Denver D-12 flotation machine was used for flotation test. The flotation cell is 1.2 liters in volume.

Table 1 Chemical analysis of dolomitic phosphate ore sample

Components	Wt. %
P_2O_5	21.70
MgO	5.01
CaO	37.96
Acid insoluble	19.13
SiO_2	16.82
$Fe_2O_3+Al_2O_3$	3.14
F	1.67
CO_2	10.72

Table 2 Size distribution and chemical analysis of crushed dolomtic phosphate ore sample

Size (mm)	Wt.%	Grade, %		Recovery, %	
		P_2O_5	MgO	P_2O_5	MgO
-12+10	20.00	17.37	6.48	15.67	26.16
10+5	38.64	21.54	5.36	37.54	41.80
-5+2	20.74	25.25	3.90	23.62	16.33
-2+1	6.32	25.50	3.70	7.27	4.72
-1	14.30	24.67	3.81	15.91	11.00
Total	100.00	22.17	4.95	100.00	100.00

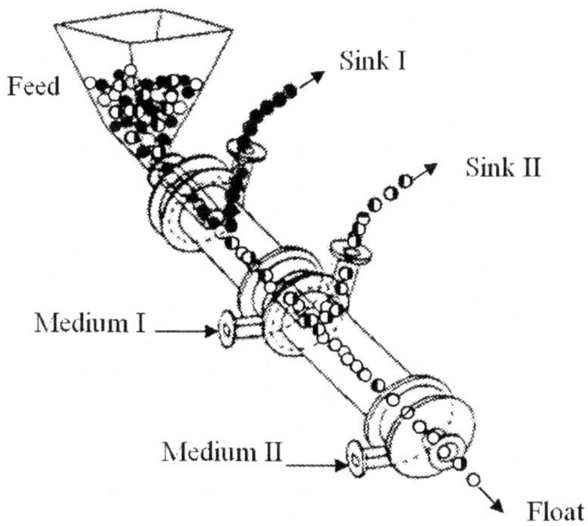

Figure 1 Schematic diagram of Tri-Flo dense medium separator

Figure 2 The Tri-Flo dense medium separator installed in the pilot plant: (1) first stage outlet, (2) second stage outlet, (3) first stage medium inlet, (4) second stage medium inlet, and (5) float outlet

Process Description

The ore sample was jaw crushed to -12 mm. The crushed ore was then screened at 1 mm. The -12+1 mm size fraction was treated in Tri-Flo DMS at controlled medium density. The experiment was carried out in two flowsheets. In flowsheet I, plus 1 mm coarse fraction was treated using Tri-Flo separator, the sink I and sink II were combined as coarse particle concentrate and minus 1 mm fraction was ground and subjected to dolomite flotation using phosphoric acid as pH modifier and oxidized paraffin soap as collector. In flowsheet II, sink I was taken as the high grade coarse concentrate, and sink II as middlings and reported to grinding and reverse flotation with minus 1 mm portion of the feed to obtain a fine concentrate. In order to ensure high P_2O_5 recovery of the separation process, the medium density in the second separation compartment should be lower than that in the first one. Because there was only one-density medium preparation unit in the pilot plant, the experiment was conducted in an alternative approach. The float was retreated in the Tri-Flo DMS using lower density medium. Both sink products of the re-treatment step were combined with the sink II of the first step as composite middlings, and the float was rejected as final tailings. The composite middlings were combined with -1 mm fraction of the feed and subjected to grinding and reverse flotation to produce a fine concentrate. Flotation tests were carried out in Denver D-12 machine and 1.2 liter flotation cell. The Tri-Flo DMS beneficiation circuit is shown in Figure 4.

RESULTS AND DISCUSSION

Float-Sink Test

The float-sink test of the crushed ore in size range of -12+1 mm was carried out and the results are plotted in Figure 5. The test results indicate that, for this dolomitic phosphate ore, the gravity separation can be used to obtain an acceptable phosphate concentrate and/or rejecting some tailings by proper control of separation densities. From Figure 5, it can be observed that, at the relative density of 2.95, the sink product contains about 35% P_2O_5 and 0.9% MgO with 77.9% P_2O_5 recovery. At the relative density of 2.9, the float contains about 7% P_2O_5 and 9.8% MgO. About 70% of total MgO can be rejected into the float with 10% P_2O_5 loss.

Dense Medium Separation

The dense medium separation tests with Tri-Flo separator were conducted in batches using 1.5 tons of sample for each test. The ore sample was fed to the screen and then entered the Tri-Flo DMS at the feed rate of 10 t/h. The most concerned parameters in the dense medium separation include relative medium density for separation, viscosity and stability of dense medium slurry, medium inlet pressures, and back pressures. The pilot plant tests were carried out to determine the operation conditions including relative medium density for separation and the first stage inlet pressure.

Medium density effect. Float-sink test results indicated that the separation of gangue minerals from phosphate could be achieved by proper control of separation density. During pilot plant test, the separation densities varied from 2.85 to 2.97. The results are plotted in Figure 6. The effect of the separation density on sink I was investigated. Figure 6 shows that at the relative density of 2.97, a phosphate concentrate containing 34% P_2O_5 and 1.16% MgO could be obtained at 61.6% P_2O_5 recovery for the first sink product. At this relative density, more than 85% MgO in the feed were rejected into final float as tailings.

Medium Inlet Pressure Effect

The Tri-Flo DMS used in the pilot plant test is a two stage dynamic dense medium separator. Medium inlet pressure is very critical for the separation. The product quality can be controlled by adjusting the medium inlet pressures as well as the back pressures. In this study, the medium inlet pressure of the first stage was varied by fixing the medium inlet pressure of the second stage at 0.15 MPa. Figure 7 shows that the medium inlet pressure had a significant effect on MgO and P_2O_5 grade as well as P_2O_5 recovery. At low medium inlet pressure, the

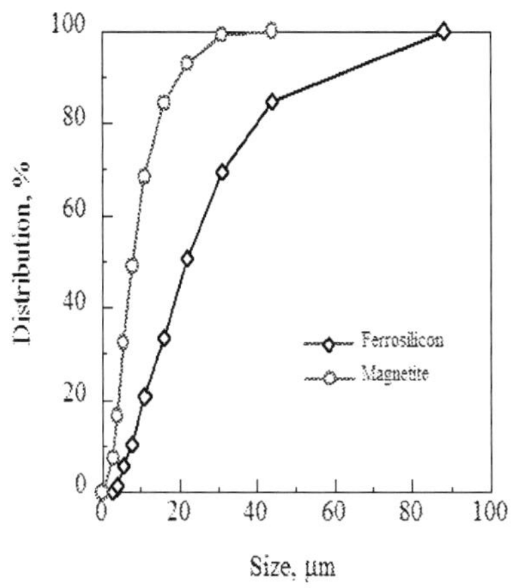

Figure 3 Size distribution of ferrosilicon and magnetite used for the tests

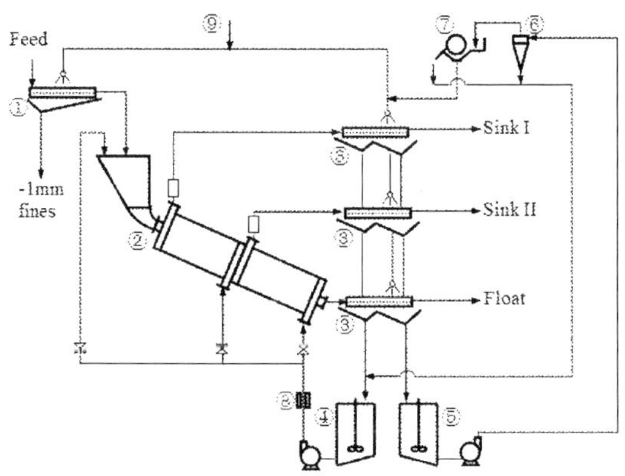

Figure 4 Schematic diagram of pilot-plant test flowsheet for processing dolomitic phosphate ore with Tri-Flo dense medium separator: (1) feed screen, (2) Tri-Flo separator, (3) drain and rinse screens, (4) dense medium sump, (5) dilute medium sump, (6) classifying cyclones, (7) magnetic separator, (8) nuclear density meter, and (9) spray water

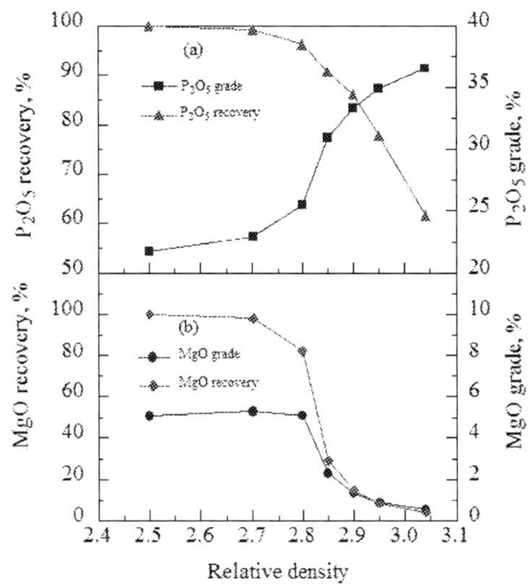

Figure 5 Float-sink test results of dolomitic phosphate ore sample: (a) P_2O_5 grade and recovery as a function of relative density and (b) MgO grade and recovery as a function of relative density

sink I could be collected as a high quality phosphate concentrate which contained 35.6% P_2O_5 and 0.9% MgO. About 95.6% MgO could be rejected into the combined product of the sink II and float. The P_2O_5 recovery in the sink I product increased by increasing the medium inlet pressure of the first stage. But the P_2O_5 grade suffered due to the high carbonate impurities in the product. Between 0.18 to 0.2 MPa, the effect of the medium inlet pressure of the first stage remained constant for this pilot-plant test.

FLOWSHEETS FOR DENSE MEDIUM SEPARATION AND FLOTATION

Flowsheet I

In this flowsheet, the dolomitic phosphate ore in size range of -12+1 mm was processed by Tri-Flo DMS to produce a coarse phosphate concentrate by rejecting low density gangue minerals such as silica and dolomite. The minus 1 mm size fraction was ground to 62% minus 74 µm (-200 mesh) and then subjected to dolomite flotation. In dolomite flotation, phosphoric acid was used as pH modifier to maintain the flotation slurry at about pH 5.5, and 2.0 kg/t of oxidized paraffin soap was used as dolomite collector. The flowsheet and the results are given in Figure 8. Figure 8 shows that the Tri-Flo DMS rejected 44.5% of the total weight of the feed as tailings which contained 9.8% P_2O_5 and 8.6%

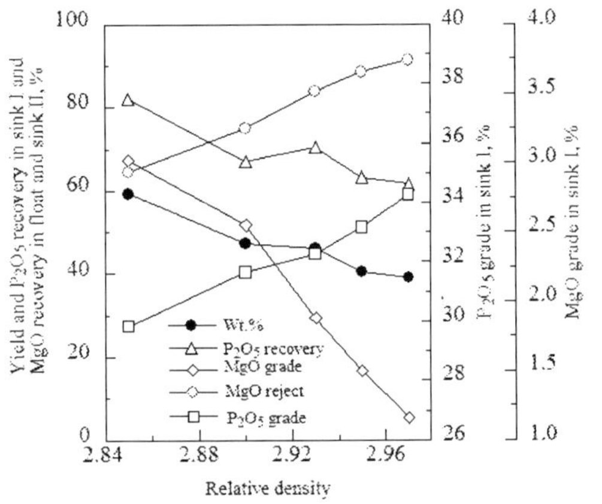

Figure 6 Effect of dense medium density on the first stage separation

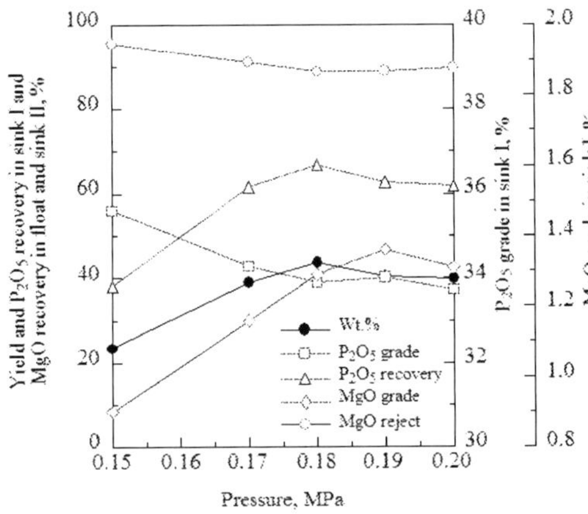

Figure 7 Effect of medium inlet pressure on the first stage separation

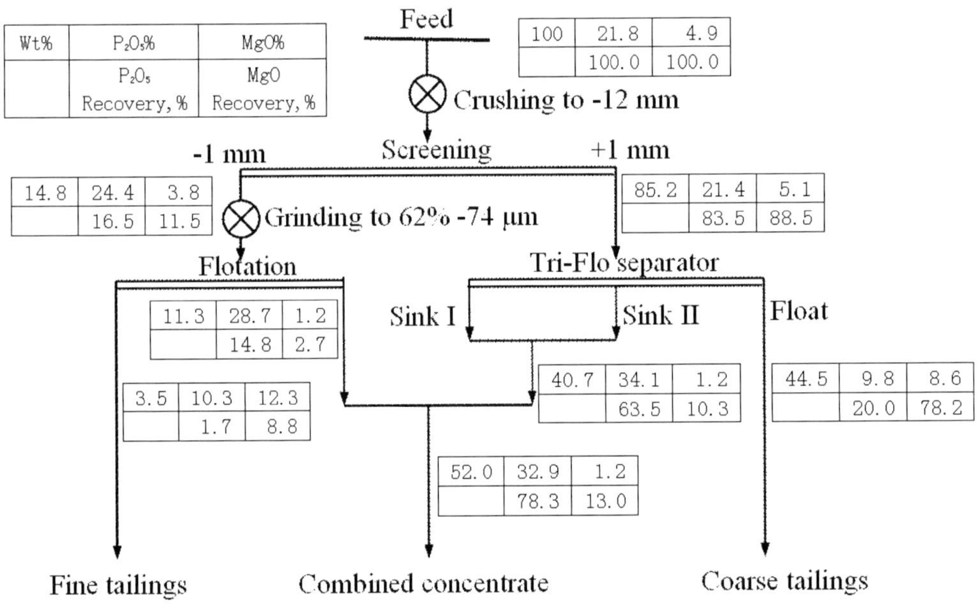

Figure 8 Mass balance flowchart for flowsheet I

MgO at 20% P_2O_5 and 78.2% MgO recoveries, respectively. The combined product of sink I and sink II of the Tri-Flo DMS contained 34% P_2O_5 and 1.2% MgO. The flotation concentrate contained 28.7% P_2O_5 and 1.2% MgO. The final composite phosphate concentrate of 32.8% P_2O_5 and 1.2% MgO could be obtained at the overall P_2O_5 recovery of 78%.

Flowsheet II

In order to produce the phosphate concentrate at low MgO grade and high P_2O_5 recovery, an

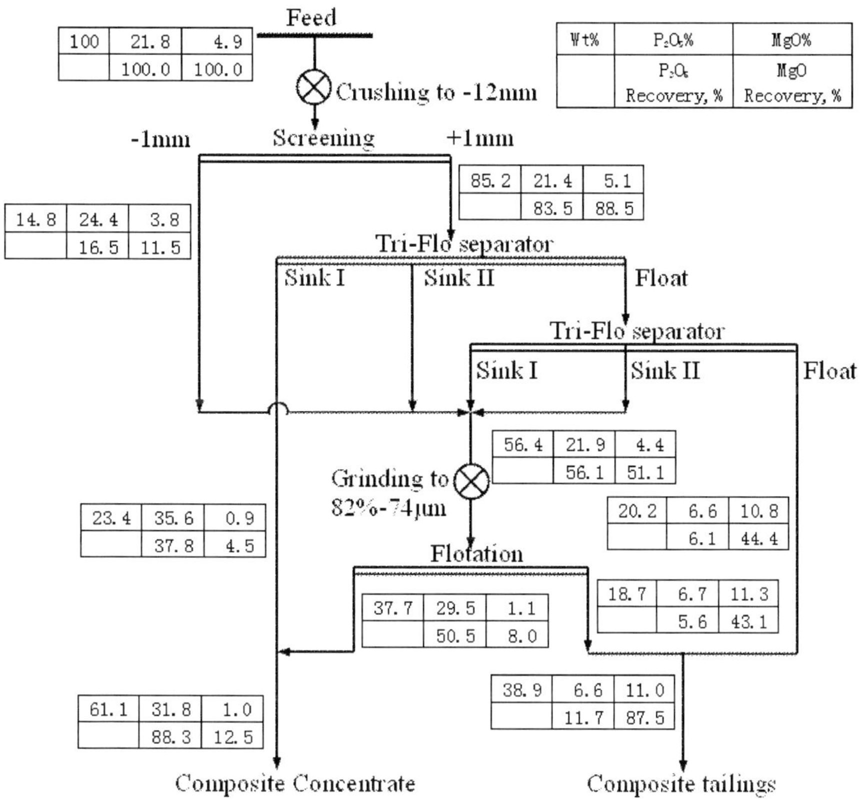

Figure 9 Mass balance flowchart for flowsheet II

alternative flowsheet was tested. The minus 12 plus 1 mm fraction was processed by Tri-Flo DMS. In this test, the medium inlet pressures were set up at 0.13 MPa for the first stage and 0.2 MPa for the second stage. The relative density of the medium was 2.95. The sink I was collected as the coarse concentrate. The float was reprocessed with Tri-Flo DMS at the medium density of 2.8 and the medium inlet pressure was set at 0.19 MPa for both stages. The objective of the reprocessing was to ensure a high P_2O_5 recovery by rejecting low P_2O_5 tailings. Both sink products of the reprocessing were combined with the sink II of the first step separation and minus 1 mm fine fraction and subjected to a grinding and flotation process to produce a flotation concentrate. For a separator equipped with two separate medium feeding units, these two processing steps can be accomplished in one single step of Tri-Flo separation by controlling the medium density and the medium inlet pressure for each stage. However, in this pilot plant unit, the production of high quality concentrate and low grade tailings were conducted in separate steps as shown in Figure 9. From this flowsheet, a composite phosphate concentrate containing 31.8% P_2O_5 and 1.0% MgO could be obtained at the overall P_2O_5 recovery of 88.3%.

CONCLUSIONS

From these pilot-plant test results it can be concluded that Tri-Flo dense medium separator can be used for processing coarse size range, high dolomite phosphate ores if the liberation of carbonate impurities from phosphate ore is not a major problem. By appropriate control of the medium density and medium inlet pressure, Tri-Flo dense medium separator can be used to process coarse size feed to produce high quality phosphate concentrate at low P_2O_5 loss in tailings.

The combination of dense media separation with flotation is a feasible flowsheet for the high dolomite ore tested.

REFERENCES

Ferrara, G., Machiavelli, G., Bevilacqua, P., and Meloy, T.P., 1994, Tri-Flo: A Multistage High-Sharpness DMS Process with New Applications. *Minerals and Metallurgical Processing*, Vol. 11, No. 2, pp. 63–73.

Fu, R., and Fan, X., 1993, Production Practice of Dense Medium Separation of Yichang Phosphate Ores. *Chemical Mines Technology*, Vol. 23, No. 2, pp. 29–32.

Lawver, J.E., Snow, R.E., Wiegel, R.L., and Hwang, C.L., 1983, Beneficiation of Dolomitic Florida Phosphate Reserves. In *Proceedings, XIV International Mineral Processing Congress*, p IV. 20.1–IV. 20.22, Toronto, Ontario, Canada.

Wills, B.A., 1997, Chapter 11 Dense Medium Separation (DMS). *Mineral Processing Technology*, 6th ed., Butterworth Heinemann (Elsevier, St. Louis, Mo.), Oxford, UK.

CHAPTER 18

Novel Microorganisms in Apatite and Calcite Flotation

M.S. Miettinen,* M. Rättö,† and R.W. Smith‡

ABSTRACT

Laboratory scale flotation experiments were performed on apatite and calcite ores using bacteria, (*Staphylococcus carnosus*, *Bacillus subtilis*, *Bacillus licheniformis*, *Pseudomonas stutzerii*, and *Bacillus firmus*) as flotation collectors or as flotation modifiers. In some cases the results were compared to that obtained with conventional collectors. Batch rougher flotation was performed in both an Outokumpu cell and a flotation column. Basic interaction mechanisms between minerals and microorganisms were studied using zeta potential measurements, hydrophobicity measurements, confocal laser scanning microscopy, TOF-SIM measurements and single mineral microflotation tests.

INTRODUCTION

Although commercial application of microorganisms as mineral surface modifiers, flotation collectors, depressants or activators is not presently done and the economic viability of such use is unknown, the promise of using such naturally occurring substances in the place of toxic chemical reagents is attractive. In general it has been found that, although microorganisms can selectively function as flocculants, flotation collectors or modifiers for single mineral or for artificial, simple mixed systems, selectivity is much more difficult in actual ores (Zheng, 1997; Zheng et al., 1997; Zheng et al., 1998). In the flotation of semisoluble salts, for example, the very slight solubility of these minerals places metal ions derived from the minerals into solution. The metal ions can adsorb not only onto all minerals present rendering all the surfaces more alike, but also adsorb onto the bacteria, further complicating the systems (Zheng et al., 2001).

In the present experimentation, bench scale laboratory tests were performed using microorganisms as collectors. Materials floated were apatite, calcite and quartz and an apatite ore and a flotation tailing containing calcite. The results were sometimes compared to those obtained with a conventional collector. In addition, in some cases, the organisms were studied as flotation modifiers when used with a conventional collector. Both an Outokumpu cell and a flotation column were used in the experimentation.

*Geological Survey of Finland, Mineral Processing, Outokumpu, Finland.
†VTT, Espoo, Finland.
‡Metallurgical and Materials Engineering, University of Nevada, Reno, Nevada.

MATERIALS

Minerals, Commercial Ores and Tailings

Natural minerals that were used were separated from handpicked samples received from Siilinjärvi (apatite) and Lappeenranta (calcite), Finland. Arkansas quartz was purchased from Wards Natural Science Establishment (USA). Prior to fine grinding, the apatite sample was washed with dilute nitric acid to remove traces of calcite and dolomite. The minerals were dry-ground using an agate mill followed by screening to obtain suitable mineral fractions for flotation and flocculation tests. A particle size fraction of –90, + 32 µm was used for the flotation tests and a –20 µm size fraction was used in the flocculation and zeta potential studies. The specific surface areas, as measured by a flowsorb analyzer, of the –90, + 32 µm fractions of apatite, calcite and quartz were 0.241, 0.292, and 0.204 M^2/gm, respectively.

The physical characteristics of the apatite ore after grinding were P_2O_5 5.0%, CO_2 7.2% and the particle size 75.7% -75 µm. The physical characteristics of the calcite flotation tailing were calcite 42.0% and the particle size 36.0% -38 µm, 61% -75 µm.

Chemicals and Microorganisms

Chemicals used in the research for pH and ionic strength control were of reagent grade. Methyl isobutyl carbinol (MIBC) was used as frother for single mineral tests. Atrac 886 was used as frother for apatite ore flotation and Dowfroth 1012 as frother for calcite tailing flotation. An amphoteric collector and a cationic collector were used in some of the flotation tests on the industrial materials.

The bacteria used were the most promising microorganisms found in a study of the potential 12 different microorganisms as flotation collectors and flocculants (Miettinen et al., 2004).

A ferrooxidans E-991375, *S. carnosus* E-91446, and *B. firmus* E-97061 were from VTT Culture Collection. The bacterial cultivation and quantity of cell production for some of the bacteria are shown in Table 1. The bacterial cells were separated from the culture media by centrifugation, washed twice with water and suspended in water. Part of the *S. carnousus* suspension was freeze-dried and used in comparative flotation studies.

METHODS AND RESULTS

Zeta Potential Studies

A Coulter Delsa 440 instrument was used for zeta potential measurements. Mineral suspensions (-20 µm) were conditioned at a selected pH value for 10 min prior to measurement. A 1×10^{-3} M $NaNO_3$ solution was used for ionic strength control. Figures 1–3 show results as a function of pH obtained on apatite, calcite and quartz, respectively. The effect of 120 mg/L of *B. subtilis*, *S. carnosus*, *B. licheniformis*, *P. stutzerii*, and *B. firmus* on the zeta potential of the minerals is demonstrated.

It is to be noted that in the open atmosphere apatite is positively charged from acidic pH values up to about pH 10. In the presence of bacteria the mineral generally becomes more negatively charged. The effect of *B. subtilis* and *S. carnosus* are similar. It appears to matter little whether a cell suspension is added or freeze dried cells are added. *B. firmus* has the least effect and at quite basic pH values actually causes the mineral's surface to become slightly more basic.

In the absence of bacteria and in the open atmosphere, calcite is positively charged at all pH values studied. In the open atmosphere in the absence of bacteria or foreign ions, calcite is more positively charged than apatite. Both *B. subtilis* and *S. carnosus* reverse charge on the mineral. In the presence of *B. firmus*, calcite becomes less positively charged, but the charge is not reversed.

In the absence of bacteria, the quartz studied is highly negatively charged. *B. subtilis* and *C. carnosus* have little affect on the zeta potential of quartz, but *B. firmus* is able to create a much more basic surface on quartz and at a pH more basic than about pH 10 is able to reverse charge on the mineral. Thus, from this figure, it appears that *B. firmus* is one of the few

Table 1 Cell cultivation

Bacterial Strain	Cultivation					Final Cell Suspension		
	Culture Medium	Volume (l)	Time (hr)	Dry weight (g/l)	Cells/ml	Volume (l)	Dry weight (g/l)	Cells/ml
A. ferrooxidans	9K Medium	4	96		8.4×10^7	0.016	3.5	2.8×10^{10}
S. carnosus E-91446	YM Broth (Difco)	200	18	4.8	2.1×10^{10}	4	168	7.7×10^{11}
B. firmus E-97061	BHI (Difco)	200	17	1.9	3.0×10^9	2	63	1.2×10^{11}
B. subtilus ATCC 6051	Zheng et al. (2001)	20	16	8.5	4.4×10^8	?	?	?

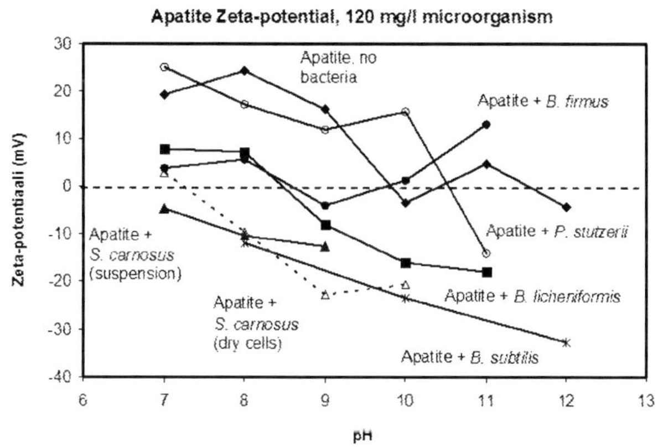

Figure 1 Zeta potential of apatite as a function of pH in the presence and absence of bacteria

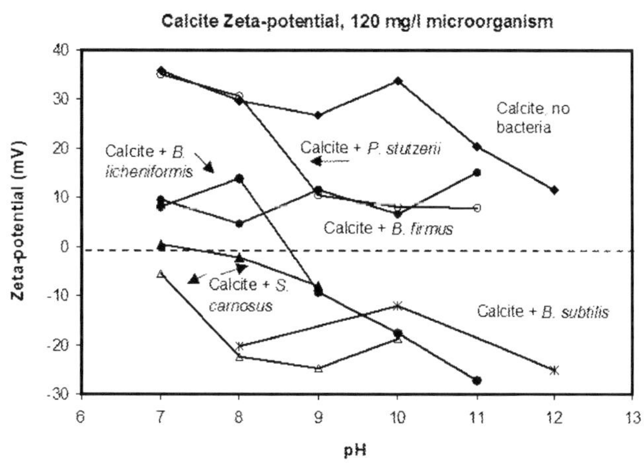

Figure 2 Zeta potential of calcite as a function of pH in the presence and absence of bacteria

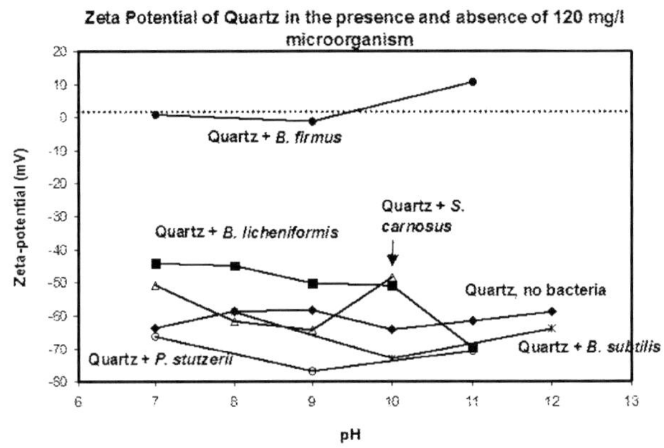

Figure 3 Zeta potential of quartz as a function of pH in the presence and absence of bacteria

Table 2 Protein concentrations in flotation water, S. *carnosus* added, 1,620 mg/L batch rougher flotation of apatite ore, pH 11

Flotation Stage	Conditioning Method	Protein (mg/l)	Protein (% of Cells Added)
Feed	Propeller Rod Stirring	45	2.8
Pulp Water after Flotation	Propeller + Flotation Stage	29	1.8
Feed	Drum Mixing	38	2.3
Pulp Water after Flotation	Drum + Flotation Stage	30	1.9

positively charged bacteria near neutrality in the absence of foreign multivalent cations.

Other Experimentation

The pulp water of ground apatite ore after flotation contained the following elements as determined by ICP analysis (mg/L): Ca 92.5, K 75.8, Mg 20.2, Na 171.2, S 138.5, P 0.3, and Cu 1.2.

Soluble protein concentrations in water from two flotation experiments with different conditioning conditions and using *S. carnosus* were measured using the method of Lowry et al. (1951). Only very low amounts of soluble protein were determined indicating that there was only minor cell breakage during conditioning and flotation. The total protein concentrations in flotation water both before and after flotation are shown in Table 2.

The protein concentrations in the pulp water both before and after flotation were very low, indicating attachment of bacteria to the minerals. Slightly lower concentrations after flotation indicates that the flotation cell agitation resulted in some additional cell attachment to minerals.

Confocal laser scanning microscopy was used to roughly determine attachment of cells to mineral particles. A protein binding dye was used to visualize the organisms on the mineral surfaces. The experimentation indicated that in high intensity conditioning (as in the Outokumpu flotation cell) all the particles are well coated with bacteria. After gentle conditioning as with the propeller rod stirrer or the drum mixer, the bacteria favored attachment to smaller sized particles.

TOF-SIM was used to determine that, indeed, varying amounts of protein are present on the minerals' surfaces.

Single Mineral Flotation

Single mineral microflotation was carried out in a modified Hallimond tube (70 mL) using nitrogen as the flotation gas and a flow rate of 8 cm^3/min. A 1 gm mineral sample (-90, + 32 mm) was added to 75 mL of distilled water and conditioned for 5 min with the microorganism at a specified pH. A frother, MIBC (13 mg/L), was then added and flotation carried out for 5 min. HNO_3 and NaOH were used for pH adjustment. Bacteria were added either as water suspensions or as dried cells. In the former case the suspensions contained between 5×10^{10} and 7×10^{11} cells 7 mL. Solids content was typically between 2–6 wt %. For addition of dry cells a similar wt. of bacteria were added.

Figure 4 depicts the single mineral Hallimond tube flotation recovery of apatite, calcite and quartz at pH 9.0 as a function of microorganism concentration using *S. carnosus* as collector. Figure 5 shows the flotation recovery of the minerals as a function of pH using a concentration of *S. carnosus* of 30 mg/L. This combination of about pH 9 and about 30 mg/L *S. carnosus* appears to be the best combination for selective separation. Figure 6 shows data similar to figure 4 using *B. firmus* as collector. Figure 7 shows data similar to Figure 5 using *B. firmus* as collector.

Batch Rougher Flotation of Industrial Ores/Waste

Ore/waste product was ground, if needed, in a stainless steel grinding mill. Initial flotation tests were conducted in a 2 dm^3 Outokumpu cell. Typically, 0.5 kg of material was conditioned for 2 min at 1,300 rpm rotor speed prior to addition of the microorganisms. Bacteria were added and the slurry was conditioned for one of the following:

- 5 min in the flotation cell
- 10 min in a vessel with a propeller rod stirrer
- 20 min in a drum

After conditioning, 12 to 20 g/t of the frother was added to the slurry in the flotation cell 1 min before the introduction of either nitrogen or air (2 dm^3/min). The flotation time was 7 min. HCl and NaOH were used for pH adjustment as required. Atrac 886 was used as frother for the apatite ore flotation and Dowfroth 1012 was used for calcite tailing flotation. Table 3 lists the Outokumpu and column flotation studies performed.

Batch Rougher Flotation of a Finnish Apatite Ore

Initial tests using the Outokumpu cell examined the effect of conditioning type, particle size and pulp density on flotation of the apatite ore. The conditioning tests indicated that intense conditioning in the flotation cell resulted in higher recoveries of P_2O_5 and CO_2 and a higher grade of CO_2.

Figure 8 illustrates the effect of particle size on recovery and grade of both P_2O_5 and CO_2. The best recoveries of both P_2O_5 and CO_2 were obtained with a particle size of 53% less than 38 mm. Flotation using microorganisms appears to require smaller mineral particle sizes than flotation using conventional collectors. Perhaps the bacteria do not render the larger minerals sufficiently hydrophobic for effective flotation.

The effect of pulp density on flotation is shown in Figure 9. A pulp density of 23% gave the best recoveries. A pulp density of 40% gave the best recoveries.

Column Flotation Compared with Outokumpu Cell Flotation

Figure 10 shows the comparative flotation of the apatite ore at pH 10–11 using *S. carnosus*, *B. firmus*, and collector A3. For flotation using *S. carnosus*, column flotation was compared to Outokumpu cell flotation. It is apparent that the amphoteric collector A3 gave superior results as compared to the bacteria. Also evident is the superior column flotation for both P_2O_5 and CO_2. This suggests that the fine bubble size and

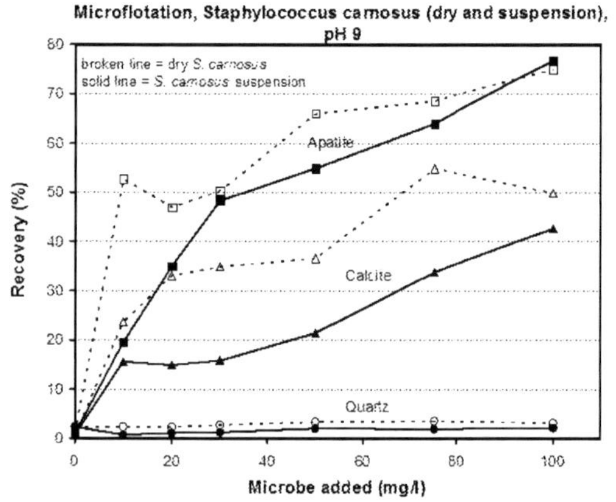

Figure 4 Flotation recovery of the minerals as a function of S. *carnosus* concentration, pH 9

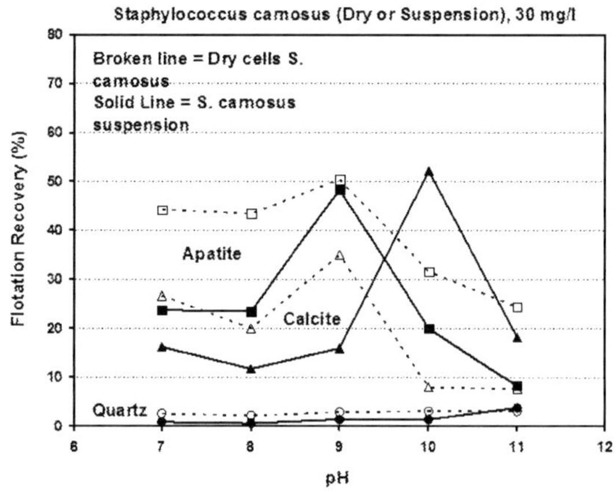

Figure 5 Flotation recovery of the minerals as a function of pH, 30 mg/L *S. carnosus*

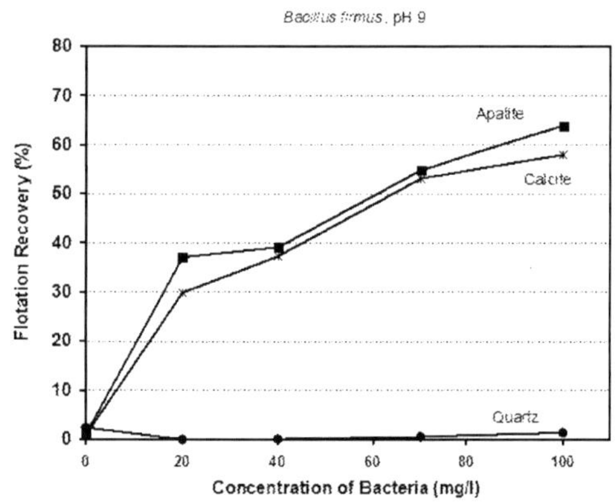

Figure 6 Flotation recovery of the minerals as a function of *B. firmus* concentration, pH 9

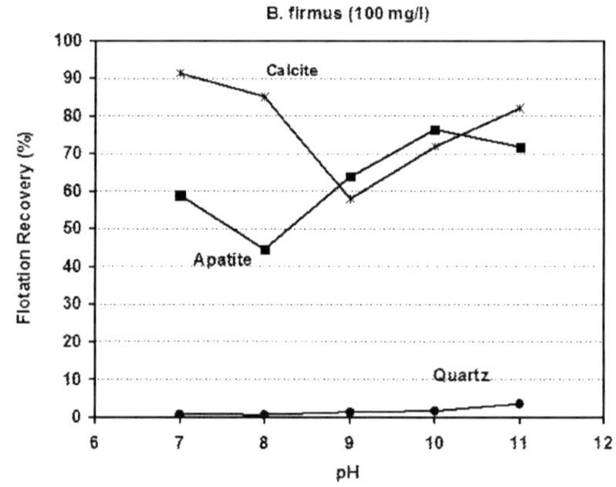

Figure 7 Flotation recovery of the minerals as a function of pH, 30 mg/L *B. firmus*

Table 3 Outokumpu and column flotation tests performed

Microorganism or Collector	Apatite Ore		Calcite Tailing	
	Outokumpu Cell	Column	Outokumpu Cell	Column
S. carnosus	X	X	X	X
B. subtilus + A3	X			
B. firmus	X		X	
Flotinor SM15			X	
A3	X			

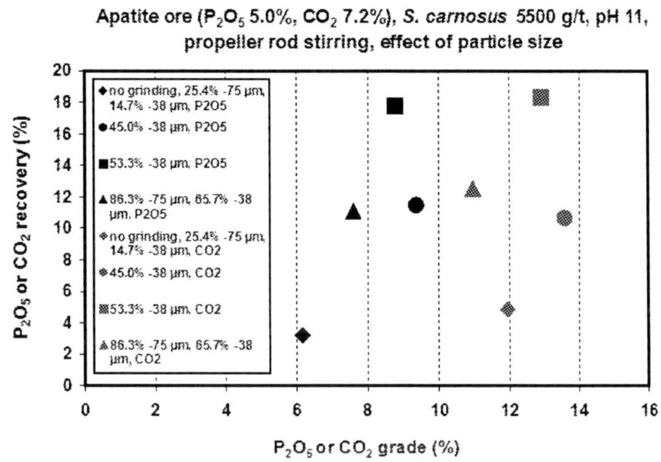

Figure 8 Effect of particle size on grade and recovery of both apatite and calcite

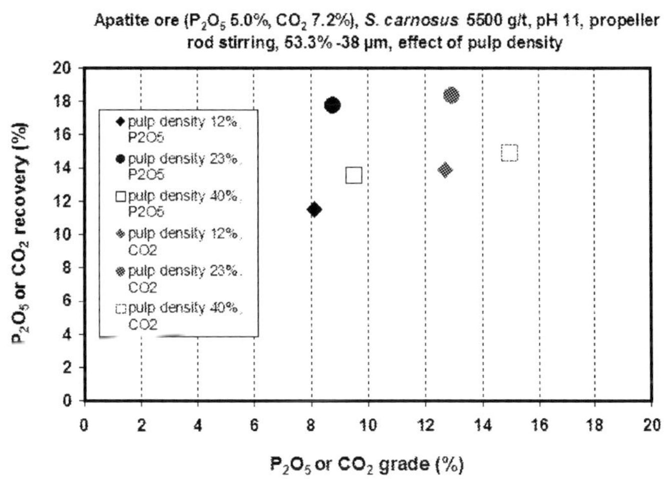

Figure 9 Effect of pulp density on recovery and grade of both apatite and calcite

the more quiescent conditions of the flotation column promote attachment.

Figure 11 shows comparative Outokumpu cell flotation of the apatite ore at pH 10 using A3 (200 g/t) as collector with varying amounts of *B. subtilis* as modifier. In the absence of bacteria, A3 functioned as an excellent collector for P_2O_5 and a very poor collector for CO_2. *B. subtilis* functioned as a strong depressant for P_2O_5 (apatite) with increasing amounts of bacteria increasing the depression. However, the microorganism functioned as an activator for CO_2 (calcite), again with increasing concentrations of the bacteria increasing the effect.

DISCUSSION AND SUMMARY

Single mineral microflotation tests in a modified Hallimond tube along with zeta potential measurements indicate the following:

1. It is difficult to correlate zeta potential data with flotation data.

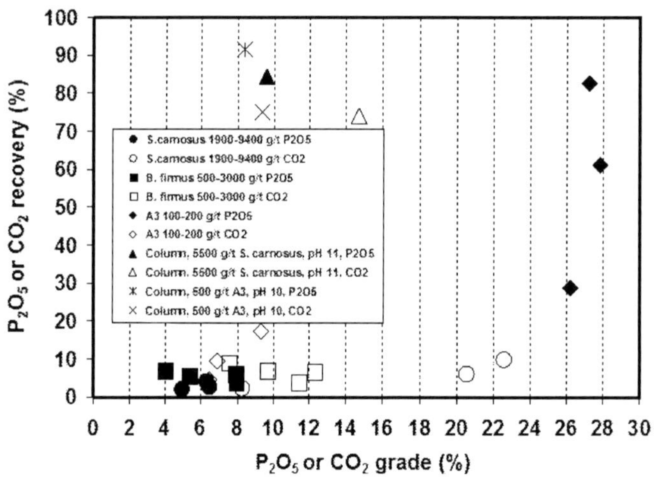

Figure 10 Flotation of the apatite ore at pH 10–11

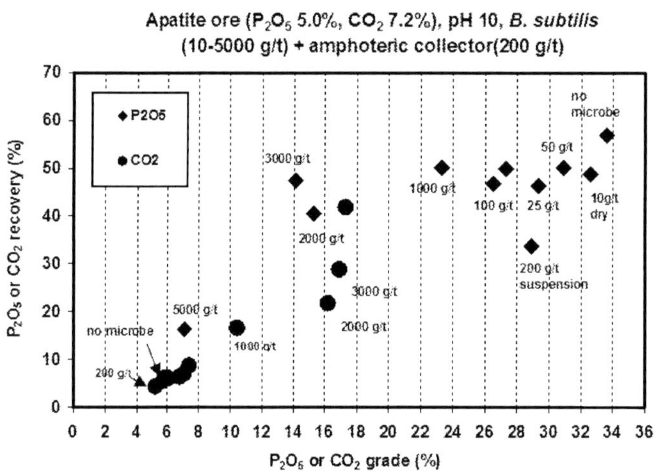

Figure 11 Comparative flotation of the apatite ore using an amphoteric (A3) collector plus *B. subtilis*

2. By adjusting pH and bacterial concentration, it is possible to set conditions, at least with *S. carnosus,* in the absence of auto activating ions, where it should be possible to float apatite selectively from calcite.

3. It is possible to substitute freeze dried cells of *S. carnosus* for cells in an aqueous suspension. However, data in Figures 4 and 5 suggest that potential selectivity is reduced when the dry cells are used.

4. Both *S. carnosus* and *B. firmus* do not function as collectors for quartz.

5. It appears that selective flotation of either mineral from the other is difficult using *B. firmus* as collector.

When flotation was attempted on the apatite ore the results were generally mediocre, conventional collectors performing better than the bacteria. It may be that the quantity of Ca(II) and Mg(II) ions in the flotation water as noted in Table 2 interfere with flotation by adsorbing on both the minerals and the bacteria making selectivity difficult. Column flotation appears to work better when bacteria are used as collector than does the Outokumpu cell. It is likely that, because of the small size of the mineral particles and the smaller bubbles in column flotation, there is better attachment of the minerals to the bubbles. Also, the loose bacterial-mineral aggregations are probably easily broken during agitation in the Outokumpu cell while the more quiescent conditions of the column enhance recovery.

ACKNOWLEDGMENTS

The financial support of the Geological Survey of Finland is gratefully acknowledged.

REFERENCES

Lowry, O., Rosenburgh, N., Farr, A., and Randall, R., 1951. *J. Biol. Chem.*, 193, 263–275.

Meittinen, M., Rättö, M., Leppinen, J., and Viikari, L., 2004. Evaluation of Potential Microorganisms in Minerals Biobeneficiation. (being prepared).

Zheng, X., Smith, R.W., Mehta, R.K., Raichur, A.M., and Misra, M., 1997. A Hydrophobic Bacterium as Modifier in Dolomitic Phosphate Rock Flotation. *Proceedings of the XX International Mineral Congress*, Vol. 4. Aachen, Germany. H. Hoberg and H. von Blottnitz, eds. pp. 537–546.

Zheng, X., Arps, P.J., and Smith, R.W., 2001. Sorption of Two Bacteria onto Dolomite and Their Effect on Dolomite Depression in Anionic Flotation Intern. *Jour. Min. Proc.*, 34, 159–172.

Zheng, X., Smith, R.W., Mehta, R.K., Misra, M., and Raichur, A.M., 1998. Anionic Flotation of Apatite from Dolomite Modified by the Presence of a Bacterium. *Miner. Metall. Proc.*, 15, 52–56.

Zheng, X., 1997. Flotation Separation of Phosphate from Dolomite Using Bacteria as Depressants. PhD Thesis, University of Nevada, Reno.

CHAPTER 19

Carbonate Separation from Sedimentary Phosphates Through Bio-Flotation

S. Farrah,* H. El-Shall,* and A.M. El-Mahdy†

ABSTRACT

This paper aims at studying the amenability of carbonate separation from a sedimentary phosphate ore through bio-flotation process. Two types of bacteria were chosen for this study. The experiments were performed using oleic acid as a collector. Flotation experiments were carried out using statistical designs for optimizing the main operating parameters. A concentrate containing 0.78% MgO and 30.15% P_2O_5 with a recovery of 92.31% was obtained from a feed containing about 2.45% MgO and 27% P_2O_5. The specification of such concentrate could not be obtained from the conventional flotation experiments, in absence of bacteria, under similar conditions. Measurements of zeta potential, adsorption of bacteria and collector as well as froth power were performed to explain the role of microorganisms in surface modification of both dolomite and phosphate, and their interaction with the collector.

INTRODUCTION

Numerous flotation studies have been carried out to separate carbonaceous gangue from sedimentary phosphate ores. However, except in China, none of these studies is applied on industrial scale due to technical or/and economic reasons (El-Shall et al., 1996; Elgillani and Abouzeid, 1993; Anazia and Hanna, 1987). The biggest difficulty arises from the fact that the physico-chemical properties of phosphatic minerals and carbonates are very similar. The surface properties of phosphate are affected not only by phosphate's own solution chemistry but also by the dissolved species from other salt-type minerals present such as calcite and dolomite leading to their surface transformation (Moudgil and Somasundaran, 1986; Anazia and Hanna, 1987).

In the mean time, bioprocessing proved to be very promising in many applications on different ores such as coal, sulfides, and gold (Somasundaran, 1998; Rao et al., 1992; Natarajan and Deo, 2001). However, very limited research work was applied to phosphate ores (Smith et al., 1991; Boice, 2000).

Through a cooperation between Central Metallurgical Research and Development Institute (Egypt) and University of Florida (USA) eight species of microorganisms were tested and just two types of bacteria were chosen for application as surface modifiers in carbonate separation from sedimentary phosphate ores (Abdel-Khalek

*Particle Engineering Research Center, University of Florida, Gainesville, Florida.
†Central Metallurgical Research and Development Institute, Helwan, Cairo, Egypt.

and Farrah, 2004). This paper aims at studying the amenability of carbonate separation from a phosphate ore, of Abu-Tartur locality, through bio-flotation process. Statistical designs were used to screen and optimize the operating parameters affecting on the flotation process. In the mean time, measurements of zeta potential, adsorption of bacteria and collector as well as froth power were also performed to explain the role of bacteria in surface modification of both dolomite and phosphate, and its interaction with the collector.

EXPERIMENTAL

Phosphate Samples

A low grade sample of Abu-Tartur phosphate ore was used in this study. It contains 24.3% P_2O_5, 1.56% MgO and 6.2% acid insoluble (A.I.). XRD of the sample indicated that apatite is the main phosphate mineral whereas dolomite (or ankerite) represents the main gangue. The sample was first primary crushed with a jaw crusher followed by roller crusher to about –2 mm. The crushed ore was then divided into patches of ~3 kg using a Jone Riffle sampler.

To prepare the feed for flotation experiments, samples were attrition-scrubbed at 55% solid for 15 minutes and then screened on a 0.25 mm screen. The oversize (-2.0 mm+0.25 mm) fraction was separated as a coarse concentrate since it is rich in apatite (~30% P_2O_5) and has minimum amount of carbonates (0.6% MgO). The undersize fraction (-0.25 mm) was further screened on a (0.044 mm) screen where the undersize fraction (-0.044 mm) was considered as phosphatic clay slimes (~31% A.I.).

The size fraction (-0.25 mm+0.044 mm) of high MgO content was taken as a flotation feed. Such intermediate size fraction (-25 mm +0.044 mm) has low content of P_2O_5 (27.07%) as a result of its contamination with MgO (~2.5%). It should be mentioned that such intermediate fraction is treated with a magnetic separation at Abu Tartur beneficiation plant (New Valley, Egypt). However, the produced non-magnetic concentrate (called fine concentrate) still has MgO higher than 1.0%.

Samples of pure single minerals of apatite and dolomite, from Ward Company, USA, were also used for studying the surface properties of the mineral-collector-microorganism systems.

Chemicals

Sodium oleate (obtained from Aldrich Chemicals) of 99% purity was used as a collector. Analytical grade NaOH and H_2SO_4 are used as pH regulators. All reagents were used without further modification.

Microorganisms

Eight different species (six types bacteria and two types of fungi) of microorganisms were isolated from the surface of Abu-Tartur phosphate ore at the Microbiology Department, University of Florida, USA. Two of them were selected for the bio-flotation experiments. The microorganisms were grown in a liquid medium containing 30% mycological agar, 10.0 g/L sucrose, 1.0 g/L ammonium nitrate and 700 mL double distilled water. The microorganisms were incubated at temperature of 25°C for two weeks.

Microorganisms counting was made through turbidity measurements using spectrophotometer at wave length 550 μm. Reading of 0.01 at this wave length is equivalent to 10^6 cells/mL. Five mL of microorganisms was centrifuged at 15,000 rpm and rinsed three times with double distilled water and thereafter diluted in 100 mL double distilled water. Samples turbidity was measured using Perkin-Elmer spectrophotometer model Lambda 3B. The results indicate that a microorganism coded as number one (identified as *corynebacterium-diphtheriae-intermedius*) has 5×10^7 cell/mL while that coded as number two (identified as *pseudomonas aeruginosa*) contains 2×10^7 cell/mL (Abdel-Khalek and Farrah, 2004). These values are used in calculating the concentration of microorganisms in the flotation experiments as well as in the adsorption and zeta potential measurements.

The selectivity of microorganisms to mineral surface was determined based on the change in size distribution of solid samples after their treatment with microorganisms. A laser particle size analyzer Fritsch model Analystte 22 was employed for measuring size analysis of samples after treatment with microorganisms. In these experiments, fixed concentration (5×10^7 cells/mL) of each microorganism was conditioned with one gram of the mineral for one hour before recording the change in size distribution.

METHODS

Evaluation of Samples and Flotation Products

Routine chemical analysis of samples was conducted using standard methods for phosphate analysis using the acid attack method. Magnesium, calcium, and iron oxides were determined by atomic absorption technique using Perkin-Elmer atomic absorption model Analyst 200. Phosphorous was determined by spectrophotometric method using Perkin-Elmer spectrophotometer model Lambda 3B. Meanwhile complete chemical analysis of the original sample was conducted using Philips x-ray fluorescence (XRF). X-ray diffraction analysis of samples was performed using a Philips Nozelco diffractometer using a Ni-filtered Co (K-alpha) radiation generated at 36 kV and 20 mA.

Flotation Experiments

Floatation experiments were performed using a micro-flotation column. The air flow rate was fixed at 30 cm^3 per min. The feed samples were conditioned with the surfactant into 100 mL round flask at the required pH at solid-liquid ratio of 20% by weight in a shaker for a period of 5 minutes. A magnetic stirrer was used during the flotation experiment to prevent particles from settling.

Adsorption Measurements

The adsorption density of collector (or microorganisms) was determined by adding one gram of ground sample of –74 μm in final volume of 100 mL of double distilled water with the definite collector or microorganisms concentration. The pH was adjusted at a certain value and the total volume was shaken for one hour. The final pH (equilibrium pH) value was then recorded. The samples were centrifuged at 15,000 rpm for 15 minutes to separate supernatant from the settled fraction. The total organic carbon (TOC) content (residual concentration) in the 40 mL of the supernatant was determined using a Phoenix 8000 total carbon analyzer. The average of three readings was taken as a measure for the residual concentration of organic carbon. The adsorbed amount was then calculated as the difference between initial and residual concentrations.

FTIR was also used for studying the adsorption of collector in presence or absence of microorganisms. The adsorption density of microorganisms in presence of collector onto the mineral surface was employed using PYE Unicam model SP 1200 infraredspectra. A single mineral of apatite or dolomite (0.2 g) was first conditioned with microorganisms for 30 min followed by the addition of collector and then further conditioning for another 30 min in final volume of 100 mL of double distilled water. The pH was adjusted at a certain value and the total volume was conditioned for one hour. The final pH (equilibrium pH) value was recorded. The samples were centrifuged at 15,000 rpm for 15 minutes. The supernatant was separated from the residual solid which was rinsed several times before drying at 30°C.

Zeta Potential Measurements

A laser zetameter Malvern Instruments model Zetasizer 2000 was employed for zeta potential measurements. A sample of 0.01 gram ground to 200 mesh was placed in 50 mL double distilled water with definite collector concentration and pH modifier having ionic strength of 10^{-2} M KCL. The pH was then adjusted to the required value. The sample was shaken for one hour. After shaking the sample, the equilibrium pH was recorded and the zeta potential of the mineral particles was measured.

Measuring Frothing Power

The frothing power of collectors and their mixtures with different microorganisms are also measured. A graduated tube of 20 cm height and diameter of 1 cm with a stopper was used for measuring the frothing power. Three milliliters of the collector (or in addition to 2 mL

Table 1 Control tests for flotation of phosphate sample with oleate collector in absence of microorganisms

Collector type	P_2O_5 %	MgO %	P_2O_5 Recovery %
Oleate (0.25 kg/t)	30.19	0.84	55.52
Oleate (> 0.25 kg/t)	Bulk flotation with 100 wt % floated		
Flotation Feed	27.07	2.24	-

Table 2 Levels of central composite design for bio-flotation of phosphate

Parameter	Symbol	(-1.68)	(-1)	(0)	(+1)	(+1.68)
pH	A	4.14	5.5	7.5	9.5	10.86
Collector dosage (kg/t)	B	1.33	1.5	1.75	2	2.17
Microorganism concentration (x 10^7 cell)	C	3.3	5	7.5	10	11.7

of microorganisms) was added and shaked for 1 min and then the froth height was recorded. The average of three readings was taken as a froth height.

RESULTS AND DISCUSSION

Biological processes have been attracting attention in mineral processing recently due to their lower operating costs and their possible applications in beneficiation of low-grade ores that are not easily concentrated using conventional processes. Development of efficient reagents and techniques to process ores has become important all over the world due to increasing demand for minerals from a dwindling supply of quality ores (Somasundaran et al., 1998).

Preliminary flotation experiments were conducted, as control tests, without addition of microorganisms, the results of which are shown in Table 1. A concentrate of 30.2% P_2O_5 and 0.85% MgO can be obtained with oleate at the expense of recovery (55.5%) while adding 0.25 kg/t oleate at pH 5.5. Addition of higher concentration (> 0.25 kg/t) of oleate resulted in bulk flotation (100 wt% floated) of the flotation feed with no selectivity.

Bio-Flotation of Phosphate Samples

The different operating parameters affecting the flotation process were first screened using a Plackett-Burman design. The statistically significant parameters were further optimized using central composite designs for each type of microorganisms. However, the flotation results using microorganisms #2 were almost the same as in its absence (Abdel-Khalek and Farrah, 2004; El-Mahdy, 2004). For this reason, the study shows only the results of applying the central composite design for optimizing the most significant parameters affecting bio-flotation of phosphate using the microorganisms #1. The studied parameters are collector dosage, concentration of microorganism coded #1, and pH. Each parameter is varied over five different levels as shown in Table 2. The other operating parameters were fixed at the following levels: conditioning the microorganism in the system for 15 min followed by conditioning the added collector for another 5 min. The variables were coded between "-" and "+"; where "-" represents the low level and "+" represents the high level of the factor. These low and high levels gives eight combinations in addition to three other experiments at their mid-point (0). This represents the normal factorial design studied before. However, six more experiments (represent –1.68 and +1.68 levels) are added to cover a wider range for each parameters as shown in Table 3. This gives 17 experiments which constitute the central composite design. The results of these experiments are depicted in Table 4 and their analysis is depicted in Figure 1.

The results in Table 4 depict the following: The grade of concentrates and their recoveries vary over a wide range with the variation of the operating conditions. For example, the P_2O_5 content varied between 27–30.1% and MgO varied from 0.78% to 2.45% while the recovery showed also a wide range of variation (20.8–100%). Experiment number 3 showed no selectivity at all. The operating conditions of experiment number 4 seems to be the optimum combinations where the best grade (0.78% MgO and 30.2% P_2O_5) with high recovery (92.6%) was obtained. Such grade and recovery cannot be obtained in conventional flotation of phosphate with oleate (Table 1), i.e., in absence of microorganisms.

Comparing the results of experiments number 3 (no selectivity) and number 4 (best selectivity) with their operating conditions showed very interesting findings related to the role of microorganisms as surface modifier. Both experiments are performed under identical conditions (pH 5.5 and 2.0 kg/t oleate) except that the concentration of microorganism coded #1 is doubled from 5×10^7 cells in experiment number 3 to 10×10^7 cells in experiment number 4. Also, these results (Table 4) show the influence of interaction between the type of microorganism and the type of collector. Such interaction seems to be synergistic in nature.

On the other hand, the statistical analysis for the results (Table 4) is depicted in Figure 1. Such analysis show the net effect of each studied parameter as follows: better grade (of lower % MgO and higher % P_2O_5) was obtained at lower pH (~5.5) rather than higher pH (~10.9). This illustrates that in reverse flotation of carbonate with oleate, it is recommended to use low pH of ~5.5, i.e., the same pH as in conventional flotation. However, the presence of microorganisms, under these conditions, enhances the selectivity probably due to its action as surface modifier.

Zeta Potential Measurements of Different Microorganism—Mineral Systems

The zeta potential measurements have been conducted for single minerals of dolomite and apatite and each type of microorganism (coded #1 and #2) used in the flotation experiments as well as their mixtures. These measurements were performed at constant ionic strength of 1×10^{-2} M KCl. Figure 2 shows the zeta potential of dolomite and apatite. The curves of apatite and dolomite show, more or less, similar trends. The electro-negativities of the surfaces of both minerals are close to each other and increase with increasing the pH within the entire range from about 2–12. Such great similarity in the surface properties of apatite and dolomite represents one of the main reasons that hinder the selectivity of their separation by the flotation process. This behavior for apatite and dolomite as salt type minerals is in agreement with that mentioned by many authors (Somasundaran and Zhang, 1999; Houot, 1982). The curves in Figure 2 do not show an iso-electric point (i.e.p.) for apatite or dolomite although they are of considerable purity as mentioned before. Many authors (El-Mofty, 1989) attributed such phenomenon to the diluted solid-liquid ratio as has been done (0.02%) in this study.

The zeta potential of the microorganisms coded #1 and #2 was also measured at constant (1×10^{-2} M KCl) ionic strength, the results of which are depicted in Figure 3. Both microorganisms have minor electro-negativity especially in the acidic medium. The zeta potential slightly increased (by not more than 12 mv) with raising the pH from the very acidic medium (pH 2.0) to the strongly alkaline medium of pH ~12.0.

These results illustrate the hydrophobicity of these microorganisms. The zeta potential for each single mineral, of dolomite and apatite, after its surface modification with each type of microorganism (coded #1 and #2) was also conducted under similar conditions. The results

Table 3 Combinations of central composite design for bio-flotation of phosphates

Exp. #	A	B	C
1	-1	-1	-1
2	-1	-1	1
3	-1	1	-1
4	-1	1	1
5	1	-1	-1
6	1	-1	1
7	1	1	-1
8	1	1	1
9	-1.68	0	0
10	1.68	0	0
11	0	-1.68	0
12	0	1.68	0
13	0	0	-1.68
14	0	0	1.68
15	0	0	0
16	0	0	0
17	0	0	0

Table 4 Results of central composite design for bio-flotation of phosphates

Exp. #	P_2O_5 %	MgO %	P_2O_5 Recovery %
1	29.92	1.21	85.77
2	29.27	1.12	96.24
3	27.067	2.45	100
4	30.15	0.78	92.31
5	29.43	1.33	92.58
6	28.93	1.68	89.93
7	28.89	1.75	83.7
8	29.44	1.51	91.49
9	30.13	0.83	79.68
10	27.28	1.44	75.1
11	29.13	1.28	91.8
12	28.36	1.58	55.92
13	24.09	1.62	20.75
14	28.5	1.34	77.59
15	28.67	1.54	85.72

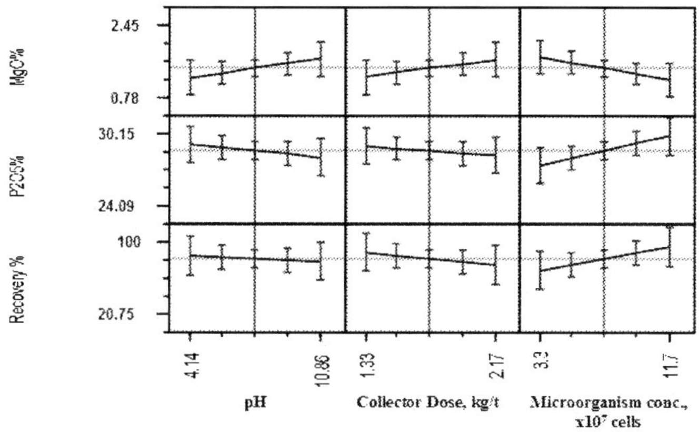

Figure 1 Statistical analysis for the central composite design

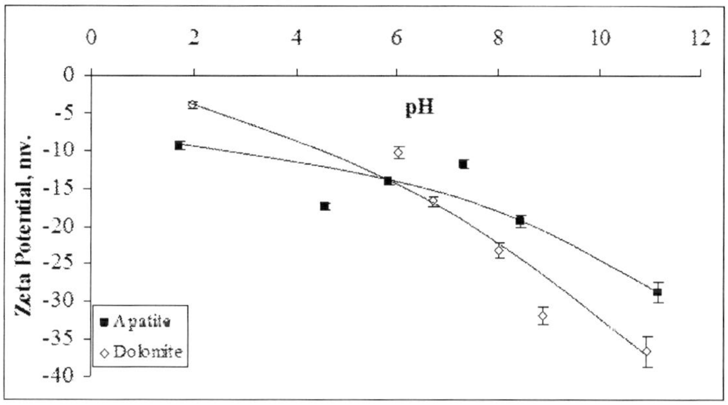

Figure 2 Zeta potential of dolomite and apatite at constant ionic strength (1×10^{-2} M KCl)

of these measurements are shown in Figures 4–5 for apatite and dolomite after their modification with microorganism coded #1 and Figures 6–7 after treatment with microorganism coded #2. The results in Figures 4 and 6 illustrate that the surface of apatite is affected by the microorganism coded #2 to a larger degree in comparison to that of microorganism coded #1. It is clear, therefore, that the microorganism coded #1, as a surface modifier, works better in the acidic pH that decreases the electro-negativity of the apatite surface, while the microorganism coded #2 works better in the alkaline region that makes the apatite surface more electronegative. Thus, in the acidic pH range, the acquired hydrophobicity for apatite can lead to a degree of aggregation and in turn form large particles (or aggregates). Such aggregates of apatite might hinder their flotation even in case of oleate adsorption (by hydrophobic interaction) due to the increasing rate of bubble-particle detachment. This might help in explaining the good selectivity achieved in flotation of phosphate with the oleate collector in the acidic pH (~5.5) while using microorganism coded #1. The influence of both microorganisms (coded

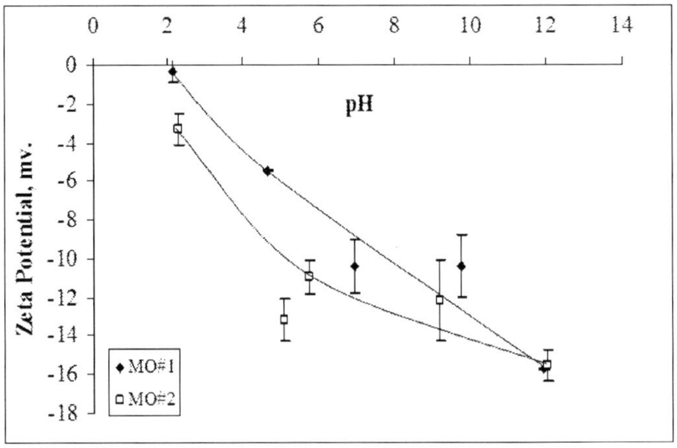

Figure 3 Zeta potential of the microorganisms coded #1 and #2 at constant ionic strength (1×10^{-2} M KCl)

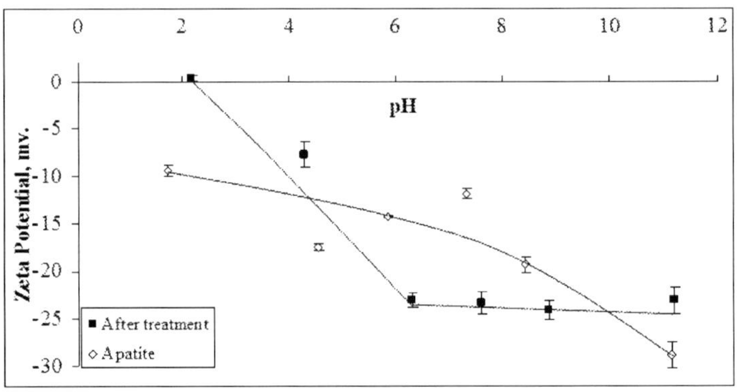

Figure 4 Zeta potential of apatite before and after surface modification with microorganism coded #1 at constant ionic strength (1×10^{-2} M KCl)

#1 and #2) on the dolomite surface shows, more or less, similar trends (Figures 5 and 7).

Adsorption Isotherms of Different Collector–Mineral Systems

Figures 8 shows the adsorption isotherms of oleate onto apatite and dolomite at pH 5.5. These results reflect the nonselectivity of adsorption process for oleate (as a fatty acid collector) onto apatite and dolomite. The isotherms show identical adsorption density over the entire range of oleate concentrations, except a very narrow region (around oleate concentration of 200 ppm) where a minor difference in adsorption density is noticed in favor of dolomite. Above 200 ppm of oleate, no selectivity in adsorption can be achieved. These results correlated very well with what was mentioned before during the flotation of phosphate with oleate at pH 5.5 (Table 1), where the selectivity (30.2% P_2O_5 and 0.85% MgO) is achieved over a very small range (0.25 kg/t) of oleate concentration at the expense of recovery (~55.5%). Above such concentration (>0.25 kg/t) bulk flotation (100 wt% floated) occurred for both apatite and dolomite and the selectivity was

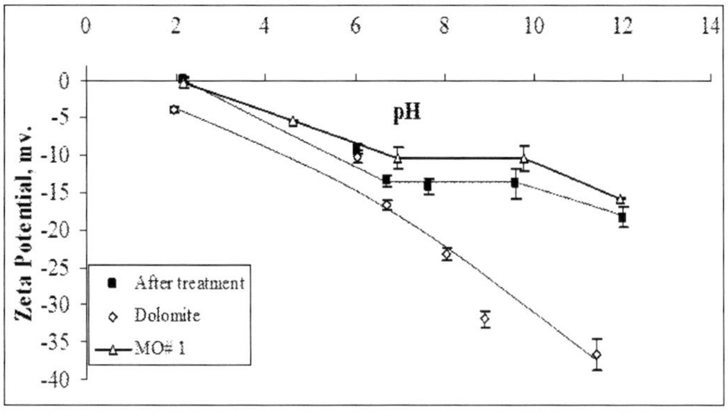

Figure 5 Zeta potential of dolomite before and after surface modification with microorganism coded #1 at constant ionic strength (1×10^{-2} M KCl)

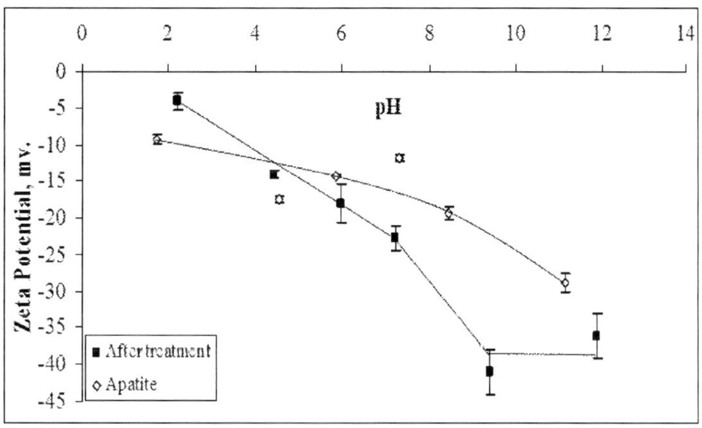

Figure 6 Zeta potential of apatite before and after surface modification with microorganism coded #2 at constant ionic strength (1×10^{-2} M KCl)

completely lost. These results for oleate adsorption (Figure 8) and its flotation behavior (Table 1) confirm the nonselectivity of oleate as one of the fatty acids collectors for phosphate flotation.

Adsorption Isotherms of Different Microorganism–Mineral Systems

The adsorption behavior of both microorganisms (coded #1 and #2) onto single minerals of apatite and dolomite was also performed. These adsorption measurements were conducted at pH 5.5 where the best flotation results were achieved with oleate. The adsorption isotherms are shown in Figures 9 and 10. The adsorption density of each microorganism onto dolomite surface is always higher than that onto apatite indicating a preferential adsorption for dolomite. However, correlation of these adsorption isotherms with the results of zeta potential (Figures 4–7) may suggest that at pH 5.5 microorganism coded #1 can be used as a surface modifier for both apatite and dolomite.

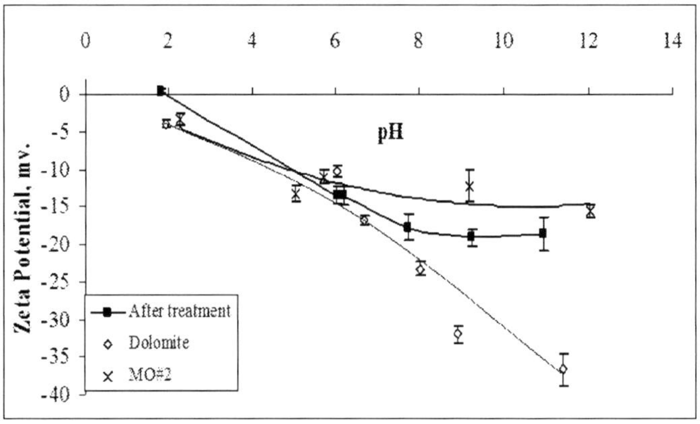

Figure 7 Zeta potential of dolomite before and after surface modification with microorganism coded #2 at constant ionic strength (1×10^{-2} M KCl)

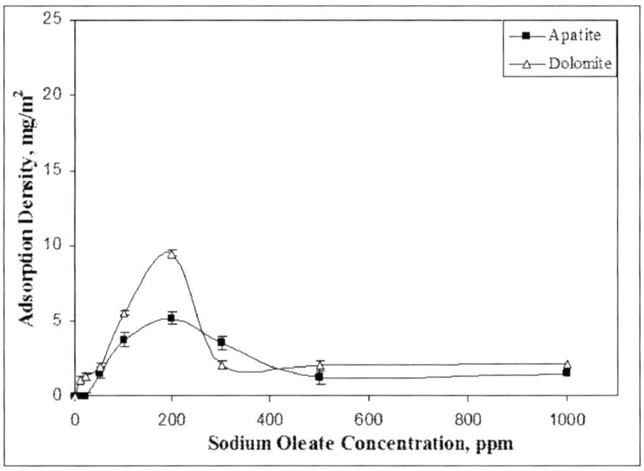

Figure 8 Adsorption isotherm of oleate collector onto apatite and dolomite at pH 5.5

Frothing Power of Different Collector–Microorganism Mixtures

In a trial to understand the role of interaction between microorganisms and collectors and in turn its effect on the selectivity of the bio-flotation process, the froth power for binary systems of collectors and their mixtures with microorganisms were measured. The results are shown in Figure 11. A sharp reduction in the froth height is recorded for oleate in presence of microorganism coded #2. Moreover, such froth height is reduced to nil in presence of the microorganism coded #1. These results show clearly that there is a strong interaction in such binary system between oleate and the microorganisms particularly coded #1. It seems that such strong interaction between oleate and the microorganism coded #1 improved the selectivity in terms of grade (0.78% MgO and 30.2% P_2O_5) and recovery (92.3%) during the bio-flotation experiments as shown before (Table

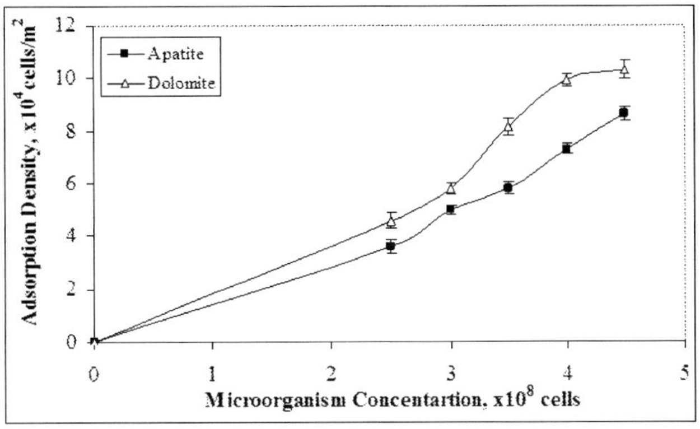

Figure 9 Adsorption isotherm of microorganism coded #1 onto apatite and dolomite at pH 5.5

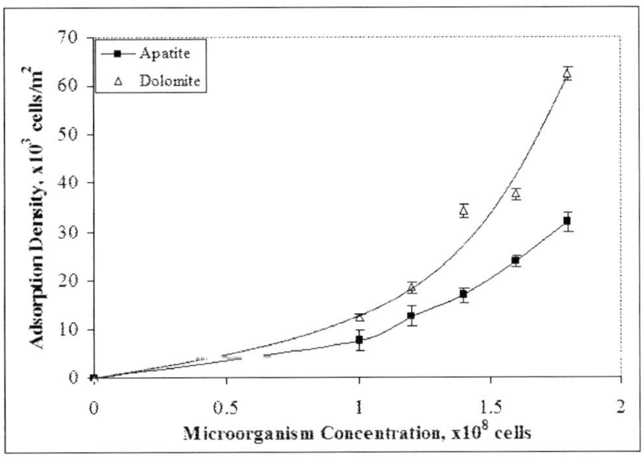

Figure 10 Adsorption isotherm of microorganism coded #2 onto apatite and dolomite at pH 5.5

4). It should be mentioned that such interaction between oleate and the microorganism coded #1 is also estimated from the analysis of the central composite design for bio-flotation shown before. Figure 12 depicts the effect of interaction between collector dosage of oleate and the concentration of microorganism coded #1 in the flotation system. This interaction is synergistic in nature that led to a significant reduction in MgO content at the higher concentration of microorganism coded #1. Also, the analysis of the central composite design for bio-flotation of dolomite illustrates that such synergistic interaction between oleate and the microorganism coded #1 becomes more favorable at the lower pHs as shown in Figure 13.

FTIR Adsorption Studies for Different Microorganisms and Collectors

Biological processes involving microbes for mineral processing were first reported by Solojenken et al. in 1976. Biological processes are attractive since microorganisms (or microbes) or microbial fat and secreted metabolites can have

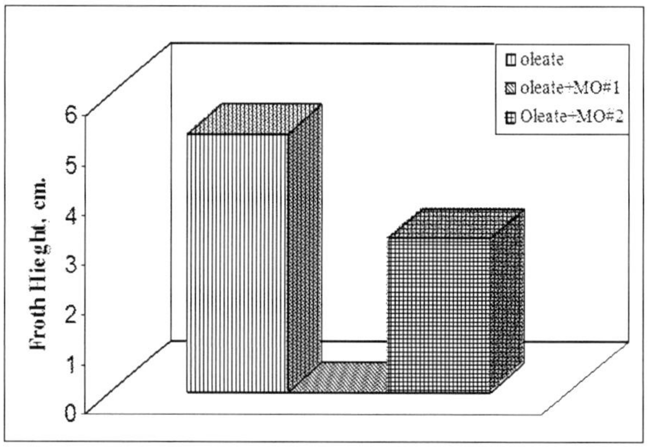

Figure 11 Frothing power of oleate and its mixtures with microorganisms (3 mL collector and 2 mL microorganism, StD. 0.4)

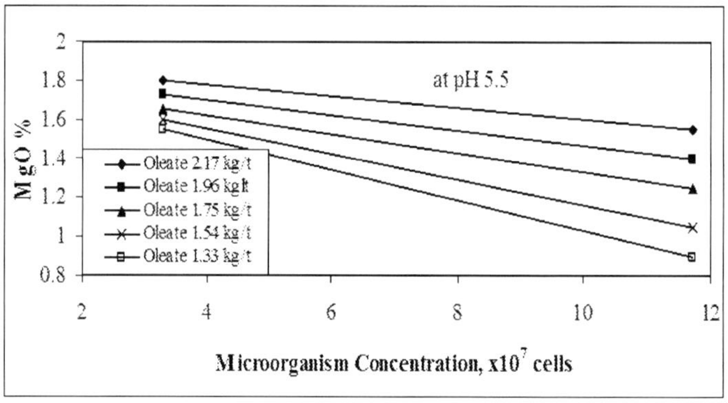

Figure 12 Effect of interaction between oleate dosage and concentration of microorganism coded #1 on the MgO content

specific interactions with minerals. Such interactions of microorganisms and their agents with minerals can be indirect, with biological products acting as surface-active agents, or direct due to microbial adhesion (or attachment) to particles bringing out surface modification. Both types of interactions can lead to alteration of mineral hydrophobicity, and in some cases cause flocculation or dispersion of mineral suspension (Somasundaran et al., 1998). Some authors (Attia and Elzeky, 1985) have suggested that microorganisms could adsorb on the mineral surface and grow. However, the bio-adsorption process is believed to be rapid enough to be completed in a matter of several minutes and not several days, or even several months, as in the case of bio-leaching. It is unlikely, therefore, that within such a short period enough bacteria metabolites are produced to affect the floatability to any measurable extent. The only other possibility for bio-flotation under these conditions is the bacterial attachment onto the minerals, which will result in a hydrophobic (hydrophilic) surface (Rao and Somasundaran, 1995). In a trail to understand the influence of such bio-modification in enhancing the

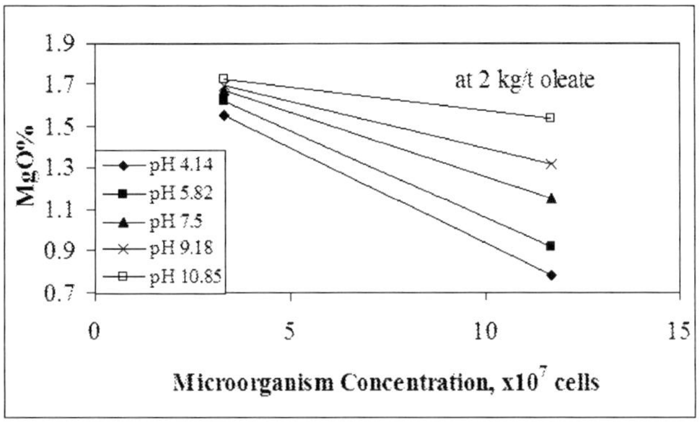

Figure 13 Effect of interaction between concentration of microorganism coded #1 and pH using oleate as collector on the MgO content

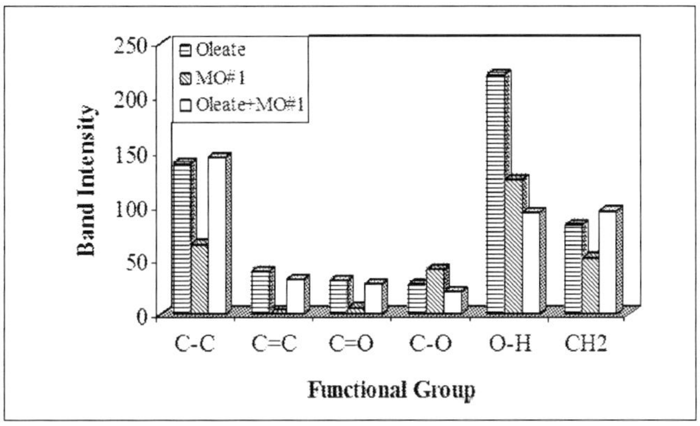

Figure 14 FTIR of oleate and microorganism coded #1 and their mixture

selectivity of the bio-flotation process and the role of interaction between collector and microorganism, FTIR measurements were conducted for both binary (microorganism-collector) and tertiary (microorganism-collector–mineral) systems. These FTIR measurements were performed using concentrations of collectors and microorganisms similar to those used at the optimum conditions for each bio-flotation system, Figures 14–15. The FTIR of the microorganisms showed the existence of O-H, C-C, CH_2, C-O, C-N and C=O bonds in decreasing order. These bonds reflect the general organic structure of microorganisms which are mainly composed of polysaccharides and lipids (protein). Polysaccharides are defined as high molecular weight carbohydrates containing many of monomeric units connected to one another by a type of covalent bonds referred as a glycosidic bond (Brock et al., 1994). Figure 16 shows a schematic for the polysaccharides structure. These polysaccharides contain many -OH and C-O groups connected to its carbon structure. Meanwhile, amino acids are the monomeric units of protein. Most amino acids consist only of carbon, hydrogen, oxygen and nitrogen. Protein contains two important functional groups, a carboxylic acid group (-COOH) and an amino

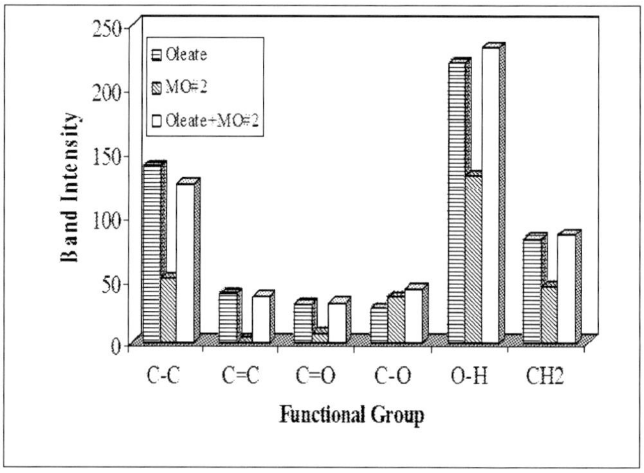

Figure 15 FTIR of oleate and microorganism coded #2 and their mixture

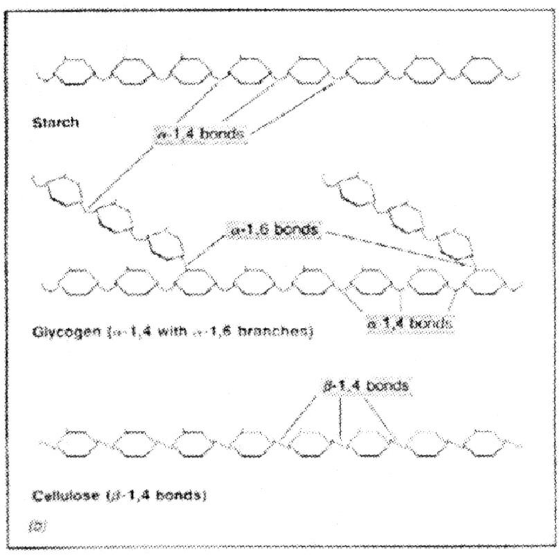

Figure 16 A schematic for the polysaccharides structure

group (-NH_2), as shown in its structure (Brock et al., 1994):

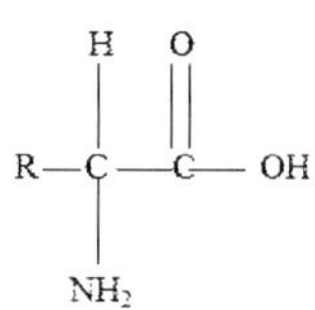

These carboxylic (-COOH) and amino (-NH_2) groups may explain the presence of C-N, C=O and OH bonds shown in the FTIR of both microorganisms. The concentrations of O-H and C-C are predominant in Figure 15 since these groups represent major constituents in the structure of both polysaccharides and protein.

Also, a very interesting phenomenon can be seen in the FTIR for a mixture of oleate with the microorganism coded #1 (Figure 14), i.e., the same combination which gave the best selectivity in bio-flotation experiments. A significant reduction in the intensity of O-H bonds is noticed after mixing oleate with the microorganism coded #1. Correlating such phenomenon with the sharp reduction in the froth height for the same (oleate and microorganism coded #1) system (Figure 11) confirms the presence of some kind of interaction between them that leads to a significant improvement in the bio-flotation experiments. This also confirms the interaction between such oleate and microorganism coded #1 system that leads to reduction of the MgO content shown before in Figure 13.

The recorded interaction between oleate and microorganism coded #1 in their binary systems is also confirmed in their tertiary systems with dolomite (or apatite). Figure 17 depicts the FTIR for adsorption of oleate and microorganism coded #1 onto single mineral of dolomite and apatite. The adsorption measurements were performed at optimum flotation pH of ~5.5. The results of FTIR (Figure 17) show that the adsorption process is interactive in nature for the system microorganism coded #1—oleate onto dolomite and apatite surfaces. The intensities of the C-C and CH_2 bonds are remarkably increased after adsorption of oleate onto the bio-modified surfaces of dolomite and apatite. It should be mentioned that these (C-C and CH_2) bonds represent the hydrocarbon structure of both oleate and microorganisms. This means that the adsorption process is synergistic in nature for the system (oleate and microorganism coded #1). However, the intensities of the O-H and C-O bonds are dramatically decreased after adsorption of oleate onto the bio-modified surface. This again confirms the presence of interaction between oleate and microorganism coded #1 as shown before in their binary systems. This may also suggest that adsorption of oleate takes place through the reaction between OH groups of the adsorbed microorganisms and the carboxylic group of oleate leading probably to their dehydroxylation with formation of water molecule. As a result of such interaction, the intensity of both C-C and CH_2 bonds increases due to oleate adsorption while that of O-H and C-O bonds decreases due to their participation in the interaction process. It seems that such interaction plays an important role in enhancing the selectivity in the bio-flotation process. However, there is a preferential adsorption for both the microorganisms and thereafter oleate onto the dolomite surface rather than apatite surface. This leads to a better selectivity that was achieved for oleate adsorption after bio-modification of the surface.

A Tentative Adsorption Mechanism for Microorganisms—Collector–Mineral System

Based on the above findings, a tentative mechanism might be suggested as described by a schematic drawing in Figure 18. Thus, microorganisms can help in the following ways:

First, bio-modification of both apatite and dolomite surfaces with microorganism coded #1 helps in covering some of the adsorption sites that causes nonselectivity of oleate adsorption.

Second, such bio-modified surfaces can also help in decreasing the high solubility of Ca and Mg ions expected at such acidic medium (5.5) and hence decreases the possibility of surface transformation that leads to nonselectivity in the dolomite-apatite system.

Third, the preferential adsorption of both microorganism coded #1 and oleate onto dolomite surface and their synergistic interaction might help in decreasing the gap of similarity with apatite (i.e., the gap in adsorption will be larger and in favor of dolomite) that leads to a significant enhancement in the selectivity of the bio-flotation process.

CONCLUSIONS

A central composite design is applied for optimizing the most significant parameters affecting bio-flotation of phosphate using oleate as collector and microorganisms coded #1. The best grade (0.78% MgO and 30.2% P_2O_5) with high recovery (92.6%) is obtained at the following

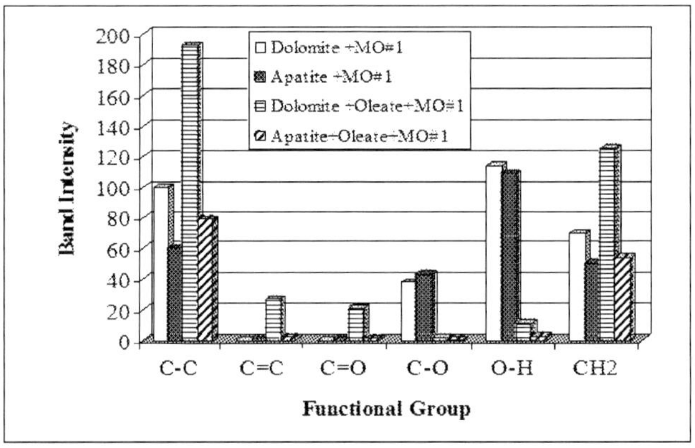

Figure 17 FTIR for adsorption of oleate and microorganism coded #1 onto dolomite

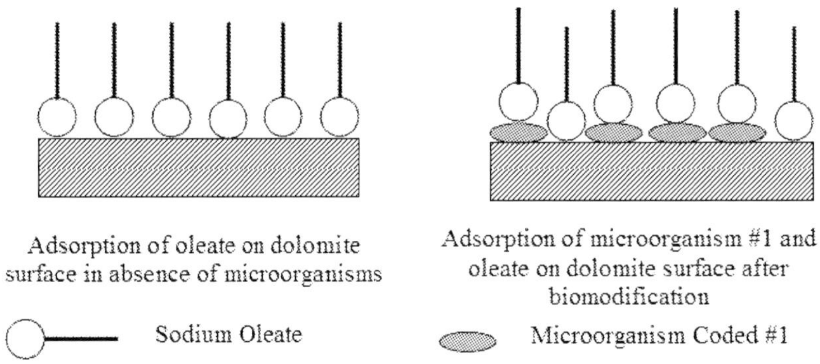

Figure 18 Adsorption of sodium oleate and microorganism #1 on dolomite surface

optimum conditions: pH ~5.5, 2.0 kg/t oleate and a concentration of microorganisms coded #1 of 1.0×10^8 cells. In absence of microorganisms, such grade and recovery are not obtained.

The frothing power (in terms of froth height) of oleate is reduced to nil in presence of the microorganism coded #1 illustrating a very strong interaction between them. It seems that such strong interaction between oleate and the microorganism coded #1 enhances the selectivity of the flotation process.

FTIR results illustrated that there is a significant reduction in the intensity of O-H bonds after mixing oleate with the microorganisms coded #1. Such interaction between oleate and microorganisms coded #1 is also confirmed in their tertiary (microorganism-collector-mineral) systems with dolomite (or apatite). Correlating such phenomenon with the sharp reduction in the froth height for the same (oleate and microorganism coded #1) system confirm the presence of some kind of interaction between them that leads to a significant improvement in the bio-flotation experiments.

The synergistic nature for the adsorption process in the microorganisms coded #1 and oleate system can help in improving the selectivity of flotation through the following mechanisms:

1. Bio-modification of both apatite and dolomite surfaces helps in covering some of the adsorption site that causes nonselectivity of oleate adsorption.
2. Bio-modification helps in decreasing the expected high solubility of Ca and Mg ions at pH 5.5 and in turn decreases the possibility of surface transformation that leads to nonselectivity in the dolomite-apatite system.
3. Preferential adsorption onto dolomite surface might help also to increase the gap with apatite leading to a significant enhancement in the selectivity of the bio-flotation process.

ACKNOWLEDGMENTS

The authors gratefully acknowledge the valuable contribution made to this paper by N.A. Abdel-Khalek of the Central Metallurgical Research and Development Institute, Helwan, Cairo, Egypt. Appreciation is also extended to M.A. Abdel-Khalek of the Central Metallurgical Research and Development Institute, and S.E. El-Mofty of the Department of Mining Engineering at Cairo University, Giza Egypt, for their contributions to this work.

REFERENCES

Abdel-Khalek, N.A., and Farrah, S., 2004, US Egypt Joint Project, Academy of Scientific Research and Technology, Egypt, March.

Anazia, I.J., and Hanna, J., 1987, *Minerals and Metallurgical Processing*, Nov., pp. 196–202.

Attia, Y.A., and Elzeky, M.A., 1985, Coal Science and Technology, Vol. 9, *Processing and Utilization of High Sulfur Coal*, pp. 673–682.

Boice, C.M., 2000, M.Sc. Thesis, University of Florida, Gainesville, Fla.

Brock, T.D., Madigan, M.T., Martinko, J.M., and Parker, J., 1994, *Biology of Microorganisms*, Prentice Hall, Englewoods Cliffs, New Jersey.

El-Gillani, D.A., and Abouzeid, A.Z., 1993, *International J. Miner. Process.*, Vol. 38, pp. 235–256.

El-Mahdy, A., 2004, M.Sc. Thesis, Mining Eng. Dept., Faculty of Eng., Cairo University, Egypt.

El-Shall, H., Zhang, P., and Snow, R., 1996, *Minerals and Metallurgical Processing*, Vol. 8, pp. 135–140.

El-Mofty, S.E., 1989, M.Sc. Thesis, Mining Eng. Dept., Faculty of Eng., Cairo University, Egypt.

Houot, R., 1982, *International J. Miner. Process.*, Vol. 9, pp. 353–384.

Moudgil, B., and Sumasundaran, P., Advances in Mineral Processing, Arbiter Symposium, SME, pp. 426–441.

Natarajan, K.A., and Deo, N., 2001, *International J. Miner. Process.*, Vol. 62, No. 1-4, pp. 143–157.

Rao, M.K.Y., Natarajan, K.A., and Sumasundaran, P., 1992, *Minerals and Metallurgical Processing*, Vol. 9, pp. 95.

Rao, M.K.Y., and Sumasundaran, P., 1995, In: *Flotation Science and Engineering*, K.A. Matis (ed.), Chapter 17, pp. 455–472.

Sumasundaran, P., Ren, Y., and Rao, M.Y., 1998, *Colloids and Surfaces*, Vol. 133, pp. 13–33.

Sumasundaran, P., and Zhang, P., 1999, In: *Beneficiation of Phosphates: Advances in Research and Practice*, Zhang, P., El-Shall, H., and Wiegel, R. (eds.), SME, Littleton, Colo., pp. 141–154.

Smith, R.W., Misra, M., and Dubel, J., 1991, *Minerals Engineering*, Vol. 4, No. 7-11, pp. 1127–1141.

Soljenken, P.M., Lyubavina, L.L., Larin, V.K., Bergelson, L.D., and Dyatlovitskaya, E.V., 1976, *Bull. Nonferrous Metallurgy*, Vol. 16, pp. 21.

CHAPTER 20

Mineral Processing of Phosphate Rock at Agrium Kapuskasing Phosphate Operations

Rawya Rifaat* and Dennis Grimm*

ABSTRACT

The Agrium Kapuskasing Phosphate Operation (KPO) in Ontario, Canada, is a challenging, highly variable ore deposit processed using grinding, desliming, magnetic separation, and flotation. The orebody is an igneous deposit consisting of iron, carbonates, silicates, and sulphide minerals. In the KPO mill, phosphate rock is ground in a SAG mill–ball mill circuit and deslimed to improve phosphate grade, and recovery and handleability. Magnetic separation is required to separate iron and titanium minerals from phosphate. Flotation is used to upgrade phosphate and calcium, while depressing silica and iron. The final concentrate is dewatered and dried using a cyclone, thickener, belt filter, and fluidizing bed dryer. The dried product is shipped to Redwater, Alberta, Canada, for acid treatment.

INTRODUCTION

In 1999, Agrium Inc. commenced processing at the Kapuskasing Phosphate Operation. In retrospect, production began without a full understanding of the ore type and milling techniques required to process this complicated ore body. Over the past five years, KPO has conducted exploration studies, mineralogy analysis, grinding surveys, laboratory and plant scale flotation, test flotation and magnetic separation test work, and dewatering and dryer circuit operational optimization. Geological and mineralogical analyses provide an indication which minerals cannot be processed in the mill and which rock types yield poor recovery and unacceptable quality. Grinding surveys and mill simulations show the positive effects of grinding equipment upgrades to increase throughput while producing a target grind size. Flotation tests show the potential of processing ore with high deleterious elements other than iron. Dewatering and drying optimization has produced high cost savings by reducing dewatering and anti-dusting reagents, limiting the consumption of natural gas. This paper describes operational strategy and equipment changes used to process this complex deposit.

GEOLOGY

The Agrium Kapuskasing Phosphate Operation, located 32 kilometres southwest of Kapuskasing, Ontario, Canada, is mined from the Cargill deposit. This deposit was identified as a magnetic anomaly in 1954 and discovered to be a phosphate deposit by International Minerals and Chemical Corporation (IMC) in 1975. From 1975 until 1997, property ownership changed

*Agrium Inc., Calgary, Alberta, Canada.

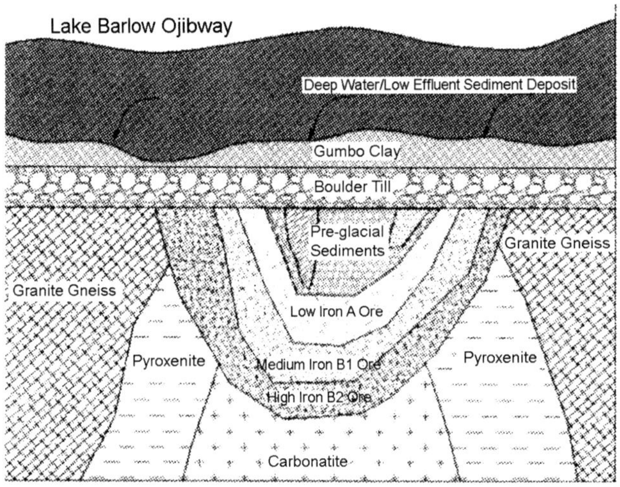

Figure 1 Cargill deposit cross section

several times until Agrium began development. The Cargill deposit is an alkalic igneous complex and consists of two weathered igneous intrusions—pyroxenite and carbonatite. During its formation, 1.8 billion to 100 million years ago, weathering of the carbonatite left behind less soluble residuum that contains concentrated amounts of apatite and iron minerals. The iron minerals are concentrated in the lower layers of the deposit because they are more prone to weathering than apatite. Woody lignite from forest moors, silica sand from weathered granite country rock, and organic rich clays from floodplain swamps filled the cavity created by weathering and make up the overburden pre-glacial sediment (PGS) that occupies the center of the ore body. Subsequent to weathering, approximately 5 million years ago, glaciation covered the deposit. At the end of the Pleistocene epoch, one million years ago, retreating glaciers deposited a layer of till and melt water. This glacial melt water formed the massive Lake Barlow Ojibway and left a layer of glacial lake (gumbo) clay up to ten meters thick on the surface of the deposit as shown in Figure 1.

The Cargill deposit is economically significant because of the presence of residuum, the final product of carbonatite weathering. Residuum is primarily a light to dark grey unconsolidated sandy material and contains up to 100% apatite (44% P_2O_5). Carbonatite is waste material because the weathering process has not enriched it. Fresh carbonatite is present in the pit as sovite and rauhaugite. Sovite, a hard white to grey color rock containing 50% or more calcite, is present along the south pit wall and is used primarily as a road building material. Rauhaugite, chemically identical to sovite but dolomite is the predominant carbonate mineral (instead of calcite), occurs along the west and north pit walls as a sandy light brown color rock.

Quality of the ore is determined by the impurity content within the residuum and classified as "A" ore, "B" ore, or marginal ore. Mostly pure residuum (high phosphate, low iron grade ore) is designated "A" ore. Low phosphate, high iron grade ore composed of residuum diluted with intermediate products of weathering, such as iron oxides and clays, is identified as "B" ore. Marginal ore has adequate phosphate content but has too many impurities for the mill to currently handle. Stockpiles are created based on ore quality and blended to produce a suitable feed grade.

MINERALOGY

The KPO ore body is variable consisting of several types of oxide, silicate, carbonate, and phosphate minerals making the grinding and beneficiation circuit design challenging. The ore, having a phosphate grade greater than 15% P_2O_5, is classified based mainly on iron grade (% Fe_2O_3), and monitored for magnesium (% MgO), aluminium (% Al_2O_3), and silica (% Si_2O). "A" ore iron content is less than 8% Fe_2O_3, "B1" ore iron content ranges from 8 to 20% Fe_2O_3, and "B2" ore ranges from 20 to 30% Fe_2O_3. "R" ore with aluminium greater than 4% Al_2O_3 is classified as "A2" ore and ore with magnesium greater than 1% MgO or 10% Si_2O_3 is identified as marginal ore. These ore types behave differently in the KPO mill due to their different mineralogical properties.

The main phosphate mineral is (fluro-)apatite ($Ca_5(PO_4)_3F$), with minor amounts of crandallite ($CaAl_3(PO_4)_2(OH)_5 \bullet (H_2O)$) and trace amounts of vivianite ($Fe_3{}^{+3}(PO_4)_2 \bullet 8(H_2O)$), monazite ($(Ce,La,Nd,Th)PO_4$), and xenotime ($YPO_4$). Apatite is the only phosphate mineral which is recoverable in the KPO beneficiation process. "A2" ore is generally composed of crandallite which is identified by high phosphate and aluminium grades and a low calcium grade. Iron minerals vary with ore type. Iron sulfides mostly occur in the "A" ore as pyrite and marcasite. Iron and titanium oxides such as hematite, magnetite, maghemite, ilmenite, and rutile, and iron hydroxides such as goethite (limonite) mostly occur in the "B" ores. Iron and magnesium carbonate minerals like dolomite, ankerite, calcite, siderite, magnesite, and kutnohorite occur mainly in the "B" ores and are identifiable by higher magnesium and calcium content. Iron is undesirable in the phosphate concentrate when it occurs as soluble iron as it causes problems in the downstream acid treatment process. Iron sulfides have a very low iron solubility, while iron oxides, hydroxides, and carbonates tend to be more soluble.

The silica content of the ore is variable and can range between less than 1% to greater than 50% Si_2O. Silica mostly occurs as quartz, but occurs in minor amounts of phyllo-silicates such as muscovite and biotite. Silica is noticeable in some "B2" ores, but has been observed in "A2" ores occurring as kaolinite.

A thorough understanding of mineralogy is important as it effects the operation and optimization of the KPO grinding and beneficiation circuits, as well as the performance of the Redwater Phosphate Operations.

MILLING PRACTICES

The main objective of the KPO is to produce concentrate meeting Redwater Phosphate Operations (RPO) size and quality specifications requirements. Size specifications are set to reduce coarse material greater than 300 microns (48 mesh) which effects reaction kinetics and reduce material finer than 37 microns (400 mesh) which causes dusting problems during offloading rail cars at Redwater. KPO's concentrate quality specifications are determined to maximize RPO's efficiency and is classified into four product types based on calcium, soluble iron, and magnesium assays. Quality specifications vary at KPO based on RPO's inventory (of different product types). Meeting product specifications can be difficult due to ore variability. The Cargill deposit is highly variable and therefore the design of the KPO mill must be adaptable to different ore types. The design and upgrade of the grinding and beneficiation circuits are intended not only to optimize production but to provide operating flexibility for varying mill feed. As explained in the geological and mineralogical section, there are several types of ores with different elemental contents. Since start-up, the operating philosophy of the KPO mill is to blend the "A" ore with the "B" ore to produce the desired feed and product grades. However, plant testing shows that the overall plant recovery increases when the ore types are run separately. To limit phosphate losses to tails, the grinding and the beneficiation circuits are operated differently depending on ore type.

Certain ore types are problematic as either phosphate recovery or product quality targets are compromised. Carbonate minerals fed into the mill produce poor quality concentrate, typically high in magnesium and soluble iron. This is due to carbonates being non-magnetic and the phosphate flotation collector is selective to calcium dominant in the carbonate minerals. High carbonate content in the final product causes foaming and scaling issues at Redwater resulting in poor efficiency at the RPO plant. Currently laboratory scale test work of a carbonate flotation stage is in progress. Phosphate minerals with low calcium content (crandallite and vivianite) affect recovery in both the KPO and RPO plants. In phosphate flotation, phosphate is collected with calcium, therefore, phosphate minerals with low calcium reports to flotation tails resulting in high phosphate losses. RPO acid addition and reaction are based on calcium content, and therefore recovery and performance of the Redwater circuit are affected when low calcium, high phosphate rock is processed. Operating equipment and parameters used at KPO to process this challenging phosphate ore body is described in Grinding and Desliming, Beneficiation, and Dewatering and Drying sections.

GRINDING AND DESLIMING

KPO grinds the ore to finer than 300 microns (48 mesh) and deslimes at 37 microns (400 mesh). A grind size finer than 300 microns is required because coarse material has a detrimental effect on process kinetics at RPO, as well as magnetic separation and phosphate flotation equipment is designed for finer grinds. The naturally present material finer than 37 microns in the ore contains elevated levels of iron, alumina, and silica, and lower grade phosphate. Furthermore, fines in the ore increase dryer natural gas consumption and flotation reagent consumption, and create offloading problems at RPO. Therefore, the fines are removed prior to beneficiation.

Currently, the grinding circuit at KPO consists of an optional toothed rolls crusher, a semi-autogenous grinding (SAG) mill operated in open circuit with a 1.83m (6ft) diameter Eimco spiral classifier (deslime classifier), a ball mill operated in closed circuit with a 2.44m (8ft) diameter Eimco spiral classifier (ball mill classifier), and twelve 25.4cm (10in.) cyclones (fines cyclones). Figure 2 shows the flow sheet for the current grinding circuit.

A belt feeder (30cm top size) or an optional rolls crusher (for coarser ore) can feed the KPO mill. The rolls crusher reduces the run of mine ore to minus 20cm and feeds the SAG mill. The SAG mill is 5.18m (17ft) in diameter and driven by a 1,500hp variable speed motor. A variable speed mill is required in this application because of the varying hardness and size of the different ore types. The "A" ore is finer but has a higher bond ball mill work index than the "B" ore. The harder "A" ore produces more efficient grinding in the SAG by preventing loading and resulting in lower SAG speeds. The soft "B" ore requires a faster operating speed to handle the load caused by limited attrition and abrasion comminution. The SAG operates with a 10% charge of 10cm (4in.) balls, and about 30% charge of balls and coarse ore.

The SAG product is minus 25.4mm, has a 76 to 80% solids content, and is classified by the deslime classifier. The deslime classifier overflow, 60 to 70% minus 150 microns (100 mesh), is screened at 1.0mm to remove organics and tramp material and is combined with the ball mill circuit product for desliming. The deslime classifier underflow reports to the 3.05m (10ft) diameter ball mill for further grinding. The ball mill is driven by a 2,000hp drive, operates in closed circuit with the ball mill classifier, and typically has a circulating load of 150 to 175%. The ball mill is charged with 3.8cm and 5cm (1½ and 2in.) balls to approximately 30% by volume. The deslime classifier overflow and ball mill classifier overflow are combined and feed the fines cyclones. The fines cyclone feed typically has a grind size of greater than 90% passing 300 microns (48 mesh) and 70% passing 150 microns (100 mesh). The fines cyclones remove minus 37 microns (400 mesh) material to increase the phosphate grade and reduce the iron, alumina, and silica grades. Fines rejection also aids in dewatering and reduces dusting issues in downstream processes.

In early 2005, the KPO mill feed will be lower grade phosphate ore resulting in an increase in

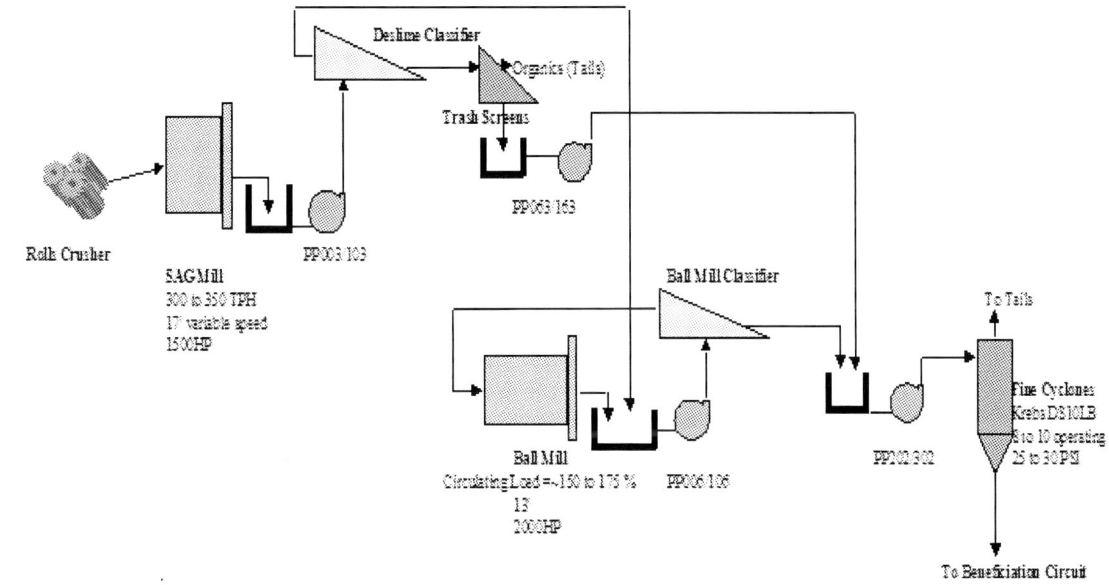

Figure 2 Current grinding flow sheet

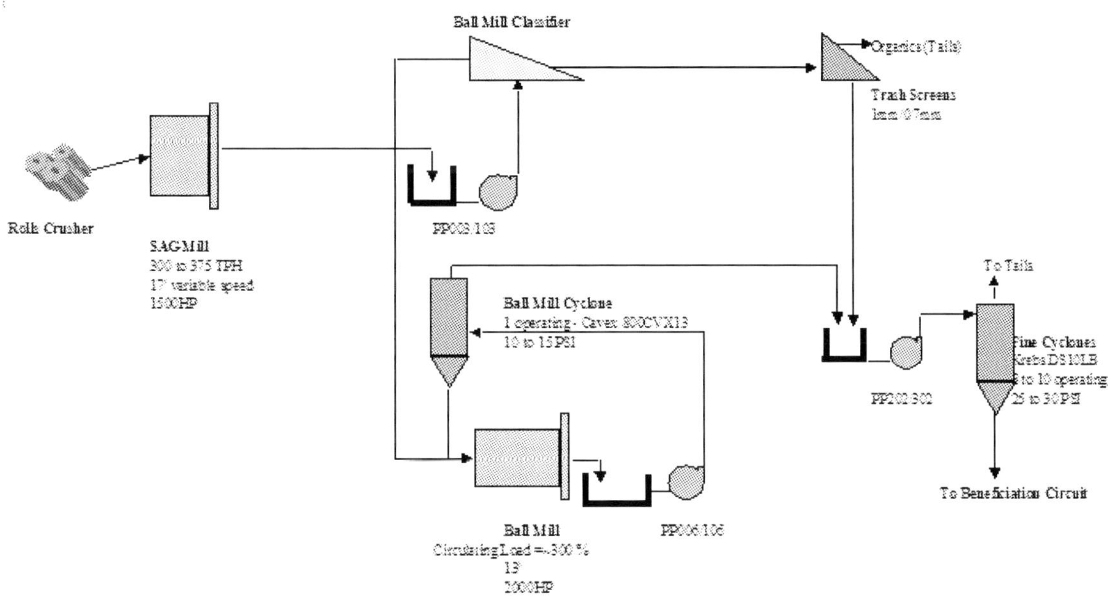

Figure 3 Upgraded grinding flow sheet

mill feed tonnage to meet target concentrate tonnage. The grinding circuit will be upgraded to handle the increased tonnage and run unblended "B" ore. Figure 3 shows the new grinding flow sheet. The ball mill classifier will replace the deslime classifier which will be decommissioned. The ball mill will operate in closed circuit with a 800mm hydrocyclone to achieve a more efficient size separation and have better operational control over the ball mill circulating load. Two additional fines cyclones will be commissioned in the fines cyclopack to handle the additional flow and maintenance requirements. To accommodate these changes the pumping and piping will be upgraded to handle the increased flow and improve availability.

BENEFICIATION

The KPO mill beneficiation circuit is designed to process variable ore grades to meet product quality specifications and maximize phosphate recovery. Currently the circuit consists of two stages of magnetic separation to remove iron, one stage of cycloning to dewater and remove extremely fine material, and a rougher stage of phosphate flotation. There is the option to bypass any or all of the last three stages depending on the feed grade. Figure 4 shows the beneficiation circuit.

The first stage of magnetic separation, low intensity magnetic separator (LIMS), removes the strongly magnetic iron from the ore and is always online. These units reject magnetite and maghematite to tails while the remaining non-magnetic product has the option to report to the second stage of magnetic separation or the second set of desliming cyclones depending on the iron feed grade. There is currently two LIMS units with the intention of installing a third unit in the near future to handle larger flows. The LIMS units operate at a fixed magnetic field of 750 gauss. The LIMS, shown in Figure 5, are four-foot diameter, horizontal, counter rotating, drum magnets in which the magnetic material is attracted to and remains upon the rotating drum to be removed by water sprays while the non-magnetic slurry reports to the discharge launder. The units are self-sufficient and require little operational control. Optimization of the LIMS consists of mechanical improvements such as an enhanced feeder system to ensure evenly distributed and consistent flow. The LIMS perform at a high phosphate recovery, typically 97% to 99% P_2O_5 recovery, as it is quite selective to strongly magnetic iron which is not associated with apatite. The LIMS circuit is always online due to the consistent performance and low losses.

The second stage of magnetic separation is fed by the LIMS non-magnetic product and removes the weakly and moderately magnetic iron. The high gradient magnetic separator (HGMS) removes the iron and titanium minerals rutile, ilmenite, goethite, hematite, and pyrite.

The HGMS unit consists of a continuous ring matrix that passes through a stationary magnetic head. Feed passes through the stationary head. The magnetization direction is induced parallel to the flow and the magnetic minerals remain on the ring matrix while the non-magnetics pass through. An underflow launder collects the non-magnetic product. The rotating ring matrix is flushed counter current to remove the magnetics and any oversize trapped in the matrix. The HGMS unit has a single head and a fixed magnetic field of 7,000 gauss. Figure 6 shows the schematics of the HGMS unit.

Ideal operation of the HGMS is controlled by rinse water, flush water, magnetic susceptibility, carousel speed, and particle size. The HGMS can be easily plugged by coarse particles. To prevent this, four derrick screens are located on the LIMS non-magnetic stream to screen at 300 microns. The oversize is pumped back to the ball mill and the undersize reports to the HGMS feed. To remedy plugged matrices, an acid wash is conducted as part of the maintenance of the unit. The acid wash removes organics as well as phyllo-silicates that pass through the derrick screens. Operating the rinse water at higher volumes will increase the fluid velocity, thus removing particles caught within the matrix. The rinse water is applied only when the matrix is in the magnetic field so it removes the non-magnetic particles. Flush water is applied when the matrix is in a low magnetic field, so it will remove

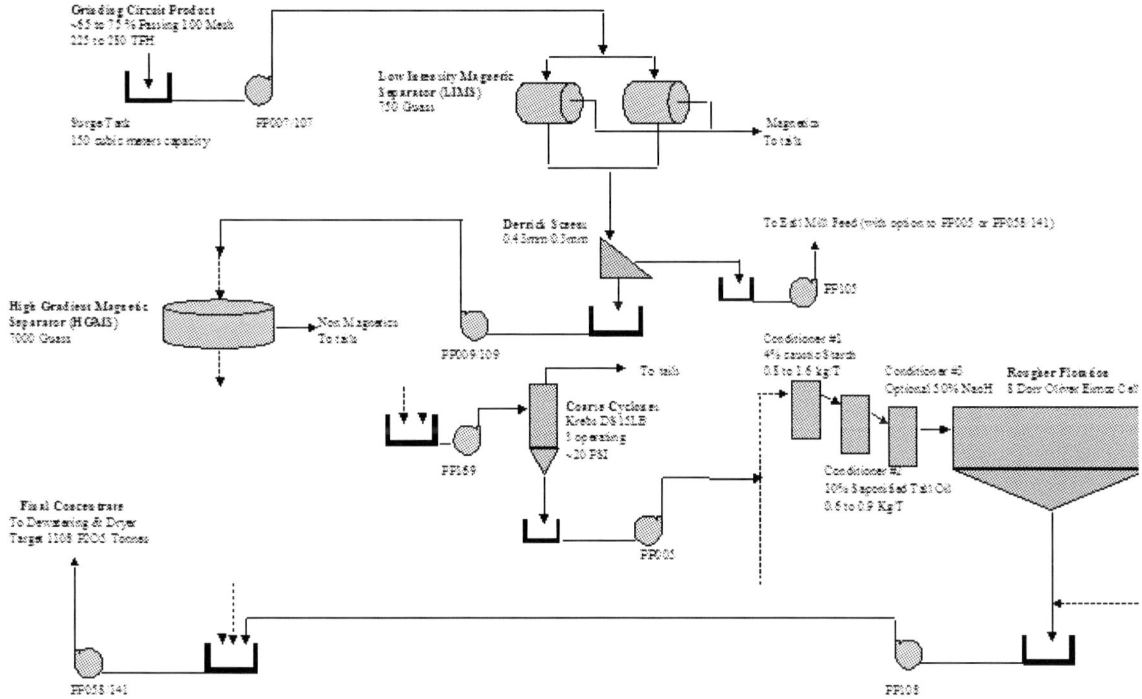

Figure 4 Beneficiation circuit

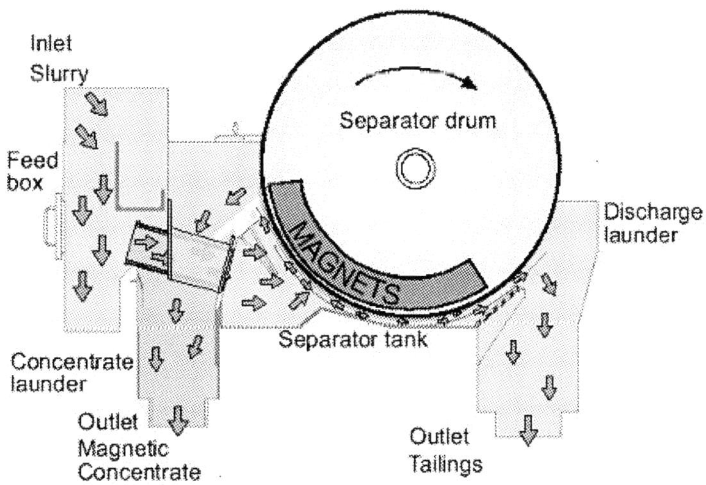

Figure 5 Low intensity magnetic separator schematic diagram

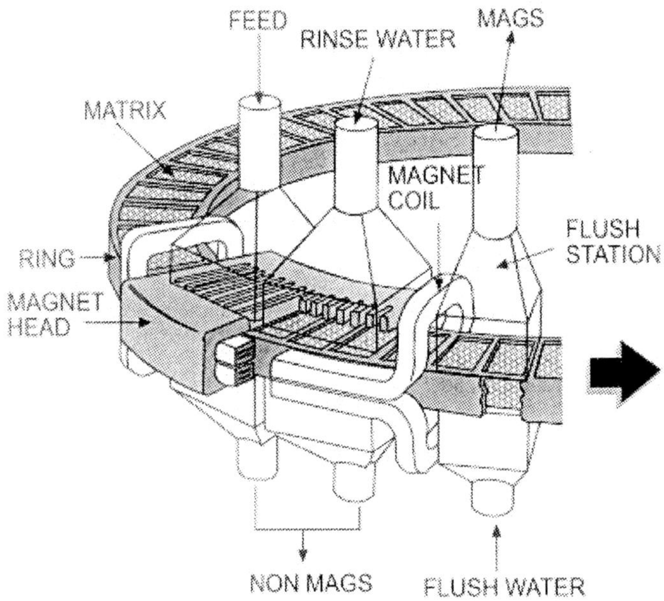

Figure 6 HGMS schematic diagram

magnetic particles from the matrix. The carousel speed affects the selectivity. If the speed is low, more iron will report to the non-magnetic product. If the speed is too high, the matrix can flood due to too much material reporting to the magnetic product indicated by poor phosphate recovery. Typical phosphate recovery in the HGMS unit is 85% to 90% P_2O_5 recovery to the non-magnetic product, and an iron reduction of 60% is expected. The mill feed iron grade determines if the HGMS is online or by-passed—typically if the mill feed iron grade is greater than 7% Fe_2O_3, the HGMS is running. The HGMS non-magnetic product feeds the second stage of desliming cyclones or the dewatering circuit as final concentrate.

The second stage of desliming cyclones, the coarse cyclones, are used to dewater the flotation feed and to remove extremely fine material. The HGMS product typically has a low density of less than 20% solids due to the large amount of rinse and flush water added. This low density has negative effects on the phosphate flotation circuit—poor reagent conditioning and sanding of coarse material. The coarse cyclones are 38cm (15in.) Krebs long body cyclones that are operated at 138kPa (20 PSI) to produce an underflow (flotation feed) density of 58 to 68% solids. The cyclone overflow reports to tails and has little phosphate losses (0.5 to 3.0% solids by mass). The phosphate recovery of the cyclones is 96 to 99%. The coarse cyclones are usually only by-passed due to maintenance as the phosphate recovery is quite high. The coarse cyclone underflow has the option to report to the flotation conditioner tanks or the dewatering circuit as final product.

Currently, at KPO the flotation circuit consists of a rougher bank to treat high iron and silica ore. The rougher cells were re-commissioned in July 2003 to facilitate the processing of lower grade phosphate ore. (Flotation was decommissioned from the KPO mill process in 2001 as magnetic separation was believed to be capable of upgrading the low grade phosphate. However, due to equipment inefficiencies and the evidence of high silica content in the ore, magnetic separation could not solely achieve required product upgrading.) During 2003, the flotation circuit ran as a plant test to prove the efficiency of phosphate flotation and determine operational and mechanical requirements for running flotation online. Early in 2004, operations commenced running the flotation circuit and throughout the year operated based on production requirements and mill feed content. Ongoing work on the flotation circuit consists of optimizing operating parameters, operating training, improving mechanical and instrumentation equipment, evaluation of cleaner phosphate flotation stages, and evaluation of stage-recovery in the rougher bank.

The rougher flotation circuit consists of three reagent conditioning tanks, eight rougher cells, sodium hydroxide storage tank, starch storage silo, starch mix and distribution tanks, and tall oil fatty acid storage, mix, and distribution tanks. The coarse cyclone underflow is pumped to the conditioner tank #1 (23.6m^3 capacity), where the slurry is conditioned with caustic starch (4% starch by volume) for a retention time of three to five minutes. This tank overflows slurry into conditioner tank #2 (22.7m^3 capacity), where the slurry is conditioned with saponified tall oil (10% tall oil by volume) for a retention time of three to five minutes. The slurry overflows into the third conditioner tank 2.3m^3 capacity), which has the option for sodium hydroxide addition to adjust pH. Conditioner tank #3 feeds slurry into the flotation cells. There are eight GL&V Dorr-Oliver 17m^3 cells in the rougher bank. The flotation concentrate reports to the dewatering circuit as final product, while the flotation tails reports to the tailings.

The operation objective of flotation is to produce concentrate which meets quality specifications while maximizing phosphate recovery and is achieved by modifying the reagent dosages, air addition rates, flotation feed density, and cell level/froth depth. The caustic starch acts as an iron depressant as well as a modifier for pH and froth. Ideal starch dosage is indicated by a pH between 10 to 11 and a stable froth with high mass content but not tight and popping. Tall oil addition controls selectivity—higher dosages yield phosphate recovery as high as 95% P_2O_5 recovery and lower grade phosphate concentrate while lower dosages yield lower phosphate recovery (70 to 80% P_2O_5 recovery) but high quality concentrate as high as 40% P_2O_5. Low volumes of air are required to produce desirable froth characteristics and collection. The cell level is adjusted to a froth depth of 10 to 15 centimeters. The flotation feed density can be adjusted by the coarse cyclone operation or water addition into the conditioning tanks. A desirable feed density is between 55 to 65% solids by mass—too low a density results in poor conditioning while too high a density results in poor phosphate recovery (probably due to viscosity characteristics of the slurry). Typical flotation recovery ranges between 75 to 95% P_2O_5 recovery depending on flotation feed grade and desired product specifications. In 2004, flotation on-stream time was about 25%. This is expected to increase in 2005 to 75 to 100% in order to handle lower grade phosphate ore.

In summary, the beneficiation circuit is designed to handle variable ore while maximizing recovery and production by having the option to run magnetics, flotation, both, or neither. All concentrate through the grinding and beneficiation circuits report to the dewatering and drying circuit prior to shipment to Redwater, Alberta.

Dewatering and Drying

The dewatering and dryer circuit at KPO is designed to produce dry phosphate product for rail shipment to RPO and recover process water from the tails for reuse. The circuit consists of a 66cm (26in.) hydrocyclone, 556.4m^3 concentrate thickener, 805.2m^3 stock tank, 100m^2 vacuum belt filter, and fluidized bed natural gas dryer. Figure 7 shows the concentrate drying and dewatering circuit. The configuration and operation of the dewatering and drying circuit depends on which ore type is being processed and which components of the beneficiation circuit are operating.

Tailings are dewatered prior to disposal and pumped in twin lines to the tailings dam. Water reclaimed from the tailings thickener,

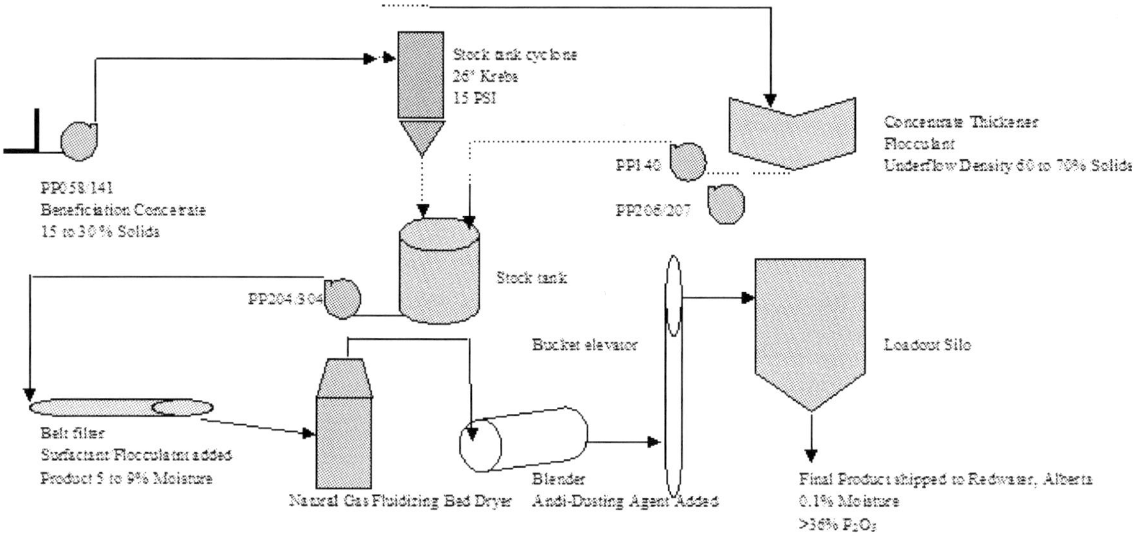

Figure 7 KPO dewatering and drying

concentrate thickener and tailings dam is reused in the mill. Mill upgrades in late 2004 will include changes to the water balance to provide a more consistent water source for the mill by increasing the volume of water returning from tailings and directing concentrate thickener overflow to the tailings dam.

Concentrate produced from magnetic separation, without flotation running, contains approximately 30 to 40% solids by weight and reports to the concentrate thickener. Under these conditions the concentrate thickener produces an underflow density of approximately 70% solids by weight. The thickener underflow is stored in an agitated stock tank prior to drying to provide a buffer between the beneficiation and dryer circuits.

Flotation concentrate is dewatered differently than non-flotation concentrate. The centrifugal concentrate thickener underflow pump is located 18.3m (60ft) from the thickener underflow well due to the thickener design. A centrifugal pump cannot pump flotation concentrate because it has an elevated pH and would settle in the thickener underflow pipe. Thus, two positive displacement (PD) pumps are used to pump flotation concentrate from the concentrate thickener to the stock tank. Although the largest available, the PD pumps are expensive to purchase, maintenance intensive, and too small to pump the full flotation concentrate flow. To reduce the flow and particle size reporting to the concentrate thickener, flotation concentrate reports first to the stock tank cyclone. To maintain the feed pressure to the stock tank cyclone, water is added to the dewatering feed pump. The underflow from the cyclone reports directly to the stock tank at 60 to 70% solids, and the cyclone overflow feeds the concentrate thickener at 10 to 15% solids. The concentrate thickener underflow is combined with the stock tank cyclone underflow in the stock tank.

Final concentrate is pumped from the stock tank to a Phillipe belt filter. Prior to filtration the concentrate is mixed with flocculent and dewatering aid in a pipe mixer. The belt filter produces dewatered product containing 5–10% moisture. Dewatered product is then fed, via a screw conveyor, into a fluidized bed natural gas dryer. Under ideal conditions the dryer uses approximately 8.5m^3 of natural gas per dry metric tonne of feed and the product moisture is

0.10%. The amount of natural gas consumed by the dryer is affected primarily by the moisture, size distribution, and tonnage of the dryer feed. Dry product is mixed with an anti-dusting agent, conveyed by bucket elevator, and loaded into 90-ton rail cars for shipment to RPO.

CONCLUSION

The Cargill orebody is a challenging and highly variable igneous ore that necessitates the use of grinding, desliming, magnetic separation, and flotation to produce a high quality phosphate concentrate. Due to the ore's inconsistency, the KPO mill has, and continues to, evolve into a highly adaptable process that can bypass any, or all, of its last three stages of beneficiation. Final product is filtered, dried, and shipped to Redwater, Alberta, via rail for acid treatment.

ACKNOWLEDGMENTS

The authors would like to thank Rene Carre, operations superintendent, and Richard Pilon, mill coordinator, for their support and contributions to this paper. The authors would also like to thank the KPO metallurgical, mining, and operations groups for their useful input.

REFERENCES

Levesque, P., Marenghi, P., and Payeur, S., Provided information and diagrams regarding the geology of the ore deposit. Agrium Inc. Kapuskasing Phosphate Operations, Kapuskasing, Ontario, Canada.

Metso Minerals, *Basics in Mineral Processing*, pp. 5:22–5:38, Metso Minerals Canada Inc., www.metsominerals.com

Vien, A., Tinney, D., Lichter, J., and Flintoff, B., *Engineering Study of the Kapuskasing Grinding Circuit*. Metso Minerals Process Technology, Kelowna, British Columbia, Canada.

CHAPTER 21

Beneficiation Studies on Sedimentary Siliceous and Calcareous Phosphate Ores in China

Shibo Zheng,* Xianjun Cao,* Zhaomin Ge,* and Wenyi Song*

ABSTRACT

Flotation is considered the most efficient method for recovering phosphate from sedimentary siliceous-calcareous phosphate resources. In this study, several flotation processes were evaluated on typical sedimentary phosphate ores from Yunnan Province, China. Those processes included direct flotation, direct-reverse flotation, reverse flotation, reverse-direct flotation and double-reverse flotation. The test results indicated that most of those processes could produce marketable phosphate concentrate with over 30% P_2O_5 and less than 1% MgO at the overall P_2O_5 recovery more than 80% from the phosphate ores containing 21.3 to 23.7% P_2O_5 and 2.8 to 4.0% MgO.

INTRODUCTION

China abounds with phosphate resources, but most of them are sedimentary siliceous and calcareous phosphate deposits described as follows:

1. Old geological age. About 90% of the phosphate deposits occurred in Cambrian or Sinian formation at the geological age of older than 500 million years. The phosphate rocks are usually in massive and very hard structure.
2. High impurities. According to the type of impurities, the phosphate rocks can be classified into three categories: siliceous, calcareous, and siliceous-calcareous (or calcareous-siliceous). About 85% of phosphate rocks in China are siliceous-calcareous. The carbonates in the ores are mainly dolomite. The MgO in the phosphate ores is quite high, usually more than 3%, in some cases exceeding 10%.
3. Middle to low P_2O_5 grade. Among the phosphate resources in China, about 6.4% of them contain more than 30% P2O5, 7.4% are in the P_2O_5 grade of 24 to 30%, 72.9% have the P_2O_5 grade ranged between 14 and 24% and the remaining with 3 to 14% P_2O_5 are in alternated sedimentary and igneous phosphate deposits. It could be seen that over 93% of the phosphate ores in China are middle to low P_2O_5 grade.
4. Fine-grained. The phosphorous deposition occurred mainly by chemical precipitation along with the precipitation of a large

*China Lianyungang Design and Research Institute (CLDRI), Lianyungang, Jiangsu, China.

amount of carbonates which are predominantly dolomite. This caused the phosphate in very fine particles intimately associated with those gangue minerals. Usually phosphate particles in the ores occur as colloidal agglomerates (known as francolite) in oölitic, pseudo-oölotic and clastic textures. Those oölites are in 0.6-0.04 mm size with inclusions of very fine-grained dolomite, calcedonite, quartz and clay minerals. In some cases, the phosphate ores must be ground to 0.05-0.01 mm for liberating impurities from phosphate.

In this study, various flotation processes, such as direct flotation, direct-reverse flotation, reverse flotation and reverse-direct flotation, and double-reverse flotation, were employed to beneficiate the typical sedimentary siliceous and calcareous phosphate ores from Yunnan Province, China, in order to produce acceptable phosphate concentrate for fertilizer manufacturing.

MATERIALS

Phosphate Ore Samples

The phosphate samples used in the study were collected from several phosphate mines in Yunnan Province, China. The samples contain 21.32 to 23.69% P_2O_5, 2.85 to 4.03% MgO and 21.30 to 25.61% SiO_2. They are typical sedimentary siliceous-carbonaceous phosphate rocks. In order to obtain the marketable phosphate concentrate, an effective beneficiation process is required to reject both carbonates and silica.

Reagents

In this study, the following reagents were used:

- **Phosphoric acid**: as pH modifier and phosphate depressant, diluted to 5% P_2O_5 solution;
- **Sulfuric acid**: as pH modifier, diluted to 5% solution
- **Soda ash**: as pH modifier, diluted to 5% solution
- **Na_2SiO_3**: as dispersant and depressant for silica, diluted to 5% solution
- **S_{711}**: a depressant for silica and carbonates, developed and manufactured by CLDRI, diluted to 2% solution
- **PA-42**: fatty acid collector for phosphate flotation, developed and manufactured by CLDRI, diluted to 2% solution
- **PA-97**: fatty acid collector for carbonate flotation, developed and manufactured by CLDRI, diluted to 2% solution
- **Fatty acid amine**: collector for silica flotation, manufactured by Arr-Maz Chemicals, diluted to 2% solution

Equipment

XMG-63 rod mill was used for grinding the phosphate sample. Flotation tests were performed with XFD-0.5L and XFD-0.75L flotation cells.

EXPERIMENT AND RESULTS

Direct Flotation Process

This is a process in which phosphate is floated, while the gangue minerals such as carbonates and silica are depressed. The flowsheet of direct flotation process is shown in Figure 1. The phosphate sample was rod-milled to a fineness of 90% passing 200 mesh. The slurry was conditioned with soda ash as pH modifier, sodium silicate and S_{711} as gangue mineral depressants and PA-42 as phosphate collector. The underflow of rougher flotation was rejected as tailings. The float of rougher flotation was subjected to two cleaning flotation steps to improve the quality of the final concentrate. The underflow of the cleaning flotation as middlings was recycled to previous operation. The float of the second cleaner flotation is the final phosphate concentrate product. The flotation results are presented in Table 1. It can be seen that the final concentrate contains 33.22% P_2O_5, 1.26% MgO and 9.38% silica. The overall P_2O_5 recovery is 81.5%.

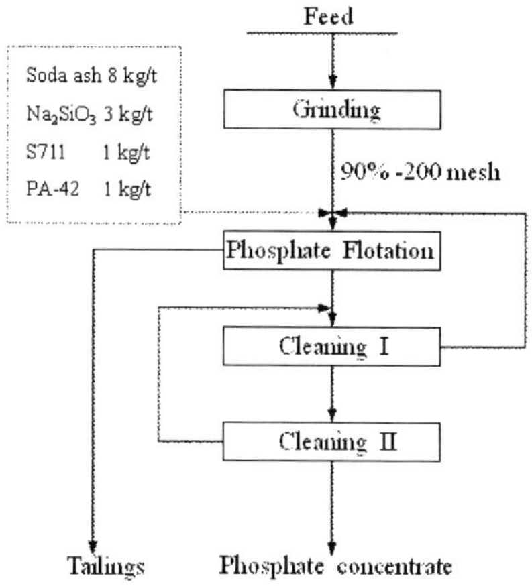

Figure 1 Flowsheet of direct flotation process

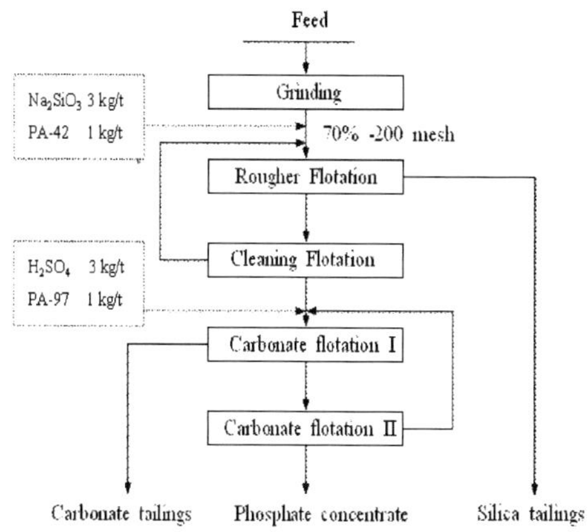

Figure 2 Flowsheet of direct-reverse flotation process

Table 1 The results of direct flotation process

Product	Wt.%	Grade, %			Recovery, %		
		P_2O_5	MgO	SiO_2	P_2O_5	MgO	SiO_2
Phosphate concentrate	53.98	33.22	1.26	9.38	81.50	23.86	19.76
Tailings	46.02	8.85	4.71	44.65	18.50	76.14	80.24
Head	100.00	22.00	2.85	25.61	100.00	100.00	100.00

Table 2 The results of direct-reverse flotation process

Product	Wt %	Grade, %			Recovery, %		
		P_2O_5	MgO	SiO_2	P_2O_5	MgO	SiO_2
Phosphate concentrate	55.82	31.70	0.78	13.60	82.97	10.92	35.63
Carbonate tailings	24.37	9.06	13.91	7.19	10.37	84.12	8.22
Silica tailings	19.81	7.18	1.03	60.38	6.66	4.96	56.15
Head	100.00	21.32	4.03	21.30	100.00	100.00	100.00

In order to effectively reject both carbonates and silica to produce a phosphate concentrate at high P_2O_5 recovery, several properties of the processes are worth to be particularly metioned. The flotation feed is very fine, 90% passing 200 mesh. The flotation slurry needs to be heated to 45°C for the improvement of flotation performance. Flotation is carried at high alkaline solution, up to pH 11. A proprietary depressant S_{711} is required to make the process work.

Direct-Reverse Flotation Process

In this process, the phosphate ore sample was ground to a fineness of 70% passing 200 mesh. The ground slurry was conditioned with sodium silicate as depressant and PA-42 as collector and subjected to direct flotation to reject silica tailings. The flotation slurry of direct flotation was maintained at 40°C. It should be noted that there was no soda ash used as pH modifier at the direct flotation stage. The rougher concentrate of direct flotation was further conditioned with sulfuric acid as pH modifier and PA-97 as carbonate collector to reject carbonate impurities. The flowsheet of direct-reverse flotation and reagent consumptions are shown in Figure 2 and the flotation results are given in Table 2. It can be seen that the direct-reverse flotation process could produce a final phosphate concentrate containing 31.7% P_2O_5 and 0.78% MgO at the overall P_2O5 recovery of about 83%.

Reverse Flotation Process

There are three main minerals in the ore sample: phosphate, silicate, and carbonates. Carbonate minerals are most fragile and easily ground because of their low hardness compared with phosphate and silicate minerals. It is expected that carbonate minerals could be liberated from other minerals with less grinding. In this case, a carbonate reverse flotation process is employed to reject the carbonate impurities only. The phosphate ore sample was ground to 50% passing 200 mesh. The mixture of sulfuric acid and phosphoric acid at a ratio of 1:1 was used as both pH modifier and phosphate depressant, and PA-97 as carbonate collector. The underflow of rougher carbonate flotation was subjected to a scanvenger carbonate flotation. The float of scanvenger carbonate flotation was recycled to rougher flotation. The froth product of rougher flotation was the carbonate tailings and the underflow of the scanvenger flotation was the final phosphate concentrate. The reagent dosages and the flowsheet are shown in Figure 3 and the chemical analysis is given in Table 3. It can be seen that it is a very simple flowsheet with less grinding. The final product contains only 0.64% MgO, but the P_2O_5 grade is quite low, only 27.5%, because of high SiO_2 content.

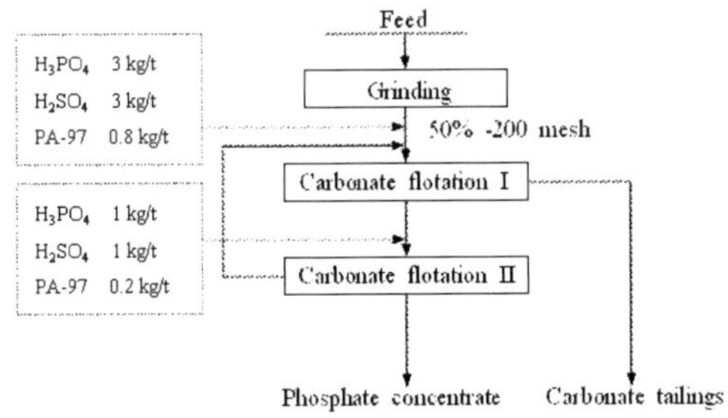

Figure 3 Flowsheet of carbonate reverse flotation process

Table 3 The results of carbonate reverse flotation process

Product	Wt %	Grade, %			Recovery, %		
		P_2O_5	MgO	SiO_2	P_2O_5	MgO	SiO_2
Phosphate concentrate	75.47	27.50	0.64	25.80	87.59	13.26	87.35
Carbonate tailings	24.53	12.00	12.80	11.50	12.41	86.74	12.65
Head	100.00	23.69	3.62	22.29	100.00	100.00	100.00

Reverse-Direct Flotation Process

As mentioned above, carbonate reverse flotation process could produce a low MgO phosphate concentrate, but the P_2O_5 grade is very low because the majority of silica impurities still remains in the underflow. In reverse-direct flotation process, carbonate flotation was first performed in acidic solution with the mixture of sulfuric acid and phosphoric acid as pH modifier and phosphate depressant and PA-97 as carbonate collector. In order to improve the quality of phosphate concentrate, the underflow of carbonate flotation was further ground to a fineness of 85% passing 200 mesh for better liberation of phosphate from silica and subjected to phosphate flotation to reject silica impurities. In phosphate flotation stage, soda ash was used as pH modifier, Na_2SiO_3 as depressant and PA-42 as phosphate collector. The slurry temperature for phosphate flotation was maintained at 40°C. The simplified flowsheet and reagent dosages are shown in Figure 4 and the results are given in Table 4. It can be seen that a phosphate concentrate analyzed P_2O_5 32.88%, 0.76% MgO and 11.30% SiO_2 could be obtained at the P_2O_5 recovery of 85.27%.

Double-Reverse Flotation Process

In the double-reverse flotation process, phosphate sample was ground to 80% passing 200 mesh. The slurry was conditioned with the mixture of sulfuric acid and phosphoric acid as both pH modifier and phosphate depressant and PA-97 as carbonate collector, and subjected to carbonate flotation. A scanvenger carbonate flotation was employed to enhance the removal of carbonate impurities. The underflow of carbonate flotation was sized at 400 mesh. Because the -400 mesh fraction had low MgO content and high P_2O_5 grade, it could be used as phosphate concentrate directly without further processing. The +400 mesh fraction was subjected to amine flotation to remove silica impurities. The underflow of amine flotation was combined with -400 mesh fines to form a composite phosphate concentrate. The flowsheet of double-

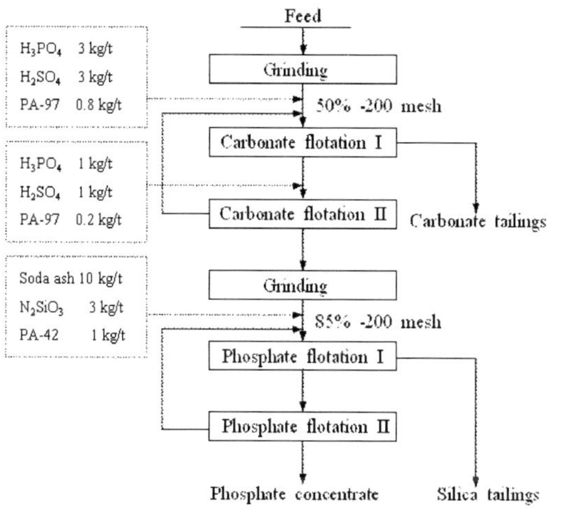

Figure 4 Flowsheet of reverse-direct flotation process

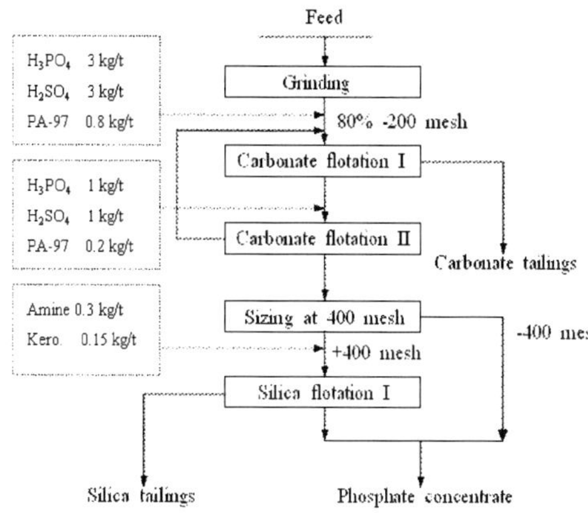

Figure 5 Double-reverse flotation process

Table 4 The results of reverse-direct flotation process

Product	Wt.%	Grade, %			Recovery, %		
		P_2O_5	MgO	SiO_2	P_2O_5	MgO	SiO_2
Phosphate concentrate	60.92	32.88	0.76	11.30	85.27	13.29	31.89
Carbonate tailings	24.13	11.94	12.27	12.24	11.96	85.55	13.68
Silica tailings	14.95	4.35	0.29	78.55	2.77	1.16	54.53
Head	100.00	21.32	4.03	21.30	100.00	100.00	100.00

Table 5 The results of double-reverse flotation process

Product	Wt.%	Grade, %			Recovery, %		
		P_2O_5	MgO	SiO_2	P_2O_5	MgO	SiO_2
Phosphate concentrate	67.82	30.42	0.69	17.21	86.79	13.66	53.63
Carbonate tailings	23.55	10.03	12.50	13.56	9.93	85.47	14.66
Silica tailings	8.63	9.00	0.32	80.01	3.28	0.87	31.71
Head	100.00	21.32	3.44	21.76	100.00	100.00	100.00

reverse flotation process is shown in Figure 5 and test results are given in Table 5. With the double-reverse flotation process, a composite phosphate concentrate containing 30.42% P_2O_5 and 0.69% MgO could be achieved at the overall P_2O_5 recovery of 86.79%.

SUMMARY

From the laboratory evaluation results of different flotation processes on typical sedimentary siliceous-calcareous phosphate ores, it could be concluded that all five processes could be used to beneficiate phosphate ores with both silicates and carbonates impurities. Direct flotation and carbonate reverse flotation processes have the simplest flowsheet structure. However, the feed for direct flotation needs more grinding than other processes and the MgO content of the concentrate produced by direct flotation is higher than 1%. The product from carbonate reverse flotation gives the lowest MgO content, but the P_2O_5 grade is also very low because the majority of silica impurities remains in the product. The other three processes, direct-reverse flotation, reverse-direct flotation and double reverse flotation, consist of procedures to reject both carbonate and silicate impurities and the final phosphate concentrate containing more than 30% P_2O_5 and less than 1% MgO could be achieved, and the overall P_2O_5 recoveries are higher than 83%.

SECTION 6

Hydrometallurgical Processing

CHAPTER 22

In-situ Mining of Phosphate Ores

H. El-Shall* and R. Stana‡

ABSTRACT

Presently, the mining of Florida phosphate requires the movement of over 100 tons of materials (overburden, sand, clay) for every ton of phosphate concentrate recovered. Not only is this energy intensive, but it also causes significant stress on the environment. In 2003, the Department of Energy solicited proposals for innovative mining ideas that could significantly improve the efficiency of mining.

An award was made to the University of Florida Engineering Research Center to evaluate the in-situ mining of phosphates using an aqueous CO_2 solution. Tests were carried out in a 6-inch diameter column, 6 feet long at pressures up to 40 psi. Initial results demonstrate that initially the MgO is leached from the ore and then the phosphate. While the tests are continuing, so far they have not demonstrated P_2O_5 concentrations that are economically attractive.

INTRODUCTION

The Florida phosphate industry is the largest mining industry in Florida, typically mining over 30 million tons of phosphate rock. Most of the rock is used to produce phosphate fertilizer, providing one of the three major essential crop nutrients. Most of the phosphate fertilizer is exported, mainly to China and India.

The industry is currently under numerous economic and environmental pressures and well could disappear in the not too distant future. While some of these pressures relate to the conversion of the phosphate rock to fertilizer, they are small compared to those involving the mining of the phosphate rock.

Mining of the phosphate rock requires the movement of several hundred millions tons of overburden and phosphate matrix and disturbance of several thousand acres of land each year. The overburden and matrix are removed by electrically driven draglines with up to 65 cubic yard buckets. Because of pit walls collapsing, generally wet pit conditions, and inability to precisely see the top or bottom of the matrix, only about 80% of the matrix is recovered.

In addition, the unmined phosphate deposits are at a deeper depth and contain significantly more impurities that require additional treatments to be useful. The most significant contaminant is dolomite, which is a mixture of calcium and magnesium carbonate.

The mined matrix is typically slurried with water and pumped to the beneficiation plant through several (up to 10) miles of pipeline. The

*Department of Materials Science and Engineering, University of Florida, Gainesville, Florida.
†R. Squared S Inc., Lakeland Florida.

pumped rock is very erosive and quickly wears the pipe. While the pipe is typically rotated (most of wear is on the bottom) every few months, it has a typical life of only 2 to 3 years.

At the beneficiation plant, the matrix is washed to remove the clay. This generates over 30 million tons of waste clay each year, which is stored as slurry in impounded areas. These waste clay ponds not only tie up the land (40% of that mined) but a large quantity of water. They are also potentially environmental liabilities, as the storage dams can break (two occurrences in the last 10 years). Even after reclamation, the clay settling ponds have limited usefulness, as they will not support much weight. The waste clay also contains about 30% of the phosphate in the matrix.

Finally, the mining operations are moving further and further away from the fertilizer production facilities, increasing the transportation costs for the rock.

Overall, the mining/beneficiation/transportation process is very energy intensive, environmentally unfriendly and requires the use of large quantities of water. Electrical consumptions are typically 90–100 KWh per ton and are at least 20% of the total mining/beneficiation cost. It is the largest expense in the mining operation.

INNOVATIVE MINING CONCEPT

An ideal solution to the above problem would be to develop a mining process that would not require the movement of anything but the phosphate minerals, not generate any clay disposal ponds, and would generate a high purity soluble phosphate. It is known that CO_2 in solution (carbonic acid) will dissolve both dolomite and phosphate rock.[1] In fact, this occurs in nature as the CO_2 in the air goes into solution in the ground water and very slowly dissolves the limestone, dolomite and phosphate rock in the ground. The calcium and magnesium go into solution as the bicarbonates. While the reaction is relatively fast for dolomite, it is much slower for phosphate rock.

However, these reactions could be sped up significantly by increasing the CO_2 in solution. This can be achieved by contacting the leach water with pure CO_2 at the hydrostatic pressure of the formation (30–40 PSI absolute). If CO_2 saturated water is contacted with the phosphate matrix, by introducing it through solution mining well complex, it will quickly solubilize the calcium and magnesium as the bicarbonates. Since a bottom layer of hard clay or solid limestone typically bound the phosphate matrix, the solution will be confined to the matrix zone. Once the solution is recovered from the ground, the calcium and magnesium can either be precipitated as the carbonates and the CO_2 recovered by heating the solution, or precipitated as the sulfates by the addition of sulfuric acid and the CO_2 recovered. Once the calcium and magnesium carbonates have been removed from the formation, the CO_2 saturated water will begin to dissolve the phosphate. While the concentration of phosphate will be quite low (<1%), it may be increased by refortifying the solution with CO_2 and reinjecting it into the ground to dissolve additional phosphate. The solubilized calcium can also be precipitated by the addition of sulfuric acid at this step or later in the process at a higher concentration.

Since the pH of the solution will be about 5, very little else will go into solution. When the matrix has been depleted, the solution mining well field can be restored by simply injecting a small amount of caustic into the well field.

Once the phosphate solution reaches about 1%, it can be economically concentrated to greater than 5% by using the reverse osmosis (RO) process. The use of RO has been demonstrated to work at the 500,000 gal/day scale under similar conditions in the ongoing clean up of the Piney Point phosphoric acid plant near Tampa, Florida.[2] Further concentration of the phosphate solution can then be accomplished in a phosphoric acid evaporator.

The concentrated acid can then be used to produce the typical phosphate fertilizer products.

In 2003, the Department of Energy solicited proposals for innovative mining ideas that could significantly improve the efficiency of mining. An award was made to the University of Florida Engineering Research Center to evaluate the in-situ mining of phosphates using an aqueous CO_2 solution.

EXPERIMENTAL

Four samples of phosphate matrix were obtained from both Cargill and IMC Phosphates. The characterization of these ore samples is given in Tables 1–2.

A test system was built as shown in Figure 1. The system consisted of a CO_2 cylinder, a commercially available soda fountain carbonator and a 6-inch diameter by 6-feet long plastic column filled with matrix, several pressure gauges and a pressure regulator.

The column was initially packed with 90.3 pounds of the IMC high MgO matrix, and a solution of deionized water saturated in CO_2 was fed in downflow at pressures up to 60 PSI. However, the process of filling the column disturbed the clay fraction in the matrix and this plugged off the column in the first few hours reducing the flow to less than 20 mL/min. The feed was then changed to upflow and the feed rate controlled by adjusting the back pressure on the column.

Figures 2–4 show the quantity of calcium, magnesium, and phosphorous in the leachate vs. time. As can be seen, after 300 hours, the MgO in the leachate began to drop and the P_2O_5 began to increase. This phase of the leaching was stopped after 433 hrs and the system modified to recycle the leachate after refortification with CO_2. Figures 5–7 show the results.

As can be seen, there was no significant increase in the P_2O_5 content of the leachate, indicating that the solution was at equilibrium with the matrix. At this point, an attempt was made to collect the leachate and precipitate the calcium before CO_2 refortification and recycling back to further increase the P_2O_5. However, a mass balance calculation showed that the matrix had been essentially depleted of P_2O_5.

The column was emptied and refilled with the IMC low MgO matrix. Figures 8 and 9 show the results for the first 264 hrs of leaching.

Table 1 Matrix solids content

Material	Solid%
Cargill High MgO	79
Cargill Low MgO	80
IMC Low MgO	79
IMC High MgO	75

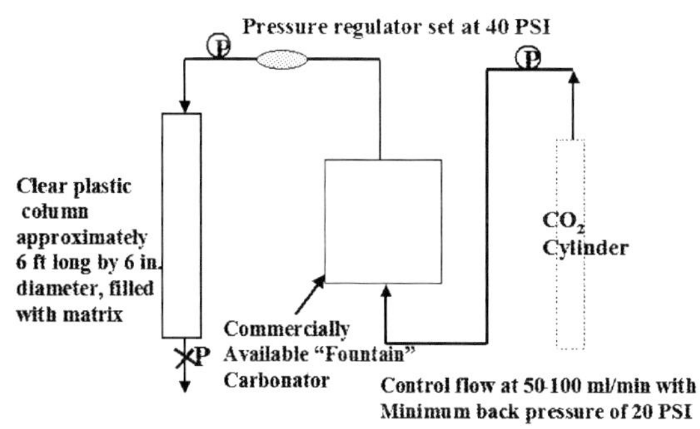

Figure 1 Experimental setup

At this point it was calculated that most of the MgO had been removed and the liquid was recirculated back through the column after CO_2 refortification. As in the first test, no significant increase in P_2O_5 content was observed. Presently, more than 100 gallons of leachate is being collected and the calcium will be precipitated and the clarified leachate refortified with CO_2 in an attempt to increase the P_2O_5 content to the objective 1%.

If the P_2O_5 content of the leachate can be increased to 1% or above, it will then be concentrated to about 27% P_2O_5, and clarified. The clarified acid will then be analyzed for iron, aluminum, magnesium, and fluorine content to determine its compatibility with the current process for making high-grade phosphate fertilizers.

Table 2 Chemical and size analyses of four samples

Cargill High MgO

Screen size, mm	%Wt	MgO %	P_2O_5 %	CaO %	Al_2O_3 %	Fe_2O_3 %
+19	3.25	3.14	3.14	14.22	0.19	0.37
-19+2	8.38	11.79	11.79	20.48	0.50	0.48
-2+1	3.56	14.86	14.86	23.58	0.40	0.60
-1+0.42	42.54	8.11	8.11	12.18	0.13	0.37
-0.42+0.1	26.44	7.18	7.18	10.92	0.15	0.32
-0.1	15.82	2.45	2.45	6.63	5.68	1.53
Feed	100	0.92	7.23	12.11	1.19	0.52
Calc.Feed	100.00	0.96	7.36	12.14	1.06	0.56

Cargill Low MgO

Screen size, mm	%Wt	MgO %	P_2O_5 %	CaO %	Al_2O_3 %	Fe_2O_3 %
+19	3.29	0.11	17.29	26.37	0.72	0.20
-19+2	11.50	0.18	14.47	22.51	0.92	0.45
-2+1	7.26	0.17	15.39	23.70	0.80	0.50
-1+0.42	26.30	0.05	4.6	6.79	0.22	0.19
-0.42+0.1	29.36	0.05	4.38	6.50	0.24	0.20
-0.1	22.29	0.48	3.21	4.58	5.14	0.75
Feed	100	0.21	6.32	10.04	1.39	0.39
Calc.Feed	100.00	0.17	6.56	9.89	1.46	0.37

IMC Low MgO

Screen size, mm	%Wt	MgO %	P_2O_5 %	CaO %	Al_2O_3 %	Fe_2O_3 %
+19	1.09	0.25	14.27	22.53	0.46	1.36
-19+2	30.99	0.22	13.59	20.89	0.38	2.68
-2+1	17.23	0.17	13.8	21.01	0.45	5.47
-1+0.42	8.51	0.62	12.35	19.01	0.36	3.88
-0.42+0.1	12.53	0.13	9.64	14.75	0.5	1.89
-0.1	29.63	0.57	5.22	8.23	4.50	3.81
Feed	100	0.31	10.31	16.18	1.73	3.15
Calc.Feed	100	0.34	10.55	16.25	1.63	3.48

IMC High MgO

Screen size, mm	%Wt	MgO %	P_2O_5 %	CaO %	Al_2O_3 %	Fe_2O_3 %
+19	0	0	0	0	0	0
-19+2	2.34	0.54	15.81	22.03	3.31	0.43
-2+1	2.12	0.32	15.17	20.70	3.08	0.42
-1+0.42	47.47	0.12	3.95	5.36	0.02	0.10
-0.42+0.1	34.55	0.04	2.86	3.78	0.34	0.20
-0.1	13.53	2.27	1.28	3.61	2.94	1.50
Feed	100	0.36	3.88	6.01	0.75	0.29
Calc.Feed	100	0.4	3.73	5.29	0.67	0.34

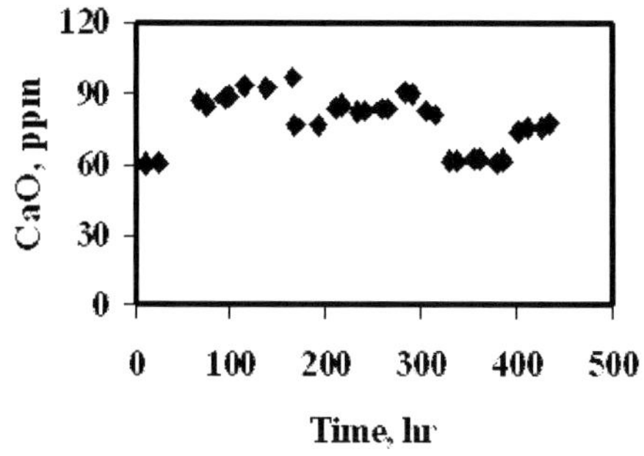

Figure 2 CaO in leachate vs. time (50 mL/min, 20 psi for first 265 hrs, 150 mL/min, 40 psi after)

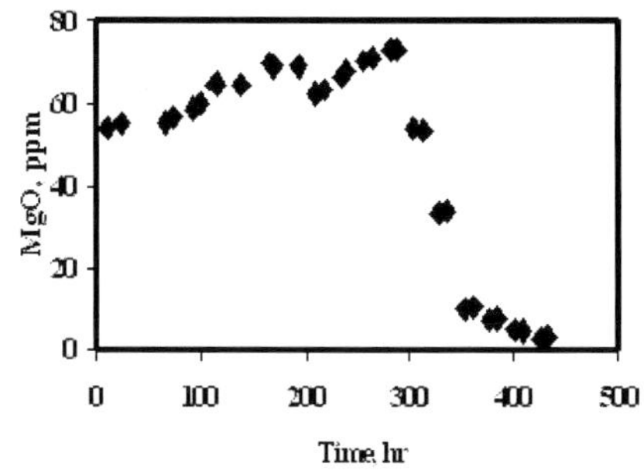

Figure 3 MgO in leachate vs. time (50 mL/min, 20 psi for first 265 hrs, 150 mL/min, 40 psi after)

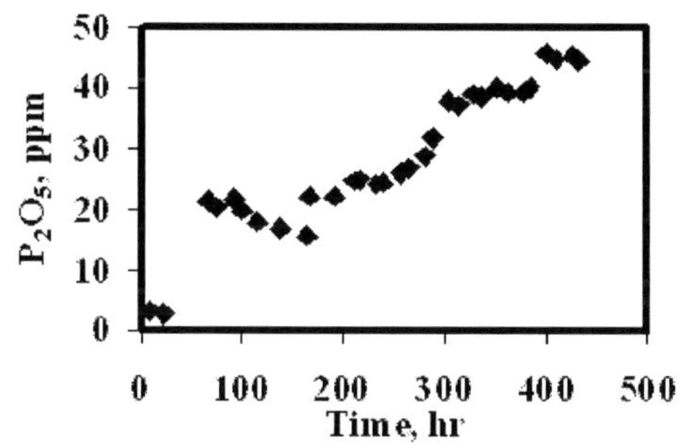

Figure 4 P_2O_5 in leachate vs. time (50 mL/min, 20 psi for first 265 hrs, 150 mL/min, 40 psi after)

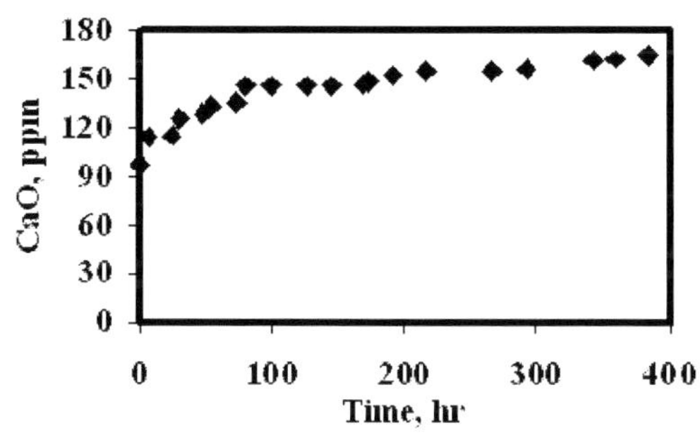

Figure 5 CaO in leachate vs. time leachate recycled at 150 mL/min after refortification with CO_2 at 40 psi

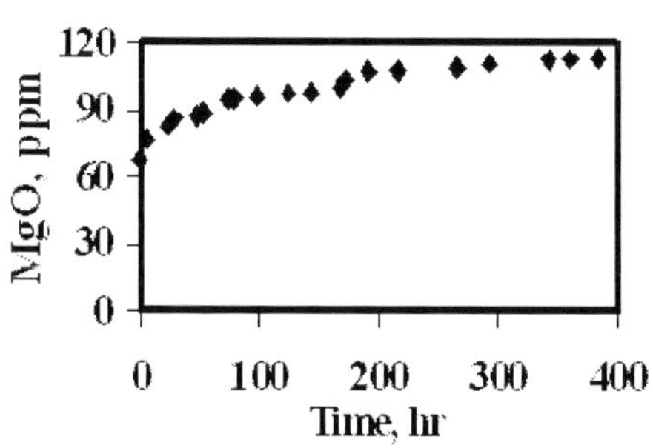

Figure 6 MgO in leachate vs. time leachate recycled at 150 mL/min after refortification with CO_2 at 40 psi

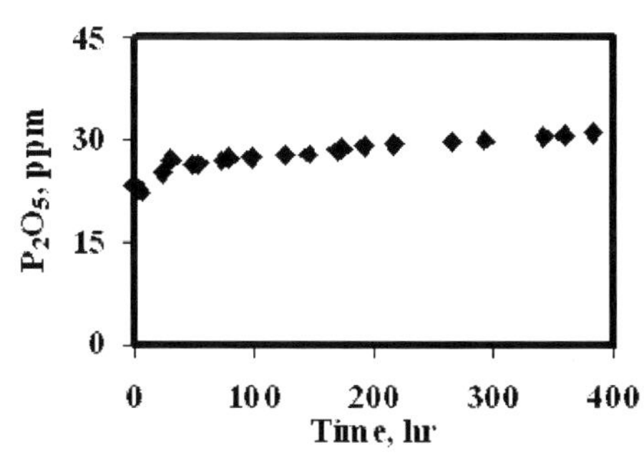

Figure 7 P_2O_5 in leachate vs. time leachate recycled at 150 mL/min after refortification with CO_2 at 40 psi

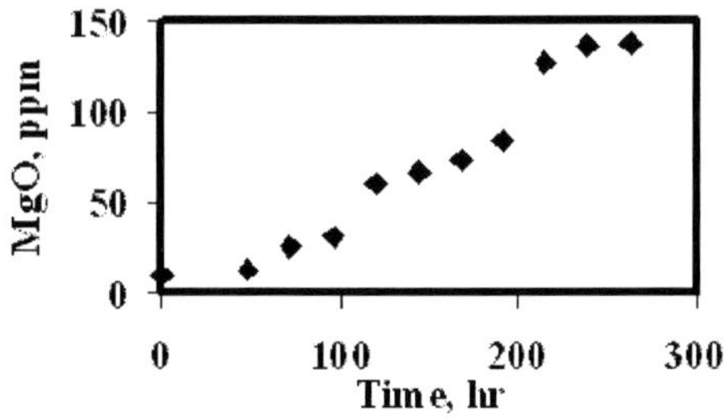

Figure 8 MgO in leachate vs. time (55 mL/min @ 40 psi feed pressure)

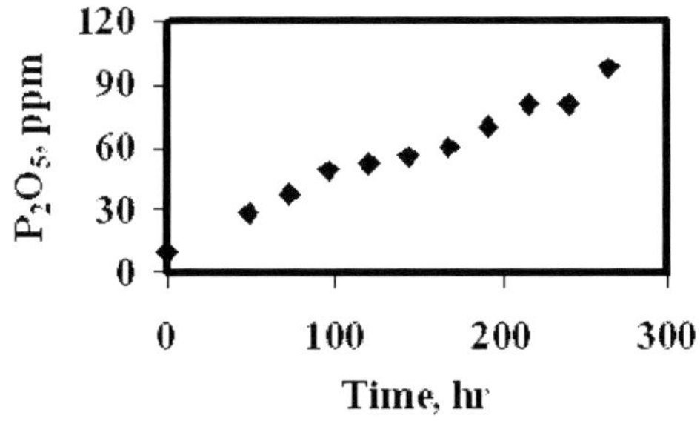

Figure 9 P_2O_5 in leachate vs. time (55 mL/min @ 40 psi feed pressure)

AKNOWLEDGMENTS

The authors wish to thank the United States Department of Energy for their financial support of this work. The authors also wish to thank IMC Phosphates and Cargill for providing the phosphate matrix samples.

Appreciation is also extended to A. El-Midany of the Engineering Research Center for Particle Science and Technology, University of Florida, Gainesville, for his contribution to this paper.

REFERENCES

1. S.I. Abu-Eishah, M.J.J. Anabtawi, and S.L. Isaac, 2004, Upgrading of carbonaceous phosphate rocks by direct carbonation with CO_2-water solutions, *Chemical Engineering and Processing*, Volume 43, Issue 8, 1085–1094.

2. Astley, Vaughn V., Reverse Osmosis of Phosphate Plant Water by Use of Novel Pretreatment Technology, *Beneficiation of Phosphates IV*, December 5–9, 2004.

CHAPTER 23

Study on Making Phosphoric Acid from Seabed Phosphorite by the Wet Process

Shifu Shen* and Yuexin Han*

ABSTRACT

It is important to study the seabed phosphorites, which are of great abundance in its reserve and may substitute for land phosphorites. A CLD04 sample, a kind of seabed phosphorite, showed a reactivity of 99.90% and a counteractant quotiety of –0.76, which was better than land apatite in reactivity. Tests making phosphoric acid from the CLD04 sample by wet process indicated a high decomposition rate and a high dissolution rate and recovery of P_2O_5, and produced better crystals of phosphogypsum with the reaction temperature of 65±2 degrees C, reaction time of 6.5 h, liquid-to-solid ratio of 2.5:1, stirring rate of 350 r/min, finished crude phosphoric acid's concentration of 25.0%~28.0%, and H_2SO_4 excess of 3%.

INTRODUCTION

The seabed is rich in solid mineral resources, such as multi-metalline tubercular, rich-cobalt crust, grit, abyssal multi-metalline sulfid, and phosphorite, which is a potential substitute for land phosphorite with its ample reserve[1,2]. It has been reported that seabed phosphorite is mainly distributed in the continental shelf sections. About 30 billion tons of the seabed phosphorite is concentrated in four huge oceanic zones: East Atlantic zone, West Atlantic zone, California zone, and Peru-Chile zone. In addition, plenty of seabed phosphorite also is located in the seamount and seadesk districts[1,2]. Generally speaking, the seamount phosphorite exists from 300 to 400 meters deep, while crust phosphorite spreads below 500 meters.

With the worldwide exploration and research in ocean resources, it is predicted that the rich-cobalt crust may be utilized earlier than the abyssal multimetal tubercular layer in commercial exploitation. The tailings of the rich-cobalt crust contain phosphorite. For the sake of comprehensive utilization of ocean resources, the study of ocean resources is of particular significance.

The international research on application of seabed phosphorite was carried out mainly in the US, the former Soviet Union, and South Africa. It was reported that seabed phosphorite contained a uranium of 0.005%–0.030%, which is higher than the land phosphorite, and the feasibility of uranium extraction from seabed phosphorite was studied[3-5]. In 1973, South Africa examined the performance of ground phosphorite as a fertilizer, showing equivalent performance as commercial superphosphate[1]. Judd

*Northeastern University, Shenyang, China.

et al. concentrated seabed phosphorite from 1.0%–1.5% P_2O_5 to 29%–31%[6]. Using phosphorite from the Japanese ocean, Samigullina et al. produced phosphoric acid by the wet process[7]. However, no studies were found on the development of seabed phosphorites in China. The objective of this study is to make phosphoric acid dihydrate from seabed phosphorites by the wet process and examine the production conditions.

MATERIAL AND METHODS

Material Composition and Mineral Analysis

The phosphorite sample CLD04 used in this study was an ocean rich-cobalt crust rock from China Ocean Association DY105-11 station. The sample was buff lumpish in color. The sample was ground and kept for testing. The chemical composition of CLD04 sample from spectroscopic and chemical analyses is shown in Table 1. In addition, rcsidual after hydrochloric acid dissolution accounted for 0.1%, and the ignition loss was 12.55%, mainly CO_2. The main mineral was fluori-carbon apatite according to X-ray diffraction (PANalytical Model X'pert), shown in Figure 1. The infrared spectrum of CLD04 is presented in Figure 2.

There were no diagnostic absorption peaks for organics such as alkyl, aryl and so on, which indicated that the concentration of organic carbon was less than 1.0×10^{-6}%. Thus the 2.55% carbon is inorganic carbon. Also there was no diagnostic absorption peak for hydroxyl around 3,430 cm^{-1} and 1,639 cm^{-1}, which indicated no hydroxyl content in CLD04. The absorption peak for PO_4^{3-} appeared at 1,045.34 cm^{-1}, 964.97 cm^{-1}, and 578.13 cm^{-1}, and the absorption apex for CO_3^{2-} occurred at 1,045.34 cm^{-1}, 865.97 cm^{-1}, and 712.91 cm^{-1}. According to published IR profiles for fluorine apatite[8], the main mineral of CLD04 was fluoral-carbon apatite without hydroxyl.

Based on *China Phosphorite, A Quality Standard for Wet Processing Phosphoric Acid* (ZBD51001-86), the concentration of P_2O_5 met the P_2O_5 standard for wet processing with MgO/P_2O_5=1.7% qualifying for Grade I phosphorite, and R_2O_3 ($Al_2O_3+Fe_2O_3$) =4.39 qualifying for Grade III phosphorite.

Analyses

The concentrations of P_2O_5 in phosphorite, phosphogypsum, and phosphoric acid was determined by phosphomoly adic acid-quinoline titration. The crystalline pattern of calcium phosphate was determined by SEM, and the crystal size of calcium phosphate was measured with a laser particle analyzer in an alcohol medium. The formulas used for the calculation were[9]:

The decomposition percentage of phosphorite K_f:

$$K_f = \frac{0.7}{Q_{CaO}/Q_{SO_3}} \times 100\% \tag{1}$$

where Q_{CaO} and Q_{SO_3} is the content of CaO and SO_3 in phosphogypsum, respectively.

Dissolution percentage K_t, i.e., percentage of P_2O_5 converted to phosphoric acid solution to that of phosphorite:

$$K_t = \left[1 - \frac{[P_2O_{5(d)}]}{[P_2O_5]} P\right] \times 100\% \tag{2}$$

Therein, $[P_2O_{5(d)}]$ is the content of the water-insoluble P_2O_5, %; P is the ratio, kg/kg phosphorite: $[P_2O_5]$ is the content of P_2O_5 in phosphorite, %.

Recovery K_w:

$$K_w = K_t \times K_l \tag{3}$$

EXPERIMENTAL RESULTS AND DISCUSSION

Reactive Capability of Phosphorite

Besides the chemical composition, the reactivity is also an important parameter to measure the quality of phosphorite for production of phosphoric acid by H_2SO_4 dissolution. The reaction rate between phosphorite and H_2SO_4 mainly depends on reaction temperature, the

Table 1 Chemical composition of CLD04 sample

Component	SiO_2	Al_2O_3	P_2O_5	Na_2O	Fe_2O_3	MgO
Content / %	0.84	1.04	28.90	1.62	0.23	0.50
Component	S	Co	CaO	F	Cl	C
Content / %	0.70	0.005	48.86	3.96	0.38	2.55

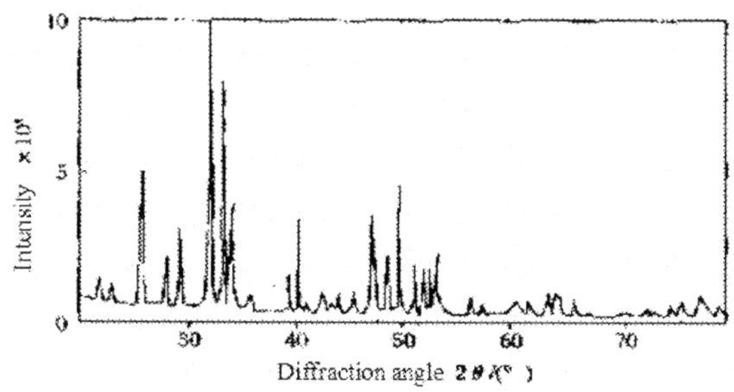

Figure 1 XRD of CLD sample

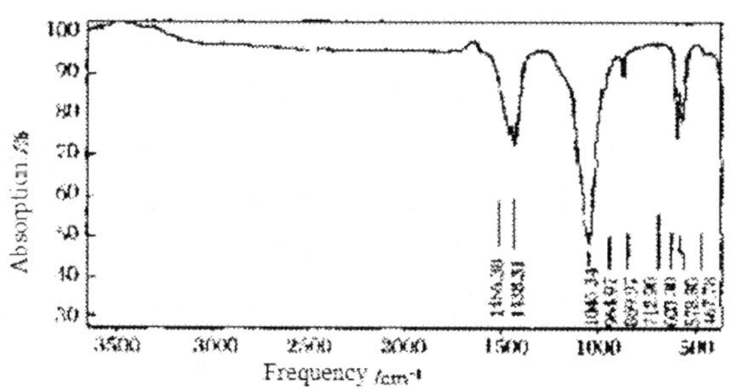

Figure 2 Infrared spectrum of CLD sample

concentration of H^+, the spread of surface liquid film, and effective surface of phosphorite particle[8]. In the wet process, SO_4^{2-} is present in solution, which can form a solid film (also called passive film) and reduce the rate that the H^+ transfers from solution to the particle surface, then slow down or stop the process. Therefore, reactivity and counter-actant quotiety are used to evaluate the reaction performance. Reactivity represents the degree at which phosphorite is penetrated by acid, and counter-actant quotiety or called counter-passivation represents the ability of phosphorite counteract passive film when it is decomposed by acid. Consequently, the reactivity of phosphorite must be considered during making phosphoric acid by wet process[9,10].

Experiment Methods

CLD04 samples were ground and particles with sises of -0.074 +0.065 mm were selected. The reactivity of particles was determined according to the following methods[9-11]:

1. **Reactivity**: The percent decomposition of phosphorite in the phosphoric acid (AR) was measured with 30% P_2O_5 at reaction temperature of 60 degrees C, stirring rate of 300 r/min (with 90 mm stirring pedal). The higher the percent decomposition, the higher the reactivity of phosphorite. The reaction equation used is

$$Ca_5F\ (PO_4)_3+2H_3PO_4=HF+5CaHPO_4 \tag{4}$$

2. **Counteractant Quotiety**: the percent decomposition of phosphorite in the mixture of H_3PO_4(AR) and H_2SO_4(AR) was measured with a reaction temperature of 60 degrees C, reaction time of 2 min, stirring rate of 300 r/min. The ratios of H_3PO_4 and H_2SO_4 were the following: H_3PO_4 with 30% P_2O_5+2% H_2SO_4, H_3PO_4 with 30% P_2O_5+4% H_2SO_4, and H_3PO_4 with 30% P_2O_5+6% H_2SO_4. Taking the concentration of H_2SO_4 in the mixed acids as the independent variable x, and the percent decomposition of the P_2O_5 in phosphorite as dependent variable y, the data were processed by the linear equation y=a+bx. B was defined counteractant quotiety, and the greater the b value, the stronger was the counteractant ability.

Experiment Results

The reactivity and counteractant ability of CLD04 determined in this study are listed in

Table 2 Comparison between reactivity of CLD04 sample and several land phosphorites

Core sample	Phosphorite Type	Content of P_2O_5 in the Core sample / %	Reactivity of 30% P_2O_5	Decomposition rate of phosphorite / %			Counteract-ant quotiety (b)
				30% P_2O_5 +2%H_2SO_4	30%P_2O_5 +4%H_2SO_4	30%P_2O_5 +6%H_2SO_4	
Morocco	sedimentary phosphates	30.46	99.70	98.16	83.97	70.12	-7.06
Wenfu 2#	sedimentary phosphates	35.28	99.70	94.88	67.91	52.40	-13.46
Kunyang	sedimentary phosphates	31.74	99.76	99.34	55.12	37.74	-11.56
Kairang-3	sedimentary phosphates	33.52	99.87	99.31	56.40	41.20	-10.94
CLD04	—	28.90	99.90	98.94	97.12	95.75	-0.76
Kela	Igneous rock	39.36	55.53	62..36	63.52	49.32	-0.87

Table 2, as well as values for several land phosphorites[10].

The results showed that the reactivity of the CLD04 sample was basically equivalent to that of Morocco phosphorite with an extraordinary higher reactivity, which was representative of the higher reactivity phosphorite used in the international phosphorus industry. The counteractant quotiety was comparable to that of Koro phosphorite, which was representative in high counteractant quotiety. The integrated reactivity was better than that of land phosphorite. According to literature, this might be caused by the higher content of CO_2[9] that is released during wet process digestion which could weaken the binding energy between passive film and CLD04 particles, or even make the film separate from the particle and form a loose and permeable thin $CaSO_4$ crystal coating, favoring the penetration of H^+ into phosphorite particle for decomposition.

Wet Process

Based on the different degrees of hydration, the process for making H_3PO_4 by wet H_2SO_4 dissolution can be divided into dihydrate, semi-hydrate, anhydrate, dihydrate–semi-hydrate recrystallization, and semi-hydrate–dihydrate recrystallization processes. Because the dihydrate process does not require the phosphorite to be of high quality and it is a mature technology, the dihydrate process was used. The basic reaction of the dihydrate process is

$$Ca_5F(PO_4)_3+5H_2SO_4 \rightarrow 3H_3PO_4+HF+5(CaSO_4 \cdot 0.5H_2O) + \text{solution} \quad (5)$$

$$CaSO_4 \cdot 0.5H_2O+1.5H_2O=CaSO_4 \cdot 2H_2O \quad (6)$$

Reaction (5) between phosphorite and sulphuric acid is actually carried out in two steps:

$$Ca_5F(PO_4)_3+7H_3PO_4=HF+5Ca(H_2PO_4)_2 \quad (7)$$

$$Ca(H_2PO_4)_2+H_2SO_4=2H_3PO_4+CaSO_4 \quad (8)$$

The experimental process and equipment are shown in Figure 3.

The effect of return phosphoric acid[9], whose concentration and quantity varies with different technological conditions, is to reduce slurry viscosity, control and stabilize the crystallization of calcium sulphate and prevent crystal breakdown and small-size crystals. In order to accurately simulate the production of crude phosphoric

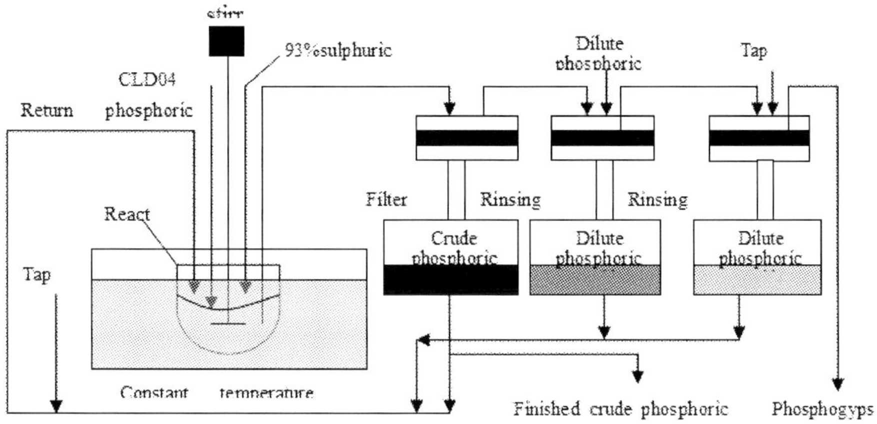

Figure 3 The technology and experimental equipment used to make phosphoric acid

acid and the reaction system, 30% sulphuric acid was used to decompose a small amount of CLD04 sample and prepare a certain amount of phosphoric acid to make the return phosphoric acid before the dihydrate flow experiment.

When the experiment started, a calculated amount of return phosphoric acid was added into the reactor while mixing. At the same time, a certain amount of CLD04 sample and sulphuric acid were added in at a precalculated rate. After being mixed for a fixed time, the slurry was transferred to the filtration funnel and filtered at a fixed vacuum. The filtrate was the finished crude phosphoric acid. The filter cake was moved to the rinsing funnel 1 and rinsed with a fixed amount of dilute sulphuric acid and filtrated. Then the filter cake was moved to the rinsing funnel 2 and rinsed with a fixed amount of tap water. Because of the limited quantity of the CLD04 sample, only a few tests could be run.

Tests were run to determine the effect of reaction time, reaction temperature, stirring rate, liquid-to-solid ratios, and the concentration of finished crude phosphoric acid on the percent decomposition of phosphorite, percent dissolution of P_2O_5, recovery of P_2O_5, crystalline pattern and crystal size of phosphogypsum.

Effects of Reaction Time (t)

Effects of different reaction times on the percent decomposition of CLD04 sample, dissolution rate of P_2O_5, recovery of P_2O_5, crystal shape and particle diameter of phosphogypsum are shown in Table 3. The condition for the tests are a temperature of 65±2 degrees C, stirring rate of 300 r/min, liquid-to-solid ratio of 2.5:1, finished crude phosphoric acid concentration of 25% and H_2SO_4 excess of 3%.

The reactivity of CLD04 sample was very good. When it was ground fine enough, 95% of the phosphorite decomposed[9] in five minutes generally with the presence of excessive H^+. Thus, reactive time had little effect on the percent decomposition of the CLD04 sample and percent dissolution of P_2O_5. However, it had a big effect on the crystal size of phosphogypsum and recovery of P_2O_5.

In the process, crystallization of calcium sulphate follows the crystal growth ΔL law[9] of Mark-keppel:

$$\Delta L = ft + L_0 \tag{9}$$

Where, ΔL is characteristic crystalline size at the age of t (it is the diameter for the clustered crystal and the length for sheet or acicular crystal), μm; L_0 is characteristic size of crystal nucleus

Table 3 Results of tests at different reaction times

code	t/h	$K_{decomposition}$/%	$K_{dissolution}$/%	$K_{refluent}$/%
1	2.5	98.1	98.33	90.4
2	3.5	98.7	98.54	91.2
3	4.5	99.5	98.92	92.3
4	5.5	99.4	98.90	94.1
5	6.5	99.4	98.94	96.4

code	Crystal shape of phosphogypsum	Mineral morphology of phosphogypsum	Average particle diameter of phosphogypsum/μm
1	rhombus	$CaSO_4 \cdot 2H_2O$	30.66
2	rhombus	$CaSO_4 \cdot 2H_2O$	32.01
3	rhombus	$CaSO_4 \cdot 2H_2O$	37.48
4	rhombus	$CaSO_4 \cdot 2H_2O$	38.13
5	rhombus	$CaSO_4 \cdot 2H_2O$	44.58

(can be ignored), μm; t is time, min or h; f is growth rate, μm/min or mm/h.

In a given system, growth rate f is a constant if the supersaturation, temperature, stirring rate and chemical component of liquid phase are constant. So a longer reaction time leads to an increase in the particle diameter of the phosphogypsum crystal, reducing the rinsing loss of phosphogypsum and increasing the recovery of P_2O_5. A reasonable reaction time of the experiment is considered to be 6.5 h.

Effects of Reactive Temperature (T)

The effect of reaction temperature on decomposition rate of the CLD04 sample, dissolution rate of P_2O_5, recovery of P_2O_5 and phosphogypsum crystal size and shape are shown in Table 4, with a reaction time of 5.5 h, stirring rate of 300 r/min, liquid-to-solid ratio of 2.5:1, finished crude phosphoric acid concentration of 25% and H_2SO_4 excess of 3.0%.

When the reaction temperature is raised, the liquid phase viscosity of the slurry decreased, which is beneficial to the diffusion of H^+, and the dissolution rate of the CLD04 sample increased. Meanwhile, an increase in the reaction temperature is also helpful to the diffusion of calcium and the growth of the calcium sulphate crystal, increasing the crystal size of phosphogypsum, lowering the rinsing loss of phosphogypsum, and improving the percent decomposition of CLD04 sample and recovery of P_2O_5. When the reaction temperature is too high (75 degrees C), the amount of calcium sulphate in the beginning of reaction is too much and supersaturation of solution increases quickly. A large quantity of crystal nuclei of calcium sulphate is produced. The diameter of the crystal particle is reduced. When temperature gets higher (80 degrees C), a certain amount of hemi calcium sulphate crystal, which has a smaller particle diameter and is not stable, formed, leading to an increase in rinsing loss of phosphogypsum. A reasonable reaction temperature in the experiment is at 65±2 degrees C.

Effects of Liquid-to-Solid Ratio (R)

The effects of different liquid-to-solid ratios on the percent decomposition of the CLD04 sample, percent dissolution of P_2O_5, recovery of P_2O_5 and crystal size and shape of phosphogypsum are shown in Table 5. The tests were conducted under the conditions of the reaction time of 5.5 h, stirring rate of 300 r/min, reaction

Table 4 Results of tests at different reaction temperatures

code	T/℃	$K_{decomposition}$/ %	$K_{dissolution}$ /%	$K_{refluent}$/%	Crystal shape of phosphogypsum	Mineral morphology of phosphogypsum	Average particle diameter of phosphogypsum /μm
1	55±2	98.2	96.2	89.2	rhombus	$CaSO_4{\cdot}2H_2O$	28.98
2	65±2	99.4	98.9	94.1	rhombus	$CaSO_4{\cdot}2H_2O$	38.13
3	75±2	99.5	98.6	86.3	Rhombus,small amount of acicular crystal	$CaSO_4{\cdot}2H_2O$	35.62
4	80±2	99.7	98.5	93.6	Sheet crystal is main, small amount of acicular and rhombus crystal	$CaSO_4{\cdot}2H_2O$	39.37

Table 5 Results of tests at different liquid-to-solid ratios

code	R	$K_{decomposition}$/%	$K_{dissolution}$/%	$K_{refluent}$/%	Crystal shape of phosphogypsum	Mineral morphology of phosphogypsum	Average particle diameter of phosphogypsum/μm
1	2.0:1	93.3	92.3	87.3	Acicular,rhombus and sheet crystal, acicular is the main	$CaSO_4{\cdot}2H_2O$	48.79
2	2.3:1	99.2	98.5	89.2	rhombus	$CaSO_4{\cdot}2H_2O$	33.40
3	2.5:1	99.4	98.9	94.1	rhombus	$CaSO_4{\cdot}2H_2O$	38.13
4	2.7:1	99.4	98.9	91.2	rhombus	$CaSO_4{\cdot}2H_2O$	25.17

temperature of 65±2 degrees C, finished crude phosphoric acid concentration of 25% and H_2SO_4 excess of 3%.

When the liquid-to-solid ratio is 2.0:1, the liquid phase viscosity of the slurry is high, which is not good for diffusion of H^+ and decomposition of phosphorite. The percent dissolution of P_2O_5 of the CLD04 sample was greatly reduced. However, the liquid-to-solid ratio increases the reaction probability of SO_4^{2-} and Ca^{2+} in the slurry. Thus growth of the crystal is faster and the particle diameter of the crystals increases. When the liquid-to-solid ratio is increased to 2.7:1, the liquid phase viscosity of the slurry drops, which is beneficial to ion diffusion to the mineral particle surface and decomposition of the mineral. However, it reduces the crystal area provided for SO_4^{2-} and Ca^{2+} per unit volume of slurry, lowers the growth rate of the crystal, improves supersaturation of solution, arouses more spontaneous nucleation, forms smaller crystals and reduces the recovery of P_2O_5. Within the optimum range of the liquid-to-solid ratio, an increase in the liquid-to-solid ratio is good for improvement of decomposition of the phosphorite, dissolution of P_2O_5 and recovery of P_2O_5. But high liquid-to-solid ratio inevitably produces more return phosphoric acid, thus reducing the final filtrate production. The optimum liquid-to-solid ratio for these experiments is 2.5:1.

Table 6 Results of tests at different stirring rates

code	n/(r·min^{-1})	$K_{decomposition}$/%	$K_{dissolution}$/%	$K_{refluent}$/%	Crystal shape of phosphogypsum	Mineral morphology of phosphogypsum	Average particle diameter of phosphogypsum /μm
1	150	93.3	92.55	84.5	Acicular,rhombus and sheet crystal, acicular is the main	$CaSO_4·2H_2O$	20.68
2	250	98.0	97.34	91.3	Rhombus	$CaSO_4·2H_2O$	30.45
3	300	99.4	98.86	94.1	Rhombus	$CaSO_4·2H_2O$	38.13
4	350	99.5	98.78	95.6	Rhombus	$CaSO_4·2H_2O$	40.57
5	450	99.5	98.86	90.4	Rhombus, chippy crystal	$CaSO_4·2H_2O$	35.84

Table 7 Results of tests at different concentrations of finished crude phosphoric acid

code	Design C_{H3PO4}, P_2O_5/%	$K_{decomposition}$/%	$K_{dissolution}$/%	$K_{refluent}$/%	Crystal shape of phosphogypsum	Mineral morphology of phosphogypsum	Average particle diameter of phosphogypsum /μm
1	20.0	99.4	99.3	97.4	rhombus	$CaSO_4·2H_2O$	40.78
2	22.0	99.1	98.9	95.6	rhombus	$CaSO_4·2H_2O$	38.44
3	25.0	99.1	98.9	94.1	rhombus	$CaSO_4·2H_2O$	38.13
4	28.0	98.6	98.2	93.6	rhombus	$CaSO_4·2H_2O$	35.31
5	30.0	98.0	97.3	90.4	rhombus	$CaSO_4·2H_2O$	26.80

Effects of Stirring Rate (n)

With the reactive time of 5.5 h, reactive temperature of 65±2 degrees C, liquid-to-solid ratio of 2.5:1, finished crude phosphoric acid concentration of 25% and H_2SO_4 excess of 3.0%, tests were conducted at different stirring rates to determine the effect on the decomposition rate of the CLD04 sample, dissolution rate of P_2O_5, recovery of P_2O_5 and crystal size and shape of phosphogypsum. These results are shown in Table 6.

An increase in stirring rate is helpful to the diffusion of ions in the liquid phase and growth of the calcium sulphate crystal. However, high stirring rates make violent collisions between the calcium sulphate crystals and the stirring device. This causes the large calcium sulphate crystal to break into fragments, reducing the calcium sulphate crystal size and recovery of P_2O_5. Optimum stirring rates for these experiments is 350 r/min.

Effects of Finished Crude Phosphoric Acid

With the reactive time of 5.5 h, reactive temperature of 65±2 degrees C, liquid-to-solid ratio of 2.5:1, stirring rate of 300 r/min and H_2SO_4 excess of 3%, tests were conducted at different concentrations of finished crude phosphoric acid to determine the effect on transformation rate of CLD04 sample. Results are shown in Table 7.

The concentration of finished crude phosphoric acid is required to be as high as possible in industrial production. But with the increase of concentration of finished crude phosphoric acid, both the concentration of calcium sulphate crystals and the liquid phase viscosity increase. So the percent decomposition of the phosphorite

Table 8 Chemical compositions of crude phosphoric acid

Component of crude phosphoric acid	P_2O_5	SO_4^{2-}	Cl^-	F^-	CO_3^{2-}
Content/($g \cdot L^{-1}$)	324.89	25.31	16.3	1.58	trace
Component of crude phosphoric acid	Al	Fe	Mg	Mn	Ca
Content/($g \cdot L^{-1}$)	0.92	1.55	0.36	1.15	0.75
Component of crude phosphoric acid	As	Co	Sr	Zn	Cd
Content/($g \cdot L^{-1}$)	5.60	24.22	2.36	40.10	0.12

Figure 4 SEM of phosphogypsum

drops and the particle diameter of phosphogypsum crystal is smaller. However, dihydrate crystals will not form at high finished crude phosphoric acid strengths. The reasonable P_2O_5 concentration of finished crude phosphoric acid in the experiment is 25.0%~28.0%.

It can be concluded from the above test conditions that a high percent decomposition of the CLD04 sample, high percent dissolution and recovery of P_2O_5 and large size and good shaped phosphogypsum crystals can be obtained with the reaction temperature of 65±2 degrees C, a reaction time of 6.5 h, liquid-to-solid ratio of 2.5:1, stirring rate of 350 r/min, finished crude phosphoric acid concentration of 25.0%~28.0% and H_2SO_4 excess of 3%. SEM of the phosphogypsum under these conditions is shown in Figure 4. Chemical compositions of the crude phosphoric acid produced under the above conditions are listed in Table 8. The contents of F,

As, Pb (1.16 mg/L) and Ba (0.106 mg/L) in the phosphoric acid are very low, and there are trace elements like Sr, Co, Cd and Zn in it. The chemical composition is very complex.

CONCLUSION

The CLD04 sample contains a high P_2O_5 content and is relatively high in F, Cl, CO_2 content. The main mineral in the CLD04 sample is fluorin-carbon apatite.

The important characteristics of phosphorite are the reactivity and counteractant capability. The CLD04 sample has a reactivity of 99.90% and a counteractant quotiety of –0.76, so it is better than land apatite in reactive capability.

Wet-process phosphoric acid can be prepared by using the CLD04 sample as a raw material by using dihydrate process. Suitable operating conditions are a reaction temperature of 65±2 degrees C, reaction time of 6.5 h, liquid-to-solid ratio of 2.5:1, stirring rate of 350 r/min, finished crude phosphoric acid's concentration of 25.0%~28.0% and H_2SO_4 excess of 3%.

Chemical components of the crude phosphoric acid prepared with seabed phosphorite as the raw material is more complex. The crude phosphoric acid has trace elements like Sr, Co, Cd, and Zn. If it is to be used to produce fine chemical products, further purification is necessary.

ACKNOWLEDGMENTS

The authors extend their grateful appreciation to Chuanyao Sun, Beijing General Research Institute of Mining and Metallurgy, Beijing, China, for his contribution to this paper.

REFERENCES

1. Xu, Maizhi. *Solid Minerals of the Ocean*. Publication Press of Qingdao Ocean University, Qingdao, 1999, 41–95.
2. Anlk K. Ghosh, Ranadhir Mukhopadhyay. *Mineral Wealth of the Ocean*, A.A. Balkema, USA, 2000, 82–87.
3. Haynes, B.W., Kramer, G.W., and Jolly, J.A. Fluorine and Uranium in Phosphate Rock Processing and Waste Materials, *U.S. Report of Investigations*, 1982, 1–17.
4. Shamsuddin, M., Wadsworth, M.E. Uranium-phosphorus Chemistry and Extraction of Uranium from uraniferous phosphorites, *U.S. Report of Investigations*, 1980, 83.
5. Slansky, M. Distribution and Possibilities of Concentrating Uranium in Sedimentary Phosphates, *U.S. Report of Investigations*, 1977, 15.
6. Judd, J.C. Beneficiation and digestion offshore phosphates [J]. *Marine Georesources & Geotechnology*, 1993, 11(3):245–257.
7. Samigullina, L.I. (USSR). Production of wet-process phosphoric acid from the Sea of Japan phosphorites [J]. *Mezhvuz. Sb. Nauch. Tr. Leningr. Tekhnol. In-t*, 1985, (9):48–51.
8. Shiwen peng, Gaokui. Phosphoric Acid, *Collection of Infrared Spectra*, Science Press, Beijing, 1982, 270–271.
9. Jiang, S., and Liu, G. *Phosphate Fertilizer and Compposite Fertilizer*, Chemical Industry Press, Beijing, 1999.
10. Pan, Z. Study on making phosphoric acid from Wongfu concentrated ore by wet process, *Guozhong Chemical Engineering*, 2000(6), 1–9.
11. Zhou, G., and Zhang, Y. Reactivity and bulking comparison of typical phosphate ores, *Phosphate Fertilizer/Composite Fertilizer*, 1996(4), 13–15.
12. Zhang, K., and Zhang, L. *Crystal Growth*, Science Press, Beijing, 1981, 115.

CHAPTER 24

Struvite—A Possible Solution to the High-Mag Rock?

H. El-Shall,* B. Bandyopadhyay,† E. Abdel Aal,‡ and R. Stana‡

ABSTRACT

A study was conducted to precipitate struvite from pond water in phosphoric acid plants as a first step for treating the water before it can be further processed for release.

Struvite is a crystalline substance consisting of magnesium, ammonium and phosphorus in equal molar concentrations ($MgNH_4PO_4 \cdot 6H_2O$). The study investigated (a) the dissolution of sufficient magnesium in pond water and precipitation and separation of calcium and (b) ammoniating the resultant liquor to precipitate struvite.

The current paper gives a background of the struvite produced commercially in the world and postulates how phosphoric acid produced from phosphate rock containing high levels of magnesium can be purified using known technology and minimizes the cost of acid purification by utilizing the separated magnesium in the production of struvite.

INTRODUCTION

Declining Phosphate Quality

The phosphate minerals exist in the form of fluoroapatite, $Ca_{10}(PO_4)_6(F,OH)_2$, francolite, $Ca_{10}(PO_4)_{6-x}(CO_3)_x(F,OH)_{2+x}$ and/or hydroxyapatite, $Ca_{10}(PO_4)_{6-x}(CO3)_x(OH)_{2+x}$. The most common means of describing the quality of any phosphate mineral is by its BPL content, expressed as weight percent BPL. BPL stands for "bone phosphate of lime" or the tricalcium phosphate equivalent of the specimen.

There has been a continuous decrease in world phosphate rock quality as reserves of high-quality rock are being depleted. As the high-grade concentrates from some sources are becoming depleted, the quality of phosphate rock, principally in Florida, but also elsewhere, is decreasing. Data indicates that the phosphate content of pre-beneficiated ore is currently decreasing by around 1% per decade.

The Major Cause of the Quality Problem in Florida is Dolomite

As mining of the phosphate minerals are exhausting the resources, the operators in Florida mining districts are having to move south. Dolomite, a magnesium bearing mineral, is continuing to increase as the mining moves south of the Bone Valley in Florida.

*Materials Science and Engineering Department, Particle Engineering Research Center, University of Florida, Gainesville, Florida.

†Process Industry Consultant, Gainesville, Florida.

‡Central Metallurgical Research and Development Institute (CMRDI), Helwan, Cairo, Egypt.

The Remedies for the Dolomite Problem

Fortunately, the magnesium which is largely present as a dolomitic impurity is amenable to separation from the phosphates by both physical and chemical separation techniques. Measures successfully utilized in the industry include:

- Attrition scrubbing of dolomite
- Selective flotation of the phosphates or dolomite
- Heavy media separation
- Pond water or other dilute acid leacning

The Issue of Process Water

While many methods can be used successfully to reduce magnesium content in the phosphate product utilized for the manufacture of phosphoric acid, the mineral impurity, namely magnesium, will build up in the process water system at each facility and would eventually find its way to the product causing serious product quality problems. One of the central problems is what to do with these magnesium bearing process waters.

Before we attempt to find a solution let's look at the "nature of the beast." The process water has the following characteristics:

- It has low pH (pH $\approx$ 1.0)
- It has phosphates ($\sim$1% P_2O_5)
- It has fluorides ($\sim$1% F^-)
- It has the following metallic and nonmetallic impurities:
  - Iron
  - Aluminum
  - Magnesium
  - Silica

The conventional means of dealing with disposal of process water is a two-stage liming process consisting of the following steps:

- Remove fluoride as CaF_2 ($Ksp=3.9\times10^{-11}$)
- Remove sulfate as $CaSO_4\bullet 2H_2O$ ($Ksp=2.4\times10^{-5}$)
- Remove phosphates as $CaHPO_4$ ($Ksp=2.7\times10^{-7}$)

EXPERIMENTAL

A Possible Solution—A Lateral Approach

The approach we have developed is based on a single stage treatment process consisting of the following:

- Remove fluoride as CaF_2 ($Ksp=3.9\times10^{-11}$)
- Remove sulfate as $CaSO_4\bullet 2H_2O$ ($Ksp=2.4\times10^{-5}$)
- Precipitate phosphate and magnesium as Struvite (magnesium ammonium phosphate) ($Ksp=1\times10^{-13}$)

Germination of the Idea

A paper entitled *Phosphate Removal and Phosphate Recovery from Waste Water Treatment Plants* by Centre Européen d'Etudes des Polyphosphates (CEEP), discusses the recovery of phosphate from municipal waste water treatment facilities in Europe and in Japan. In particular, this paper covered the reactor design parameters for struvite precipitation, struvite precipitation in sewage treatment works at the Imperial College and full-scale testing of a struvite precipitation reactor at Treviso bio-P sewage works in Italy.

Scope of This Work

- Demonstrate feasibility of struvite production from phosphate wastes
- Laboratory studies
- Develop a conceptual process scope with:
  - A preliminary estimate for cost of treating process water
  - A preliminary estimate for struvite production cost
  - A suggested means of integrating struvite in fertilizer blend formulation
  - A preliminary estimate of treated water quality

Table 1 Summary of experimental conditions using IMC pond water

Test Conditions	Test #3		Test #5		Test #6
Sample Volume	500 mL		500 mL		500 mL
Dolomite Treatment	Stage 1	Stage 2	Stage 1	Stage 2	Single Stage
Amount, g	8.5	8.5	5.0	13.0	18.0
Agitation time, hrs.	3	3	3	1	3
pH	3.4	5.6	3.0	5.0	5.3
Ammonia Treatment					
Volume of NH_3, mL	16.5		12.0		12.0
Agitation time, hrs	1		1		1
Final pH	8.8		8.5		8.4

The Proposed Process

Neutralize with dolomite:

- Precipitate fluoride as CaF_2
- Add Mg to water to the desired mole ratio for struvite
- Neutralize with ammonia
- Precipitate struvite
- Do all in a single stage

Materials Used

These included a sample of IMC pond water, a sample of IMC pilot unit R-O concentrate (made by concentrating Piney Point pond water by a factor of two), agricultural grade (calcined) dolomite with 53.9% $CaCO_3$, 45.4% $MgCO_3$ and 50% less than 100 mesh, and an ammonia solution (29% NH_3).

Test Procedure and Conditions

Four tests were conducted using the conditions given in Table 1. For the two stage processes, the solids were not removed prior to the introduction of additional chemicals for the second stage treatments. However, the solids were allowed to settle for four hours and removed prior to the addition of ammonia. For test #6, the solids were not removed until after the addition of ammonia.

The analytical results on the clarified liquids for three of the tests are given in Table 2.

As can be seen, all test procedures effectively removed the calcium. However, test 6 gave the highest phosphate removal and the lowest magnesium removal. All of the tests gave treated water with nitrogen concentrations higher than the initial concentration. While none of the treated water would meet drinking water standards, the quality is significantly improved.

The precipitated solids were also analyzed and the results interpreted as to the species precipitated. For test 3, the solids removed after the addition of the dolomite consisted of 36.1% calcium fluoride, 51.7% struvite, 3.6% gypsum, and 3.0% sodium fluoscilicate. The solids precipitated after the addition of the ammonia contained primarily 86.5% struvite and 11.1% calcium phosphates.

For test 5, the solids removed after the addition of the dolomite consisted of 55.5% gypsum, 15.5% calcium fluoride, 17.4% calcium phosphates, 5.6% sodium fluoscilicate, and 4.7% struvite. The solids precipitated after the addition of the ammonia contained 90.0% struvite, 6.6% sodium fluoscilicate, and 2.3% gypsum. For test 6, the solids consisted of 54.6% struvite, 21.0% calcium fluoride, 19.5% calcium phosphates, 2.7% sodium fluoscilicate, and 2.2% gypsum.

A second set of tests were conducted using reverse osmosis (RO) concentrate from the

Table 2 Summary of analytical data

Sample ID	Pond Water	Run #3, Filtrate		Run #5, Filtrate		Run #6, Filtrate	
			%removal		%removal		%removal
Mass, g	520	460	74.4	387	92.1		
% P_2O_5	2.34	0.63	99.7	0.21	99.9	0.08	97.4
% MgO	0.075	0.0002	99.9	0.0001	99.9	0.0041	89.4
% CaO	0.189	0.0002		0.0001		0.0038	97.6
% SO_4	0.84	0.67		0.41		0.83	
% SO_2	0.31	0.02		0.03		0.02	
% F	1.40	0.08		0.14		0.07	
% N	0.106	0.413		0.252		0.191	

Table 3 Analytical results from solids precipitated from Piney Point RO concentrate

Sample ID	Expt #3, CaF_2 Phase	Expt #3	Expt #5, CaF2 Phase	Expt #3	Expt #6, Mixed Phase
$MgNH_4 .6H_2O$	56.2	86.5	4.7	90.0	54.6
CaF_2	39.2%	0.9%	15.7%	0	21.0%
$CaSO_4 .H_2O$	5.2	0.5	56.2	2.3	2.2
$CaHPO_4 .2H_2O$	3.9	0	17.6	0	0
$Ca_5(PO_4)_3 OH$	0	11.1	0	1.1	19.5
$N2_4 SiF_6$	3.3	1.1	5.6	6.6	2.7

Table 4 Summary of the reagent consumed and costs for the four tests

Sample ID	Expt #3,	Expt #4	Expt #5	Expt #6
Dolomite Usage				
Ton/1000 gal	0.142	0.141	0.150	0.150
\$/1000 gal	8.50	8.45	9.00	9.00
Ammonia Usage				
Ton/1000 gal	0.020	0.016	0.014	0.014
\$/1000 gal	5.38	4.40	3.93	3.92
Total \$/1000 gal	13.88	12.85	12.92	12.92
Total Reaction time, hrs	7	4	5	4

Table 5 Shows a possible fertilizer blend that could be made with struvite

Sample ID		N	P	K	Mg
Formulation		7	14	21	3
Ingredients needed/ton	2000 lbs	140	280	420	60
	lbs/ton blend		$8.45	$9.00	$9.00
Struvite	903.5	54	280	0	90
Muriate of Potash	700	0	0	420	0
Urea	186.5	86	0	0	0
Limestone Filler	210				
Formulated Blend		7.0%	14.0%	21.0%	4.5%

Piney Point plant. This water was produced by first passing the pond water through an IMC proprietary pretreatment process to reduce the calcium and fluorine concentration and then passing it through a reverse osmosis unit that concentrated the water by about a factor of two.

The analysis of this water was 0.90% P_2O_5, 0.084% CaO, 2.50% SO_4, 0.10% MgO, 0.304% ammonia N, 0.025% F, 0.013% SiO_2, and 0.944% Na_2O.

A series of tests were conducted using this water with the same conditions as given in Table 1. The analysis of the solids are given in Table 3.

As can be seen, total reagent costs are relatively similar at about $13/1,000 gallons of water treated (Table 4).

Table 5 shows a possible fertilizer blend that could be made with struvite.

CLOSING REMARKS

Struvite is roughly equivalent to DAP plus dolomite in nutrient content. It contains 7% ammonia, 16% MgO and 29.1% P_2O_5. A ton of struvite is equivalent (more than equivalent in ammonia) to 0.75 tons of dolomite and 0.63 tons of DAP. If a cost of $35/ton of dolomite and $150/ton of DAP is used, struvite could be worth $121/ton. Struvite may have a higher or lower value in the marketplace since it is a very slow release fertilizer depending on the quantity available and the marketplace needs. Since it is a slow release fertilizer, the P and N will be virtually 100% utilized, compared for less than 50% for DAP. However, this is only of value for crops or plants (some pasture grasses and ornamentals are candidates).

CHAPTER 25

The Evaluation of Phosphate Rocks, from Beaker to Bench Simulator

Vaughn V. Astley*

ABSTRACT

There is always a need to evaluate samples of new phosphate sources for use in existing wet process phosphoric acid plants. Sample sizes can be as small as a few grams to several tons. Three methods are described that can provide the most information from any given sample.

INTRODUCTION

The worldwide production of large tonnages of phosphoric acid relies upon the ready availability of phosphate rock. This raw material is located in many areas of the world, with major deposits in Morocco and Florida. However, there are very many other significant but smaller deposits that from time to time need to be evaluated as to their economic potential. Indeed the sourcing of rocks from diverse geographical areas and thus their use both alone and as a blend to produce phosphoric acid can indeed provide many synergistic and thus interesting benefits. In addition rocks can vary by sedimentary or igneous origin and various levels of beneficiation and impurity level, thus the evaluation of such a diverse multitude of feed stocks must be treated with care.

In each case the rock first must be evaluated by small scale acidulations in order to determine the rock reactivity, the filterability of the produced calcium sulphate, the resultant phosacid quality and even the sulphuric consumption and P_2O_5 recovery that is to be expected. There are many references to various techniques that have been used, but I will first describe the small batch test that we have developed as well as a very small semi-continuous procedure. Finally, the larger scale bench scale phosacid simulator will be described together with some of the tests that we have carried out on several ores.

BEAKER SCALE ROCK ACIDULATION TEST

Small Scale Initial Test Work

- Only about 1 kg rock needed
- Simple equipment
- One technician
- Requires about ½ day

The equipment is shown in Figure 1.

The bench scale rock acidulation test is designed to evaluate, at a small scale, the ability of phosphate rock to produce wet process phosphoric acid and to obtain a preliminary evaluation of the filtration behavior of the attack tank slurry.

*Dr. Phosphates, Lakeland, Florida.

Figure 1 Equipment for bench scale digestion

The test uses 130 grams of rock per trial, with between four and six sequential trials used to evaluate a given rock sample. In addition to the rock, starting materials for the first trial are 22% P_2O_5 phosphoric acid, a small amount of phosphoric acid plant seed gypsum (both typically from an operating phosacid plant) and sulfuric acid. A computer mass balance program is used to calculate the quantities of sulfuric acid and seed gypsum for the trial. The rock and sulfuric acid are then added concurrently to a temperature controlled, stirred reactor over a 20 minute period. If needed, water can also be added to the reactor to make up for evaporation. At the conclusion of the rock and sulfuric acid additions, the attack slurry is allowed to mix for 60 minutes and then vacuum filtered. After removal of the product phosphoric acid the filter cake is washed twice, with each of the wash filtrates collected and kept separate. For the first trial the filter washes are de-ionized water.

Six Sequential Tests to Attain Equilibrium

- Start with plant recycle acid and other materials
- After the first test, we use the recycle acid from the prior test

Samples of the product acid, two filter washes and gypsum cake are submitted for chemical analyses. For the second and subsequent trials, the 22% P_2O_5 acid, seed gypsum and first filter wash are obtained from materials produced in the previous trial.

After four to six trials the product phosphoric acid and gypsum are considered representative of the rock being evaluated. The results should give excellent information regarding how a larger simulation should perform, particularly if acid quality, filtration performance or recovery issues need attention.

Typical information measured is shown in Table 1. From this information you can calculate the parameters shown in Table 2.

Table 1 Inputs and outputs for beaker scale digestion

Inputs

Inputs	
Desired % P_2O_5 in #1 Filtrate	28.00
Desired % Solids in Slurry	40
Rock Weight (grams)	130
Rock % P_2O_5	30.29
Rock % SO_4	0.66
Rock % CaO	43.47
Rock % Insol (est, or actual)	15.45
% P_2O_5 in Heel Acid	25.33
% SO_4 in Heel Acid	2.74
Actual H_2SO_4 Concentration (%)	75
Sulfuric Acid % excess	2
Estimated Rock Digestion Efficiency (%)	95
Final estimated filter cake % moisture	25.00%
Estimated #2 Filtrate Sp. Gr.	1.164
Estimated #3 Filtrate Sp. Gr.	1.000

Outputs

Outputs	
*** Gms diluted H2SO4 needed	120.85
*** Grams heel acid needed	161.70
*** Grams recycle gypsum needed	2.66
*** Grams 1st cake wash (#3 filtrate)	152.81
*** Grams 2nd cake wash (Pond Water)	142.18
DI Water Added	65.4
Grams #1 Filtrate From Rock	133.6
Grams #1 Filtrate From Heel Acid	146.3
Total Liquid Produced (gms)	279.9
Gms. Total Solids	194.8
Wgt. Slurry	474.7
Rock/Sulfuric Addition Time (min)	40
Rock Rate (gm/min)	3.25
Sulfuric Rate (gm/min)	3.02
Sulfuric Rate (ml/min)	1.81

Table 2 Bench scale test results

	Test #4	Test #5	Test #6	Average (Tests 4-6)
Rock Weight (gms)	130.5	130.2	130.2	130.3
Average Slurry Temperature (°F)	164.4	166.2	165.7	165.4
Calculated Slurry % Solids	43.1%	46.3%	45.3%	44.9%
Defoamer Used (gms)	0.64	0.61	0.45	0.57
Product Acid Filtration Time (sec.)	**5.12**	**6.38**	**3.45**	**4.98**
Specific Gravity @ 60 °F	1.3947	1.3991	1.3940	1.3959
%P_2O_5	**30.47%**	**30.89%**	**30.58%**	**30.65%**
%SO_4	**1.45%**	**2.32%**	**1.59%**	**1.79%**
% Solids	**3.12%**	**4.16%**	**3.89%**	**3.72%**
#1 Filter Wash Time (sec.)	6.6	16.0	8.3	10.3
#2 Filter Wash Time (sec.)	7.3	11.2	6.6	8.3
Filter Cake % Moisture	17.0%	18.3%	15.3%	16.8%
Filter Cake Total %P_2O_5	1.07%	1.35%	1.22%	1.21%
Filter Cake Water Soluble %P_2O_5	0.34%	0.52%	0.09%	0.32%
Filter Cake %CaO	25.26%	24.10%	25.12%	24.83%
P_2O_5 Mass Balance (%)	100.1%	105.0%	100.5%	101.9%
Digestion Efficiency (%)	**95.9%**	**95.1%**	**93.5%**	**94.8%**
Tons 100% H_2SO_4/Ton P_2O_5	2.51	2.55	2.55	2.54

	%P_2O_5	%SO_4	%CaO	%Fe_2O_3	%Al_2O_3	%MgO	%F	%Na_2O	% Insol.	MER
Rock	30.29%	0.66%	43.47%			1.38%			15.45%	
Product Acid (Test #6)	30.58%	1.59%	2.14%	0.72%	0.35%	1.37%	2.25%	0.11%		0.0798

Limited Information, But...

- Data evaluation is based on tests 4, 5, and 6
- Filtration times are "relative" to the control rock
- Rock digestion efficiency is also a reasonably comparable benchmark
- Gypsum samples, photomicrographs will provide added insight to compare rocks
- Phosphoric acid analyses are used to predict impact on downstream products
- Additives or crystal modifiers can be tested during acidulations

VERY SMALL CONTINUOUS BEAKER-SCALE PHOSPHORIC ACID PRODUCTION TESTS

Imagine a slug of slurry circulating within a Prayon type reactor. It takes about 7 minutes for a batch of slurry to circulate:

1. First we add ~2.5 g of rock
2. After 60 sec add ~1 mL sulphuric and 4 mL of recycle acid
3. After 6 minutes remove ~6 mL of slurry

If further information is required prior to larger scale testing, or if insufficient ore is available, then a semi-continuous small scale unit can be operated. This has been operated very successfully where only small rock samples are available. This method simulates a slug of slurry circulating within a Prayon type phosacid reactor. A typical unit uses a heel of about 150 mL of agitated slurry kept at temperature in a water bath. About 2.5 grams of rock are added together with about 4 mL of recycle acid to the slurry, after about a minute a 1 mL slug of 93–98% sulphuric acid is added, and the mixture allowed to react for a further 7 minutes or so. At this time about 6 mL of slurry are removed, and the additions of rock, 22% recycle acid and sulphuric acid are repeated. The filtration of this withdrawn slurry provides both product acid and a sufficient quantity of recycle phosphoric acid to maintain operation. A very good simulation of filterability and acid quality can be obtained.

BENCH-SCALE PHOSPHORIC SIMULATOR

Phosphate rock is digested in a six compartment attack tank as shown in Figure 2. The attack slurry is re-circulated from the #6 to the #1 compartment. Attack gravity and % solids are controlled using the 22% recycle acid.

The large scale simulator that has been used by IMC consists of a rectangular tank divided into seven compartments. There are underflows between 1 and 2, 3 and 4, and 5 and 6, whereas compartments 2, 4, and 6 overflow to the next. A phosphoric gypsum slurry recycle is maintained at about 6 liters/min from compartment 6 to compartment 1. Rock is added either as a dry material or in slurry form to compartments 1 or 2, together with the 22% recycle phosphoric acid from the filtration step. Concentrated sulphuric acid is added to one or more of compartments 2 through 4. Typically the system must be heated to maintain temperature and an efficient fume extraction system should be provided. As the reactions proceed, the free sulphate level is measured, and any adjustments deemed necessary are made to the sulphuric acid metering system. After about 30 minutes sufficient slurry will have overflowed compartment 6 and accumulated in the filter feed compartment 7.

The filtration can be carried out using either a continuous small belt filter or, in our experience, a batch filtration is preferred. The filter consists of a horizontal filter mesh with filter cloth somewhat similar to a conventional tilting pan. The small scale unit we have does in fact tilt to discharge the resultant gypsum cake. About 7.5 liters of slurry are placed in the filter and a vacuum applied through the product acid receiver. As soon as cake surface is evident, the vacuum is released and the intermediate wash added. Again vacuum is applied but through the 22% receiver, and again until the cake surface is visible. A final water wash is applied and vacuum applied through week wash receiver and extended for about 20 seconds beyond cake dry. The filter is rotated to dump the gypsum and all the filtrates transferred to their respective tanks for the next filtration cycle. The individual filtration times are recorded as is the cake thickness and samples of the various streams retained. The filter is shown in Figure 3. Operating data is collected using a RS Logix 5,000 PLC. Citect is used as the HMI to control the process, log data and display process trends. This is shown in Figure 4.

Figure 2 Six compartment attack tank

Figure 3 Bench-scale tilting pan filter

Figure 4 Process control and data acquisition system

SECTION 7

Process Analysis and Control

CHAPTER 26

An Evaluation Criterion for Florida Phosphate Flotation Results

Charles Guan*

ABSTRACT

The existing methods including the economic efficiency have limitations to evaluate Florida phosphate flotation performance. A new separation efficiency,

$$E = R_{BPL} R^{a}_{BPL} / R^{m}_{BPL}$$

is proposed as the evaluation criterion, where R_{BPL} is the recovery of the flotation under evaluation, R^{a}_{BPL} is the recovery of an additional operation upgrading the concentrate into a super concentrate at the highest BPL and producing a tails having the feed grade, and R^{m}_{BPL} is the maximum recovery of the flotation process separating the feed into a tails with the lowest achievable BPL and the super concentrate. R^{a}_{BPL} serves as a correction factor for the concentrate BPL, and R^{m}_{BPL} for the feed grade. The separation efficiency has been easily applied to evaluations of the fatty acid testing results and the production data of the same plant or different plants.

INTRODUCTION

It is well known that results of flotation tests and production are given in numbers, such as weight, rate, grade, and recovery of each product. For phosphate flotation, neither recovery nor grade of concentrate can be used as a criterion to evaluate the results because they are interdependent, e.g., an increase in concentrate grade simultaneously decreases the recovery, and vice versa. The method of factorial design of experiments (DOE) has not been used widely in the beneficiation field partially because of the lack of widely accepted single criterion for statistical tests of significance—evaluation of the results (Macdonald et al., 1985). Table 1 shows 27 possible cases for comparing results of two flotation processes. It should be noted that "0" stands for the same results, "+" the better results, "–" the worse results, and "?" indicates it cannot tell. It has been estimated that for the same feed grade (Cases 1–9), about 22% of the cases could not be evaluated according to recovery and concentrate BPL. For different feed grade cases (Cases 10–27), the questionable cases account for about 56%.

Although the graphical analysis of recovery-concentrate BPL curves has been widely adopted, and the economic excellence has been suggested as the evaluation criterion (Kelly, 1991; Schulz, 1970), they have limitations to compare results of two different feed grades and

*The Mosaic Company, Mulberry, Florida.

Table 1 Possible cases for comparing results of two flotation processes

Case	Feed BPL	Conc BPL	Recovery	Performance
1	0	0	0	0
2	0	0	+	+
3	0	0	-	-
4	0	+	0	+
5	0	+	+	+
6	0	+	-	?
7	0	-	0	-
8	0	-	+	?
9	0	-	-	-
10, 19	+, -	0	0	-, +
11, 20	+, -	0	+	?, +
12, 21	+, -	0	-	-, ?
13, 22	+, -	+	0	?, +
14, 23	+, -	+	+	?, +
15, 24	+, -	+	-	?, ?
16, 25	+, -	-	0	-, ?
17, 26	+, -	-	+	?, ?
18, 27	+, -	-	-	-, ?

tell the technical excellence. For flotation plant operation, there has not been a method to evaluate its performance with changes in feed and concentrate grades and recovery. Separation efficiencies have been proposed as the evaluation criteria since 1913 (Schulz, 1970). However, these are mostly suitable for the comparison of the same feed grade results. Therefore, it is necessary to establish a suitable criterion for comparing results. Ideally, the criterion should be as follows:

- A single relative number—actual to maximum achievable
- In the range of 0 to 100%
- Represents the separation in terms of recovery, feed grade and concentrate BPL
- Physically meaningful
- Simple

SEPARATION EFFICIENCY AS EVALUATION CRITERION

Schulz (1970) reviewed various methods that were proposed for evaluating separation processes since 1913. The methods were classified into three categories:

1. Relationships that are not true efficiencies in the usual sense. The derived figures do not compare an actual separation with either an ideal or standard one.
2. Basic efficiencies that compare results with a perfect separation.
3. Relative efficiencies that compare results with those of a standard test.

Methods that attempt to compare the results of a separation with a perfect separation are mostly of the three categories. In a perfect separation process, all the valuable mineral particles would report to the concentrate, while all the waste mineral particles to the tailings. The separation would be 100% efficient. It was concluded that the technical excellence of a mineral concentration process should be expressed uniquely and quantitatively by the separation efficiency equation. For phosphate flotation, the equation can be written as

$$E_P = R_{BPL} - R_{Insol} \quad (1)$$

Where E_p is the relative separation efficiency that compares results with 100% perfect separation, R_{BPL} the BPL recovery, and $R^m{}_{Insol}$ the recovery of insol in the concentrate. Equation 1 has been applied as evaluation criterion by various researchers (Gruber, 1999; Leal Filho et al., 2002) to compare phosphate flotation results with the same feed grade and different concentrate BPLs in the same study. However, for different feed grade results, it will be problematic using Equation 1 for the same purpose since the recovery is very much feed grade dependent for phosphate flotation. Analyzing the industrial production data and various testing data indicates that a perfect separation with 100% efficiency is almost impossible to achieve, there are always recovery limits for phosphates and sand in concentrate for a given feed BPL. The limit/maximum separation efficiency of the process for the given feed grade can be expressed as

$$E_m = R^m_{BPL} - R^m_{Insol} \quad (2)$$

Where $R^m{}_{BPL}$ is the maximum concentrate BPL recovery, and $R^m{}_{Insol}$ is the minimum insol recovery in concentrate.

Instead of comparing results with the 100% efficiency of the perfect separation, the results should be compared with the limit/maximum separation efficiency. Therefore, the following equation is proposed as an evaluation criterion for phosphate flotation results:

$$E = \frac{R_{BPL} - R_{Insol}}{R^m_{BPL} - R^m_{Insol}} \times 100 \quad (3)$$

Where E is the relative separation efficiency which compares results with the maximum separation. The concentrate BPL recovery, R_{BPL}, can be calculated according to

$$R_{BPL} = Yield_{Conc} \frac{BPL_{Conc}}{BPL_{Feed}} \text{ and } Yield_{Conc} = \frac{BPL_{Feed} - BPL_{Tail}}{BPL_{Conc} - BPL_{Tail}} \times 100 \quad (4)$$

The insol recovery in concentrate, R_{Insol}, can be calculated according to

$$R_{Insol} = Yield_{Conc} \frac{Insol_{Conc}}{Insol_{Feed}} \quad (5)$$

The concentrate BPL recovery limit—the maximum BPL recovery achievable for a given BPL_{Feed}, R_{BPL} can be calculated according to

$$R^m_{BPL} = Yield_{Super\_Conc} \frac{BPL_{Super\_Conc}}{BPL_{Feed}}$$

$$Yield_{Spuer\_Conc} = \frac{BPL_{Feed} - BPL_{Tail\_Limit}}{BPL_{Super\_Conc} - BPL_{Tail\_Limit}} \times 100 \quad (6)$$

BPL_{Tail_Limit} is the lowest achievable BPL in tails, which could be taken from the historical data, e.g., 0.5%. BPL_{Super_Conc} is the super concentrate BPL, it could be the zero insol BPL, ZIBPL. The ZIBPL can be obtained according to the following equation:

$$ZIBPL = \frac{BPL_{Conc}}{100 - Insol_{Conc}} \times 100 \quad (7)$$

The minimum insol recovery in concentrate, $R^m{}_{Insol}$, can be calculated according to

$$R^m_{Insol} = Yield_{Super\_Conc} \frac{Insol_{Super\_Conc}}{Insol_{Feed}} \quad (8)$$

$Insol_{Super\_Conc}$ can be estimated according to

$$Insol_{Super\_Conc} = 100 - \frac{BPL_{Super\_Conc}}{ZIBPL} \times 100 \quad (9)$$

Equation 3 can be transferred as

$$E = \frac{\left\{\frac{BPL_{Conc}}{BPL_{Feed}} \times \frac{BPL_{Feed} - BPL_{Tail}}{BPL_{Conc} - BPL_{Tail}}\right\} \times \left\{\frac{BPL_{Super\text{-}Conc}}{BPL_{Conc}} \times \frac{BPL_{Conc} - BPL_{Feed}}{BPL_{Super\text{-}Conc} - BPL_{Feed}}\right\}}{\left\{\frac{BPL_{Super\text{-}Conc}}{BPL_{Feed}} \times \frac{BPL_{Feed} - BPL_{Tail,Limit}}{BPL_{Super\text{-}Conc} - BPL_{Tail,Limit}}\right\}}$$

or

$$E = \frac{R_{BPL} \times R^{a}_{BPL}}{R_{BPL,Limit}} \tag{10}$$

Where R^{a}_{BPL} represents recovery of additional BPL upgrading step separating concentrate (BPL_{Conc}) into a super concentrate at BPL_{Super_Conc} and a tails at BPL_{Feed}.

Equation 10 indicates the separation efficiency is the ratio of recoveries of two separation processes as shown in Figure 1. The numerator of the equation represents the overall recovery of separation process as shown in Figure 1a. The process consists of the flotation operation under evaluation and additional BPL upgrading step separating the concentrate into a super concentrate and a tails having the feed grade. The denominator of the equation stands for recovery of a flotation process as shown in Figure 1b. The process is an operation separating the feed into a tails with the lowest achievable BPL and a super concentrate.

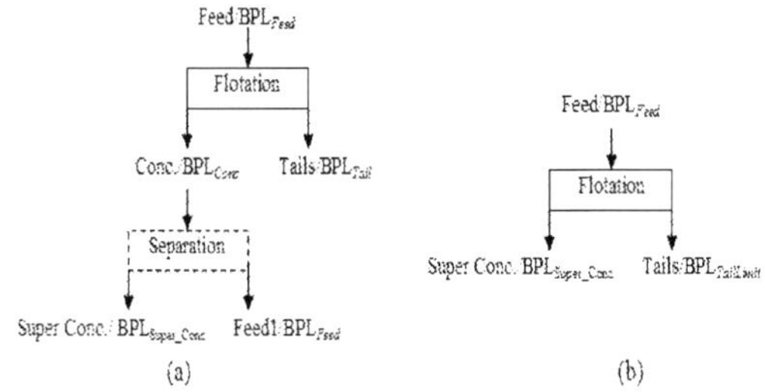

Figure 1 (a) Flotation process under evaluation with an additional separation process; (b) maximum separation efficiency flotation process

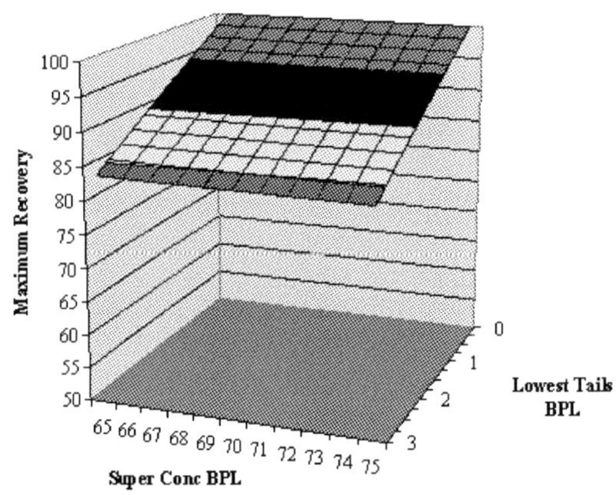

Figure 2 Maximum recovery vs. the super concentrate BPL and the lowest tails BPL

DISCUSSION

Maximum Achievable BPL Recovery

For Florida phosphate flotation, there is no liberation process involved. In nature the liberation of phosphate from sand in flotation feed is incomplete. A portion of the unliberated phosphates will never be recovered by the flotation, and always report to tails. This will result in a tails BPL limit for process to achieve, the lowest tails BPL. As the same manner, some of the unliberated sand will never be rejected and always report to the concentrate. The BPL of super concentrate having the minimum sand is considered as the highest concentrate BPL. Recovery of a process which could separate a given feed into a concentrate containing the highest achievable BPL and a tails with the lowest tails BPL would be the maximum achievable BPL recovery.

At constant feed grade and concentrate BPL, the effects of the highest achievable concentrate BPL and the lowest tails BPL on the maximum BPL recovery have been investigated. Figure 2 shows that the maximum recovery is significantly dependent on the lowest achievable tails BPL, and almost independent of the super concentrate BPL. It increases with the decrease of the lowest achievable tails BPL.

When 75% BPL is used as the super concentrate BPL, the maximum BPL recoveries at different feed grades and the lowest tails BPLs are calculated. The results are shown in Figure 3. It

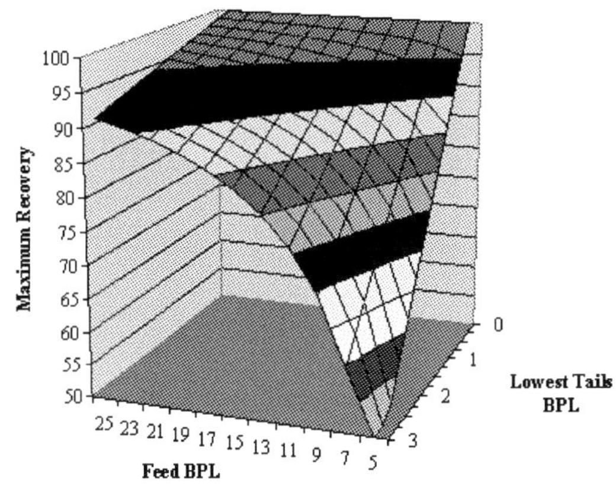

Figure 3 Maximum recovery vs. feed and the lowest tails BPLs

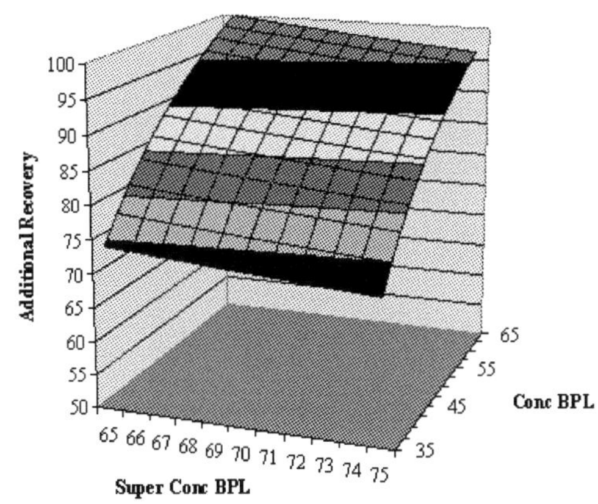

Figure 4 Additional recovery vs. the super concentrate BPL and the concentrate BPL

can be seen from Figure 3 that the maximum recovery is affected by both the lowest tails BPL and the feed grade when the super concentrate BPL is fixed. The maximum recovery decreases with the feed BPL. It is suggested that the maximum achievable BPL recovery mainly acts as a correction factor for feed grade in the separation efficiency evaluation.

Additional BPL Upgrading Recovery

Regardless of how the liberated sand particles are floated, concentrate BPL should be upgraded to the same BPL as the super concentrate BPL used in the maximum BPL recovery. If the concentrate is considered as a mixture of the super concentrate and feed material which is considered to be by-passed, the recovery due to the feed in the concentrate should be deducted. Therefore, an additional separation step is needed to remove the feed material from the concentrate to make the super concentrate BPL. The effects of the feed grade, concentrate BPL, and the super concentrate BPL on the additional BPL upgrading recovery are shown in Figures 4 and 5. As can be seen, the additional BPL upgrading recovery increases with the increase of the concentrate BPL, and decreases with the increase in the feed grade and the highest concentrate BPL. Also it appears that the additional BPL upgrading recovery plays a correction role

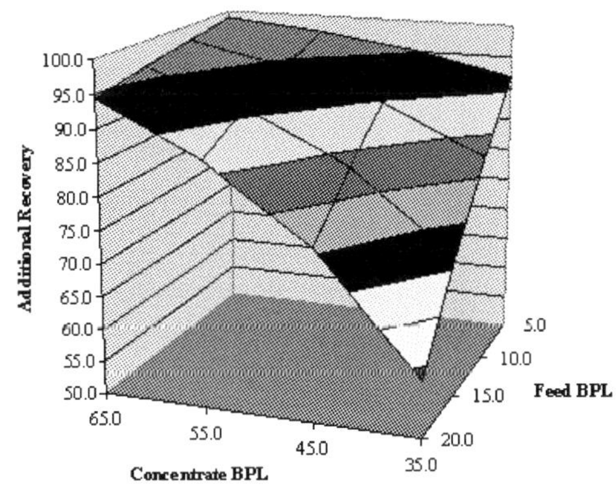

Figure 5 Additional recovery vs. feed and concentrate BPLs

for the concentrate BPL in the separation efficiency evaluation.

Applications of the Separation Efficiency

The first example of how the separation efficiency may be applied to evaluate phosphate flotation test data is shown in Figure 6. In this example, two fatty acids were tested at different dosages using 11.6% BPL plant fine flotation feed. Figure 6a shows plots of recoveries and concentrate BPLs versus the dosages for the two fatty acids, Figure 6b presents the recovery

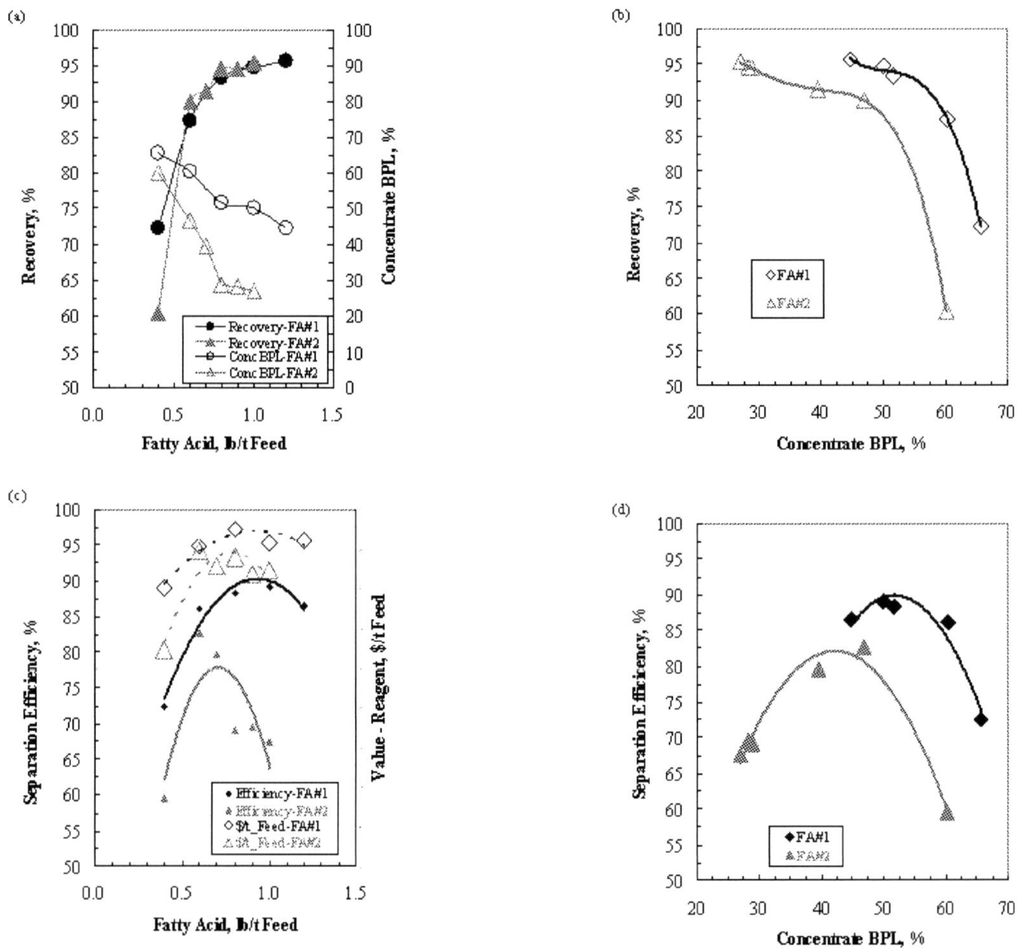

Figure 6 Comparison of lab flotation test results of two fatty acids

concentrate BPL curves. The separation efficiencies at different fatty acid dosages were calculated according to Equation 10 and are plotted in Figure 6c. Shown in Figure 6c is also economic information in terms of values of concentrate produced minus the reagent costs per ton of feed. It was assumed that amine flotation would be used to make the same BPL final concentrates. The overall concentrate recovery is calculated according to

$$R_{Overall} = R_{Rougher} \times R_{Amine} \tag{11}$$

The rougher recovery, $R_{Rougher}$, was determined from the test and the amine recovery, R_{Amine}, was projected according to

$$R_{Amine} = a + b \times BPL_{RougherCon} \tag{12}$$

The amine usage was predicted according to

$$lb / t_{Amine} = c - d \times BPL_{RougherCon} \tag{13}$$

Equations 12 and 13 were obtained by regression of the plant data. The feed was assumed to be upgraded to concentrate at 70% BPL by rougher flotation and amine flotation. It can be

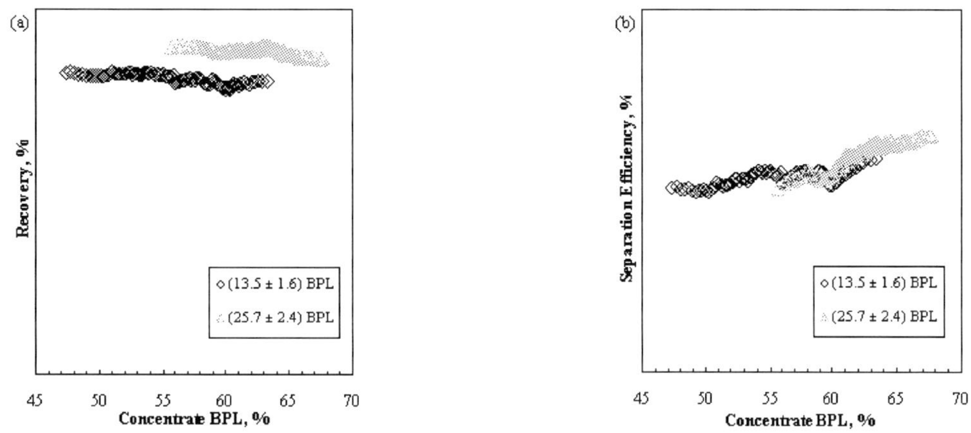

Figure 7 Fine rougher flotation performance comparison of Plant 1 at different feed grades

seen from Figure 6 that the recovery-concentrate BPL curves are good for the comparison of the two fatty acids in this example. However, it does not tell the difference among the different fatty acid dosages for the same fatty acid. On the other hand, the separation efficiency plots (Figures 6c and 6d) show there is an optimum efficiency for each fatty acid. Coincidently, there is also an optimum economic point for each fatty acid tested, and the optimum separation efficiency is achieved at or close to the test conditions of the best economic produced.

This example demonstrates that the separation efficiency could be used as an evaluation criterion for comparison of lab flotation test results of two fatty acids, as well as to determine the best flotation conditions from the tests.

The second example is to show how the separation efficiency may be applied to evaluate phosphate flotation production data at different feed grades. Figure 7a shows plots of 20 weeks rolling average Plant 1 fine rougher recoveries versus fine rougher concentrate BPLs at two different feed BPL ranges over 10 years. It can be seen from Figure 7a that changing feed grade would certainly lead to new recovery–concentrate BPL curves. It indicates that the higher the feed grade, the more upright the grade-recovery curve. It would be very unfair to say the flotation performance was better at the higher feed grade than at the lower grade due to the fact that the feed grade is usually not controllable. The separation efficiencies of the two different feed grades were calculated according to Equation 10 using 75% BPL for the highest achievable BPL in super concentrate and 0.5% BPL for the lowest achievable tails BPL. Figure 7b shows the plots of separation efficiencies versus the concentrate BPLs for the two different feed grades. It can be seen that the two efficiency-concentrate BPL curves have been brought much closer to each other. This example indicates that the separation efficiency has been fairly corrected for the feed grade effect in the evaluation, and it is possible to compare flotation performances at different feed grades.

The third example is to demonstrate how the separation efficiency may be applied to compare flotation performances of two plants. Figure 8a shows plots of 20 weeks rolling average fine rougher concentrate BPLs and recoveries versus fine rougher feed BPLs for two different plants. It is clearly shown that concentrate BPL and recovery increase with the feed BPL for both two plants. It further shows that in the 6 to 20 feed BPL range, Plant 1 produced lower BPL concentrate with higher recovery than Plant 2. The separation efficiencies were calculated using 75% BPL as the super concentrate BPL and 0.5% BPL as the lowest tails BPL, and the results are

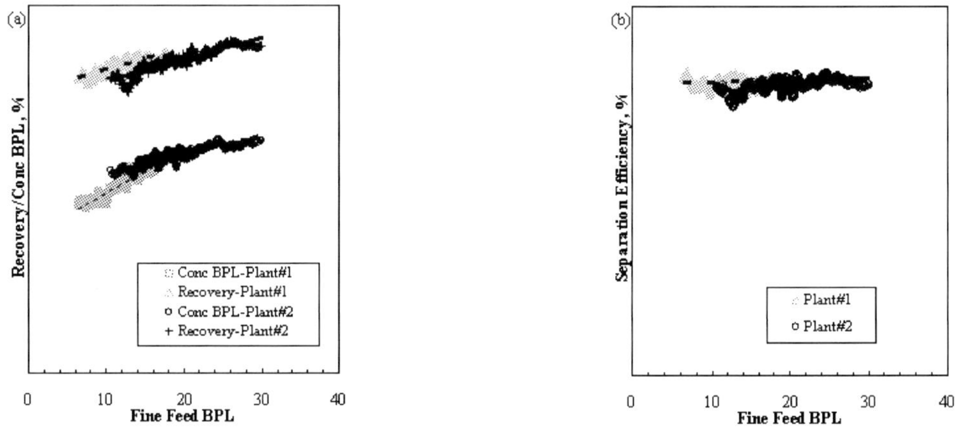

Figure 8 Fine rougher flotation performance comparison of Plant 1 and Plant 2

presented in Figure 8b. It can be seen that the separation efficiency is almost independent with the feed BPL for Plant 1 and slightly increases with the feed BPL for Plant 2. The difference between the two plants is statistically insignificant.

Other Considerations

Without a doubt, if applicable, the economic result will be the ultimate decision making criterion for operation and selection of processes. However, there are also limitations for economic efficiency as an evaluation criterion due to it is incomparable for financial results of two different feeds including feed grade, size distribution, and surface properties. Similar to the recovery, the separation efficiency derived in this study is considered as technical criterion for comparing the flotation results with different feed grades and concentrate BPLs.

CONCLUSIONS

The graphical analysis of recovery-concentrate BPL curves, the separation efficiency (R_{BPL}-R_{Insol}) proposed by others, and the economic efficiency have limitations to evaluate Florida phosphate flotation performance. The separation efficiency proposed in this study, is the only one satisfying the requirements of an evaluation criterion for Florida phosphate flotation results:

- A single relative number: It is actual to maximum achievable
- In the range of 0 to 100%
- Comprehensive of feed grade, concentrate BPL, and recovery
- Physically meaningful
- Simple to apply to evaluate test results and the plant production data

REFERENCES

Gruber, G.A., 1999, Anionic Conditioning for Phosphate Flotation. In: *Beneficiation of Phosphates: Advances in Research and Practice,* P. Zhang et al. (editors), SME, Littleton, CO, pp. 303–313.

Kelly, E.G., 1991, The Evaluation of Separation Efficiency. In: *Evaluation and Optimization of Metallurgical Performance,* D. Malhotra et al. (editors), SME, Littleton, CO, pp. 239–252.

Leal Filho, L.S., Assis, S.M., Barros, L.A.F., and Peres, A.E.C., 2002, Activation and Depression of Silicates During Anionic Flotation of Igneous Apatites. In: *Beneficiation of Phosphates: Fundamentals and Technology*, P. Zhang et al. (editors), SME, Littleton, CO, pp. 55–65.

Macdonald, R.D., Brison, R.J., and Hellyer, W.C., 1985, Flotation Test Procedures. In: *SME Mineral Processing Handbook*, N.L. Weiss (editor), AIME, New York, NY, pp. 30:96–30:104.

Schulz, N.F., 1970, Separation Efficiency: *Transaction of Society of Mining Engineers*, AIME, Vol. 247, pp. 81–87.

Yancey, H.F. and Geer, M.R., 1951, Efficiency and Sharpness of Separation in Evaluating Coal Washery Performance, *Mining Engineering*, Vol. 3, June, pp. 507–517.

CHAPTER 27

Laser-Induced Breakdown Spectroscopy as a Sensor for Radiometric Sorting of Minerals

M. Gaft,* I. Shapir-Sofer,* H. Modiano,* and M. Brestel*

ABSTRACT

Gated laser-induced breakdown spectroscopy (LIBS) enables on-line, essentially instantaneous, qualitative analyses of ore compositions on a moving belt. Measurement is based on intensity ratios of emission lines characteristic of specific ions. Thus it is a promising tool for radiometric sorting of minerals. A prototype LIBS machine was successfully tested for on-line, real-time evaluation of Mg concentration in phosphate ore (Four Corners Mine, Florida). Such a sorter may be easily adapted for use in other laser-based techniques, such as time-resolved luminescence, gated Raman spectroscopy, gated second harmonic generation, and photometric.

INTRODUCTION

In many mineral-processing installations, it is of critical importance to provide efficient mineral purification. Complicated processing steps, involving grinding to very fine particle sizes, are used in order to carry it out. However, purification of the coarse feed particles before grinding can provide significant benefit in overall energy efficiency and final product purity. When particles containing impurities are separated before grinding, they do not contaminate the feed stream and are not unnecessarily ground. Such separation can be achieved by radiometric sorting performed before the grinding stage, when the particles have sizes in the order of several centimeters.

Mineral sorting is a physical process in which particles are separated on the basis of their mineralogical composition. None of the minerals in the material treated undergoes any chemical change. The objective of any mineral-beneficiation process is to split the material being treated into two or more process streams. Valuable minerals are diverted into one stream, usually called the concentrate stream, while nonvaluable minerals are diverted into a second stream, which is called the tailings stream. As little as possible of the valuable minerals should be lost to the tailing stream. Sometimes when the mineral separations are difficult, a third stream, known as a middling stream, is introduced (Mokrousov and Lileev 1979, Salter and Wyatt 1991, Cutmore and Eberhard 2002).

Modern automated sorters are machines that examine feed lumps on an individual basis, compare the measured properties of each lump with predetermined criteria, and then separate the lumps into different products. The available sensing technologies include nuclear, optical, electrical, magnetic, acoustic, thermal, and various others; selection is specific to the ore. Most

*ITL Optronics Ltd., Emerson, New Jersey, and Petach Tikva, Israel.

of the properties used are not directly connected with mineralogical composition of accepted and rejected minerals. The only method that directly determines the major elements of separated minerals is the x-ray fluorescence (XRF) technique. Nevertheless, it is not suitable for elements with an atomic number less than 20. Furthermore, in many cases the XRF method is not selective enough, because many elements have characteristic emissions similar to those of valuable components (Ostapenko 1990).

The classical method for elemental analysis of geological samples is atomic emission spectroscopy. About three-fourths of the common elements can be distinguished by this technique, with lower limits of detection in the range of 0.5 to 5 mg/g. ICP-AES (inductively coupled plasma-atomic emission spectroscopy) offers excellent measurement precision, usually from 0.1 to 3% relative standard deviation; it is a rapid multi-elemental method, by which 30 to 40 common elements can be determined within about 2 min. The main drawback of this technique, which precludes its use in radiometric sorting, is the need for sample preparation.

LIBS has been developed for applications requiring elemental analysis of materials where conventional methods are not usable (Rusak et al. 1997). LIBS differs from ICP-AES in that the plasma is generated directly on the sample surface; no sample preparation is needed. LIBS, as an analytical technique, has the following advantages compared to ICP-AES:

- The analyses process is fast, because no sample preparation is needed.
- The capability for simultaneous multi-element determination and surface analysis remains.
- With the advent of miniaturized lasers and gated CCD detectors, field-portable instrumentation is achievable.
- Remote analysis is possible by transporting the laser beam through a fiber optic cable or by focusing the laser beam on the sample.

The main drawbacks are much lower sensitivity and reproducibility compared to ICP-AES. But even a semi-quantitative analysis may be quite adequate for radiometric sorting of minerals, where we do not need a detailed chemical composition of the minerals to find the specific lines that enable us to identify the corresponding mineral.

The purpose of this paper is to present several examples of LIBS implementation for radiometric sorting of minerals.

EXPERIMENTAL METHODS

Laboratory Setup

The excitation sources were Nd:YAG lasers with a pulse duration of ~5 to 87 ns. Lasers of both the fundamental (1,064nm) and first three harmonics (532, 355, and 266 nm) were used. LIBS emission was collected at an observation angle that was 90° from the laser incidence. An Andor iStar gated image-intensified CCD system was used at the spectrometer output to analyze spectra. The Q-switch SYNC output from the laser power supply was used to trigger the system, enabling spectral measurements in selectable time intervals. These intervals are characterized by a delay time: the time between the end of the laser pulse and the beginning of the measurement; and gate width: the time between the beginning and the end of the measurement. In essence, the detector is closed during the delay time and open for signal integration during the gate width. Afterwards the system waits for the next laser pulse and repeats the cycle. The number of pulses required was determined according to signal-to-noise ratio, dependent on emission intensity. Delay time was varied from 250 ns to 5 ms, while the gate width was usually 30 ms. A CCD array with 1,024 elements, and gratings with 300 to 1,200 lines/mm were used. Spectral resolution was from 0.5 to 2 nm.

Quartz fibers, which enable spectral measurement in the range from 200 to 900 nm, were used for delivering the emission to the monochromator. Advantages of fiber optic spectroscopy are modularity and flexibility of the system.

Industrial Sorter—Principal Scheme

A typical radiometric sorter consists of three parts: feeders, sensors, and ejectors. The feeders are universal for all types of radiometric sorting. The sensors are the identification apparatus, whose implementation is the subject of our research. We use a laser irradiation source, focusing and collecting optics, and computer software that attributes the pebble to one product or another. The computer would in turn generate a signal to ejectors, according to a preselected algorithm.

There are three main methods of pebble irradiation and survey used in radiometric sorting: (a) integral, where the irradiation and survey zone is bigger than the largest pebble, i.e., a simultaneous analysis of the whole sample, scanning is not performed; (b) passive scanning with a narrow aperture, where the irradiation and survey zone have a strip form perpendicular to the pebbles' movement; (c) active scanning, where the irradiation and survey zone is much smaller than the pebbles, and at each time interval a small part of a pebble's surface is analyzed.

LIBS sorting may not employ integral irradiation; the high energy density required to generate plasma imposes the use of an intense laser beam focused on a small area (less than 1 mm in diameter), and some scanning is called for. However, if each pebble consists of a single mineral, one laser pulse is sufficient to determine the pebble composition. For uniformly embedded ores the situation is different, and 2–5% of the pebble area has to be covered for confident identification of the useful mineral. Nonuniformly disseminated ores demand even more extensive scanning, whose extent has to be determined individually for each specific case. Contemporary lasers and detectors are quite compatible with such demands.

Field Prototype

A working field prototype, based on the same principle as the laboratory equipment, was constructed for evaluation of phosphate ores with high dolomite content. A smaller and more robust CCD spectrometer was incorporated. Such spectrometers have no moving parts and use slits of fixed width that may be selected depending on needed spectral resolution. Likewise, the gratings may be specified, usually from 300 to 1,200 lines per mm. The latest versions of such spectrometers are suitable for gated operation. An external trigger is applied to the spectrometer, and a TTL output port is available to trigger a pulsed laser. The delay before starting the measurement can be set in the software, starting from several nanoseconds. Plasma radiation is produced at the mineral's surface by a focused 1,064 nm Q-switched Nd:YAG laser, with a pulse energy up to 45 mJ at a repetition rate of 1/20 Hz. A single lens assembly is used for both focusing the laser beam and gathering plasma emission; a beam splitter diverts the returning plasma light to the fiber optic cable that carries the signal to the spectrometer and detector.

The system has sealed panels, shock absorbers, and an air conditioner, enabling operation outdoors under environmental conditions, such as dust, rain, extreme temperatures, and mechanical vibrations.

EXPERIMENTAL RESULTS

Experimental Database

We used time-resolved spectroscopy because plasma emission can be divided into two time domains. The first few hundred nanoseconds are characterized by a broad emission originating from the bremsstrahlung of the free electrons and electron-ion recombination. Weak lines show up on the strong continuum; they are identifiable as ionic lines of the plume constituents. The second time domain is characterized by an emission spectrum, in which narrow atomic lines dominate, corresponding to the elements present in the plume. The line strength is proportional to the atomic concentration. This second time regime lasts for several microseconds. Given these temporal features, it is clear that a superior identification has to be done by time-resolved spectroscopy. That is, the delay time with respect to the laser pulse has to be extended in order to take advantage of the

Table 1 LIBS lines of elements important for radiometric sorting of mine

Element	λ_{emis}
Ag I	328.1, 338.3, 405.5, 418.5, 421.0, 478.7, 487.4, 520.9, 546.6, 567.7
Al II	308.2, 309.3, 394.4, 396.2; 515.0, 560.0, 618.0
Au I	242.8, 267.8, 311.1, 389.8, 406.5, 461.2, 479.5, 596.8, 627.8
B I	249.6
Ba II	413.6, 455.4, 493.4, 614.2, 650.0
Be II	313.0
Ca I	422.7, 430.0, 445.0, 527.0, 558.0, 617.0, 656.0, 672.0,
Ca II	317.0, 370.0, 373.0, 393.3, 396.7, 644.0
Cr I	298.6, 301.8, 358.0, 359.5, 425.5, 427.5, 429.0, 520.9
Cr II	268.1, 283.5, 360.5
Cu I	324.8; 327.4, 510.6, 515.3, 521.8
F I	532.0; 603.8; 604.5
Fe I	274.0, 358.0, 374.0, 382.0
K I	766.5; 769.9
Li I	610.0, 670.8
Mg I	285.2, 383.8, 518.3
Mg II	279.5, 280.2
Mn II	257.6; 259.5, 279.7; 293.3; 403.3
NaI	589.00, 589.6
P	253.4, 255.3
PbI	280.2; 357.2; 363.9; 368.3; 405.8; 423.5
Si I	250.7, 263.1, 288.3, 391
Si II	386.0, 413.0, 504.0, 576.0, 634.0, 637.0, 657.0
Sn I	270.6, 284.0, 286.3, 301.1, 303.4
TiII	308.0, 317.0, 323.0, 334.0
Zn II	213.9, 330.3; 334.5, 412.0, 429.0, 471.0, 481.0, 636.0

informative line spectrum, while avoiding the first time domain, corresponding to the strong, unstructured continuum. Furthermore, different emissions have different decay lifetimes, which may be effectively used for selective detection.

To determine the temporal behavior of element-related emissions, plasma emission spectra of important elements were recorded at different delay times. Such information is very useful for finding a good compromise between improving signal-to-noise ratio and cutting off spectrally nondistinctive bremsstrahlung emission. We developed a database of time-resolved LIBS under 1,064, 532, 355, and 266 nm laser excitations for the spectral emission range of 200 to 900 nm. The main minerals in our library are apatite, calcite, fluorite, dolomite, borosilicates, scheelite, quartz, feldspars, micas, native gold and silver, and oxides of Mn, Fe, Ti, Al, Pb, and Sn (Table 1). The data demonstrate that in practically all cases it is possible to find emission lines that enable differentiation between accepted and rejected elements. Based on a mineral's LIBS library, a method has been developed for real time detection and content evaluation of minerals as they are conveyed on a moving belt, using the intensity ratios of the emission lines characteristic of specific elements or minerals. Because associated minerals always have different chemical compositions, of major and minor elements, the measured value of one or more of these ratios is an indicator that enables all phases to be consistently identified and assessed within a short time span, consistent with both LIBS and the moving belt system (Gaft and Nagli 2004).

LIBS POTENTIAL IN RADIOMETRIC SORTING OF MINERALS

LIBS sensing may be superior in quality to more than 20 methods that have been studied and tested in the radiometric beneficiation industry. There are several reasons for this: a powerful laser beam generates plasma on every type of mineral; LIBS gives information on virtually any element with very high sensitivity; associated minerals always have different combinations of elements, which is manifested in their breakdown spectra; and the method is not rock sensitive, thus may be used for practically any mineral association with minor changes, mainly in software. Several examples are presented below.

Phosphate with High Dolomite Content

Among the deleterious materials (Fe_2O_3, Al_2O_3, CaO, MgO) in phosphate rock, dolomite is the most troublesome. Dolomite causes higher consumption of sulfuric acid, reduces filtration capacity, and lowers the P_2O_5 recovery in fertilizer manufacturing. In Florida, the MgO in final phosphate concentrate is usually required to be less than 1%. With the depletion of the low-dolomite, easy-to-process Bone Valley siliceous phosphate deposits, phosphate mining moves further south and southeast, and the phosphate matrix becomes leaner in grade and high in dolomite. This makes it more critical to control the quality of the pebble product, particularly the MgO content. Current practices require either stockpiling of pebble product until quality control data from sampling become available, or making the shipping or discarding decision based on visual observations of rock as it is being produced. Obviously, this practice can result in the shipping of undesirable products to the chemical plant, or the discarding of acceptable pebbles (Gao et al. 2002).

A reliable instantaneous analytical method for batch sorting of phosphate rock according to MgO concentration maybe LIBS (Gaft et al. 2002). A prototype LIBS system was developed by ITL for on-line, real-time MgO content evaluation in phosphate rock. Figure 1 demonstrates typical breakdown spectra of Florida apatite and dolomite under 1,064 nm Nd:YAG laser excitation, with pulse energy of 15 mJ. Dolomite $(Ca,Mg)(CO_3)_2$ contains Mg as a major element and does not contain P and F, while phosphate $Ca_5(PO_4)_3(F,O,CO_3)$ contains major elements P and F, and does not contain Mg. These differences offer a good opportunity to differentiate between the materials using LIBS. The characteristic line of P at 254 nm is very weak, and the strongest lines in the spectra of apatite belong to Ca ions. The characteristic lines of F are relatively weak and clearly seen only after delay time of several microseconds, when the overall intensity is already low. The lines of Mg appear in dolomite spectrum at 518 nm and in the UV part of the spectrum at 280 nm (MgII) and 383 nm (MgI). The decay time of MgI emission is short compared to CaII. LIBS analyses enable us to conclude that the mostly efficient discriminative feature between dolomite and phosphate is the ratio between MgI emission at 383 nm and CaII emission at 393–397 nm after a relatively short delay of 500 ns. Under such conditions, the apatite intensity ratio ($I_{394.3}/I_{384.5}$) is 30.2, while for dolomite ($I_{394.3}/I_{384.5}$) it is 1.1 higher in apatite dolomite by 27 times.

A LIBS system was successfully tested on the main conveyer in the Four Corners Mine, Florida. Evaluation of dolomite contents was accomplished as follows. Conveyer belt is moving with 2 m/s velocity, thus each laser pulse strikes on a different piece of pebble.

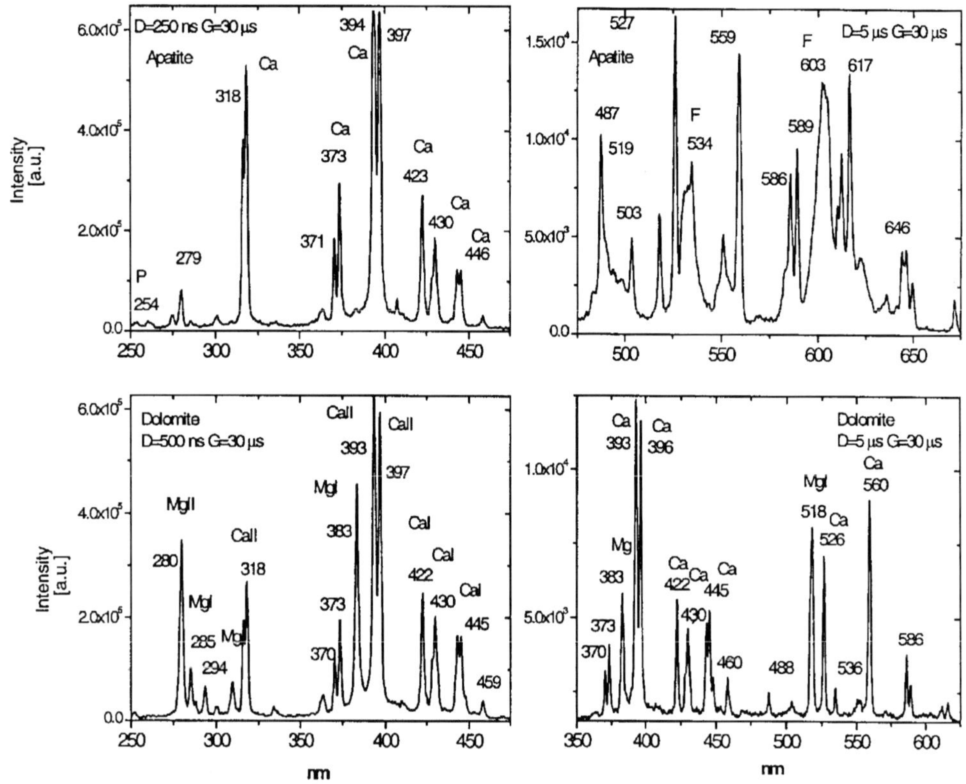

Figure 1 Breakdown spectra of Florida apatite and dolomite

According to the breakdown spectrum, specially developed software identifies the corresponding particle as apatite or dolomite. After averaging 2,000 pulses, the dolomite percentage was calculated. A set of experiments has been performed with two main aims:

- To check the system's long-term operation in real field conditions with different laser intensities and pulse repetition rates
- To analyze collected samples in the laboratory, in order to independently evaluate the dolomite content and compare it with the LIBS system results

During four days of field experiments, the rock on the conveyer changed from relatively dry to wet. It was found that in all conditions the laser intensity is adequate for satisfactory LIBS, while for relatively dry samples, it is possible to work starting from 15 mJ of laser intensity. High quality spectra have been received for the whole range of frequencies, namely from 2 to 20 Hz. At the last stage, the machine worked in "Processing" mode for 4.5 hours during three successive days, evaluating the dolomite content in the rock on the conveyer. Twenty rock samples have been collected from the conveyer during this time and analyzed in the laboratory. The impressive coincidence between LIBS results and traditional laboratory analyses was achieved (Figure 2). Thus, using on-line LIBS evaluation and corresponding ejectors, the phosphate rock with elevated MgO content may be separated from the main stream.

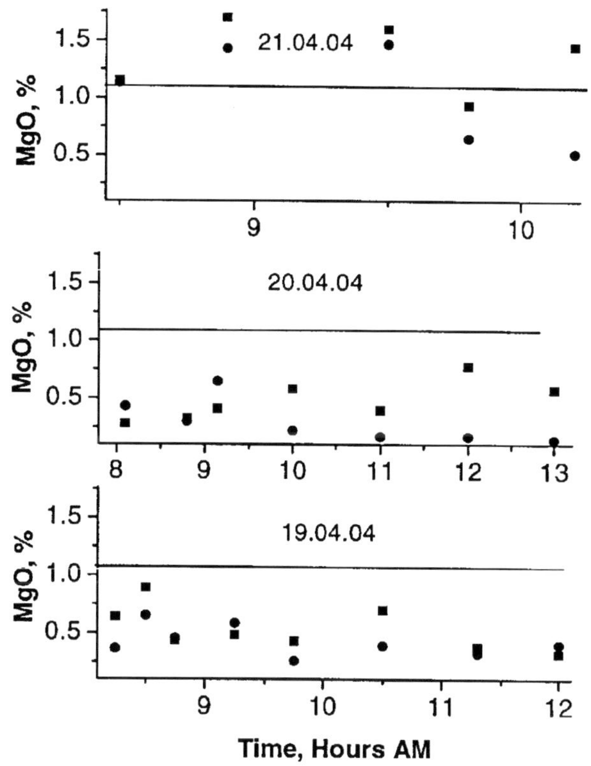

Figure 2 Comparison between LIBS (circles) and laboratory (squares) data on MgO content during beta-site test (Four Corners Mine, Florida)

Base Metals—Polymetallic Sulphides

The base metals, which include all the copper, lead, zinc, arsenic, and tin ores, presently appear to be the very large "gray area" in minerals sorting applications. There is only one known implementation: the "Precon" concentrating device, using gamma scattering in order to evaluate the sum metal content of a pebble (Wills 1992).

LIBS is a good candidate for these applications. For example, Figure 3 represents breakdown emission spectra of polymetallic sulphides with characteristic lines of Fe. In addition, the lines of Cu, Zn, and Pb are detected. The relative intensities of these elements' lines vary from sample to sample, which is evidently connected with their different contents in individual lumps.

Thus, the LIBS sensor enables us to accomplish the following tasks:

- On-line evaluation of useful Cu, Zn, Pb, and gangue Si contents for batch sorting of the rock
- Individual sorting of pebbles for concentrate (sulphides) and waste (silica, mica) using characteristic lines of Fe and Si
- Individual sorting of sulphides pebbles according to base metals concentrations to give two streams with high and low contents of useful elements

The first two tasks may be accomplished with the present version of the LIBS sensor, while the last one is more difficult. The reason is that base metals bearing minerals, such as galena PbS,

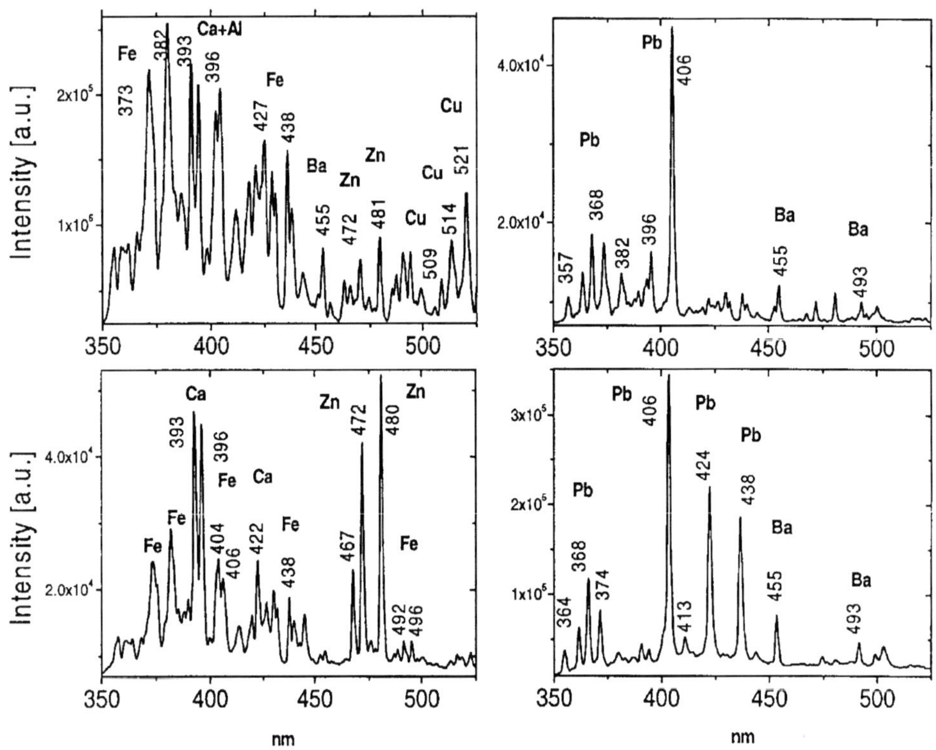

Figure 3 Breakdown spectra of polymetallic sulphides

chalcopyrite $CuFeS_2$, and sphalerite ZnS, are disseminated in massive sulphide lumps and a certain number of points has to be analyzed for proper evaluation. In such a case, 20 analyses/second may be not enough. Good correlation between averaged surface LIBS analyses and bulk laboratory data for drilling core samples (Bolger 2000) evidences that such individual sorting of sulphide pebbles according to base elements contents is feasible. The exact area which has to be analyzed for this task has to be determined individually for each specific case, but contemporary lasers, detectors, and computers appear to be quite adequate.

Dead Sea Salts

Solid and liquid sylvite KCl and liquid, containing sylvanite (sylvite KCl and halite NaCl), were analyzed in order to qualify LIBS as a tool for express on-line control of Na, Mg, and Ca impurities. The average concentrations of these elements in sylvite are 150 ppm for Na and 10 ppm for Mg and Ca.

Figure 4 shows breakdown spectra of solid and liquid Dead Sea samples. It is evident that these impurities are quite detectable. To check the sensitivity level in this case, spectra of several chemical samples from Merck with known impurity concentrations were measured. In all cases, K, Na, Mg, and Ca were detected.

Fuel Minerals

There is an ongoing need within the international coal mining and utilization industries to better characterize coal and combustion products. Areas of particular interest include improved coal characterization; the need for quantitative data on mineral distribution for better understanding of combustion behavior; and a generic analysis method for various

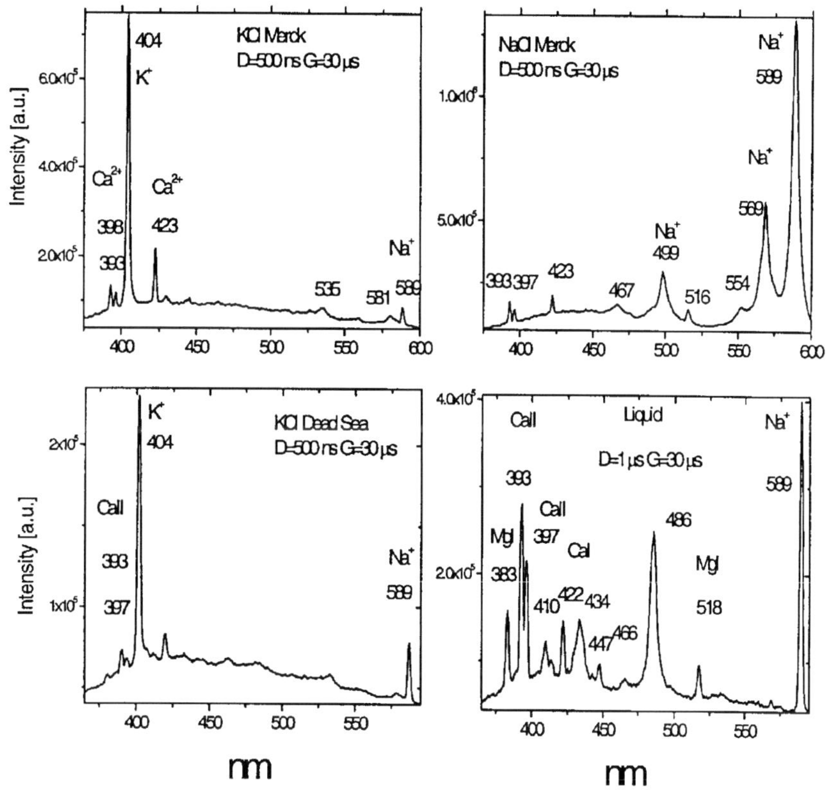

Figure 4 Breakdown spectra of Dead Sea and Merck salts

environmental waste issues surrounding the industry, including coal coal-dust analysis. Both particulates and unbroken coal can be measured by LIBS, which makes the technique attractive to mine geologists, combustion engineers, and waste and environmental scientists alike. The detectable elements in the LIBS spectra include the principal coal components (C, H, O, N) and additional inorganic components such as Al, Ca, Fe, K, Na, Mg, Si, and Ti (Figure 5). Thus, the method can discriminate the elements that belong to most common ash-forming minerals in coal, quartz, illite, kaolinite, calcite, ankerite, siderite, apatite, chamosite, gypsum, pyrite, and iron oxides.

Another possible application may be on-line LIBS analyses of carbon in ash. The excessive amount of C may indicate that the coal burning temperature has to be correlated.

COMBINATION OF SPECTROSCOPIC TECHNIQUES FOR SORTING

A LIBS sorter may be modified for use in other kinds of laser-based techniques. For example, it may, with minimal modifications, be used for photometric sorting based on laser beam reflectance.

Time-resolved luminescence may be realized by replacing the infrared laser source with an ultraviolet (UV) one, such as a third- or fourth-harmonics Nd:YAG laser. Such a sorter may be effective for diamonds, scheelite, fluorite, cassiterite (Figure 6), and many other minerals (Gorobets et al. 1997; Gaft et al. 2004).

Gated Raman extracting is made possible by adding a notch filter to the collecting optics in order to suppress the reflected laser light. To improve sensitivity, a spectrometer with an

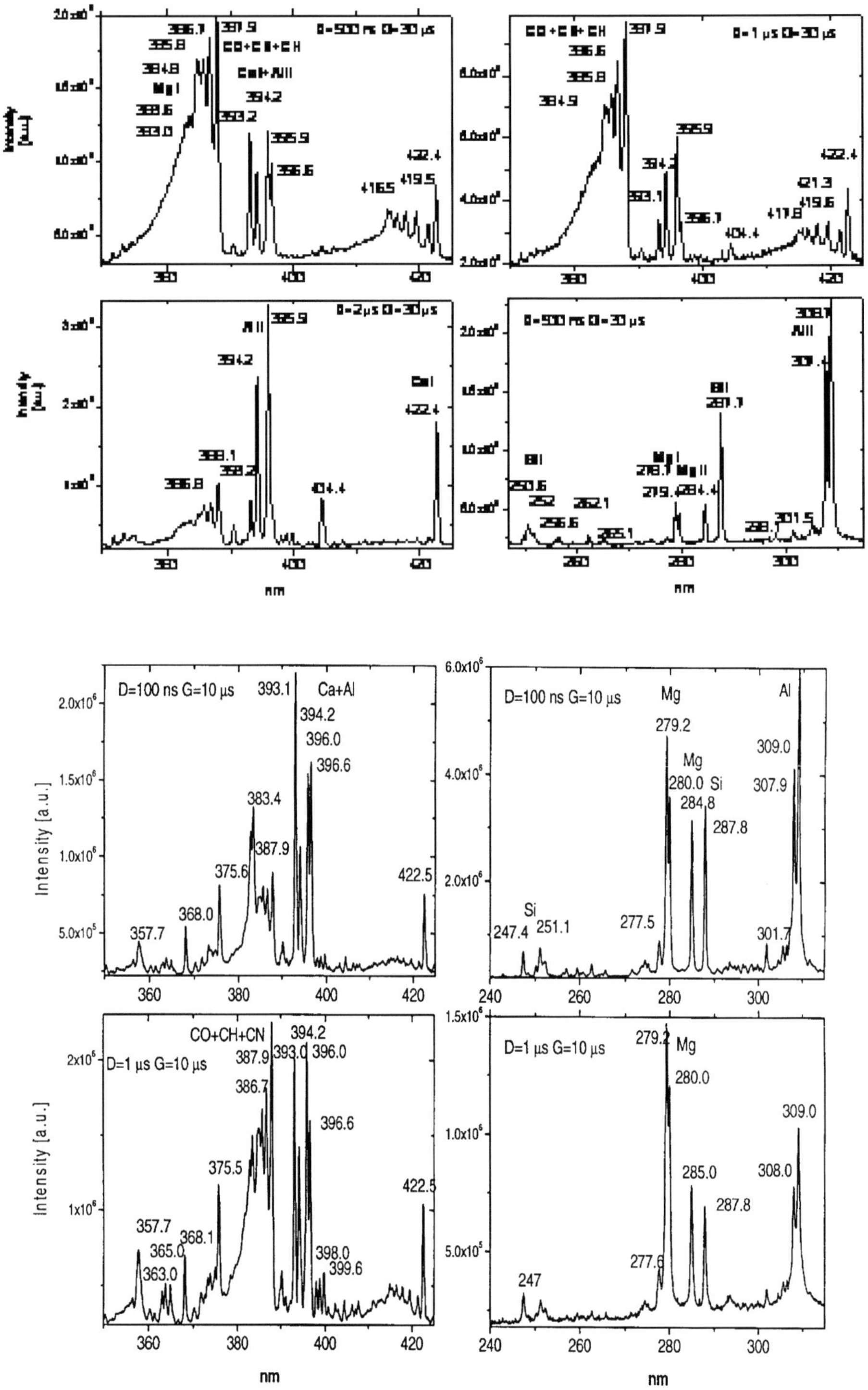

Figure 5 Breakdown spectra of coal

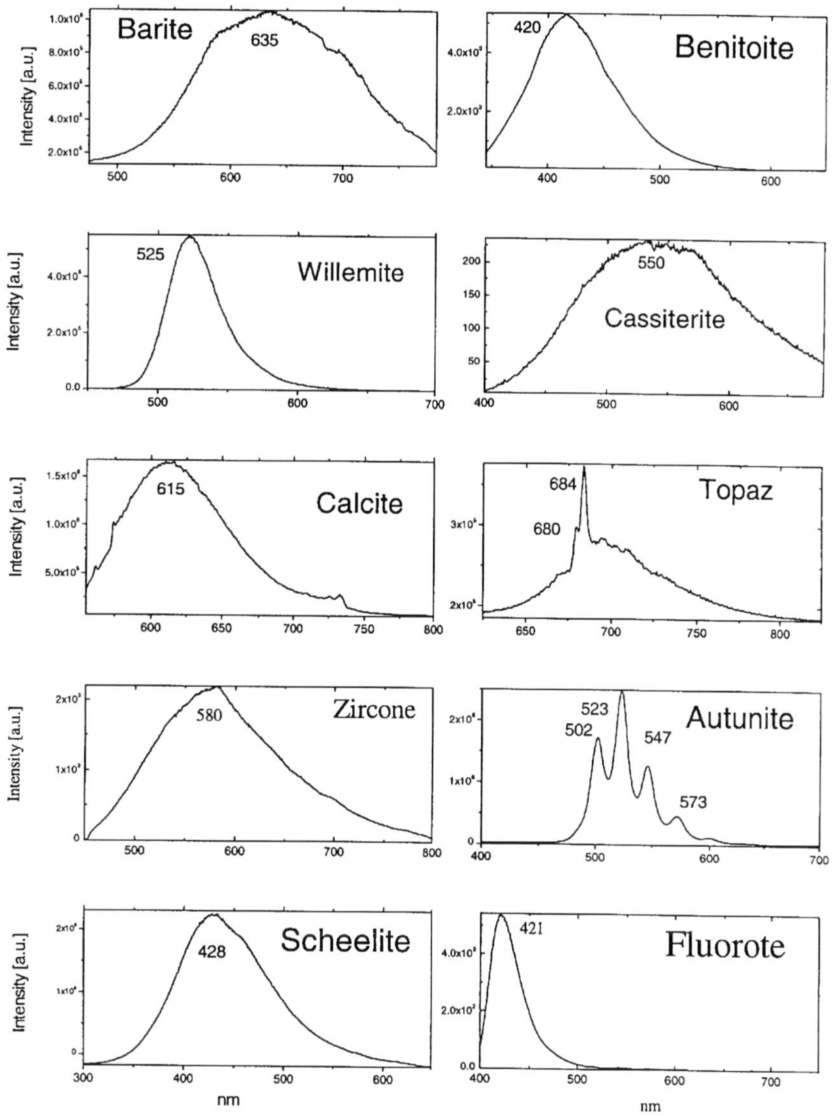

Figure 6 Laser-induced luminescence spectra of several minerals suitable for luminescence radiometric sorting

intensified CCD is needed. The second harmonic of Nd:YAG is preferred in order to remain in the wavelength range of CCD sensitivity. Nevertheless, the third and fourth UV harmonics may be potentially useful in order to excite into absorption bands and achieve resonance Raman. The most interesting potential subject for Raman sorting may be diamonds (Gudaev et al. 1997), especially in combination with laser-induced time-resolved luminescence, where both signals may be used simultaneously (Figure 7). Such a technique may be especially useful for diamonds without x-ray excited luminescence, which are not sorted by contemporary sorting machines. Other potential minerals suitable for Raman sorting may be magmatic apatite, dolomite, datolite, danburite, anatase, zircon, barite,

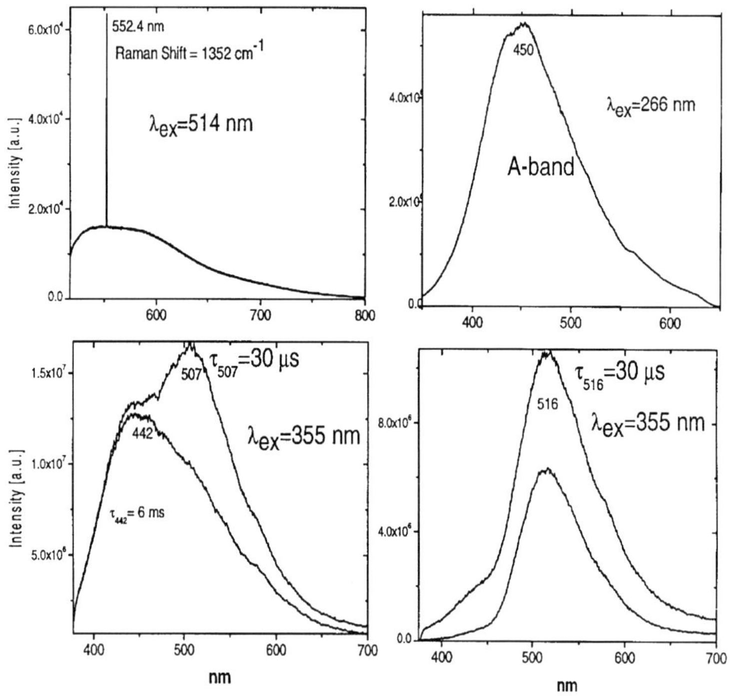

Figure 7 Raman and laser-induced luminescence spectra of diamonds, which may be used for radiometric sorting

quarts, calcite (Figure 8), and others (Gaft et al. 2004).

Such modified laser-based systems were proposed and tested for remote sensing of minerals. For example, minerals analysis at a distance of 10 meters was performed with a Raman/LIBS system. By analyzing the cation information collected in the LIBS experiment alongside the anion information provided by gated Raman scattering, a more complete image of the mineral's structure and composition can be obtained (Bishop et al. 2002, Hugh et al. 2002).

CONCLUSIONS

Laser-based time-resolved spectroscopy, mainly LIBS, luminescence, and Raman, is a very promising sensing technique for radiometric sorting of minerals. It may be used for batch and individual sorting of practically any type of minerals.

REFERENCES

Bishop, J., Lane, M., Murad, E. 2002. The power of combining multiple spectroscopic techniques for mineral identification on Mars and the necessity of lab and field study of analog materials, LPI Mars Spectral Workshop II, Lunar and Planetary Institute, Houston, TX.

Cutmore N., Eberhard, M. *Ore Sorting in 2001.*

Gaft, M., Nagli, L. 2004. Mineral Detection and Content Evaluation Method. US Patent 6,753,957 B1, June 2004.

Gaft, M., Nagli, L., Panczer, G., Reisfeld, R. 2002. Laser-induced Luminescence and Breakdown Spectroscopy Evaluation of Phosphates with High Dolomite Content. In *Beneficiation of Phosphates, Fundamentals and Technology*, P. Zhang, H. El-Shall, P. Somasundran, R. Stana, eds. SME, Littleton, Colo. 145–152.

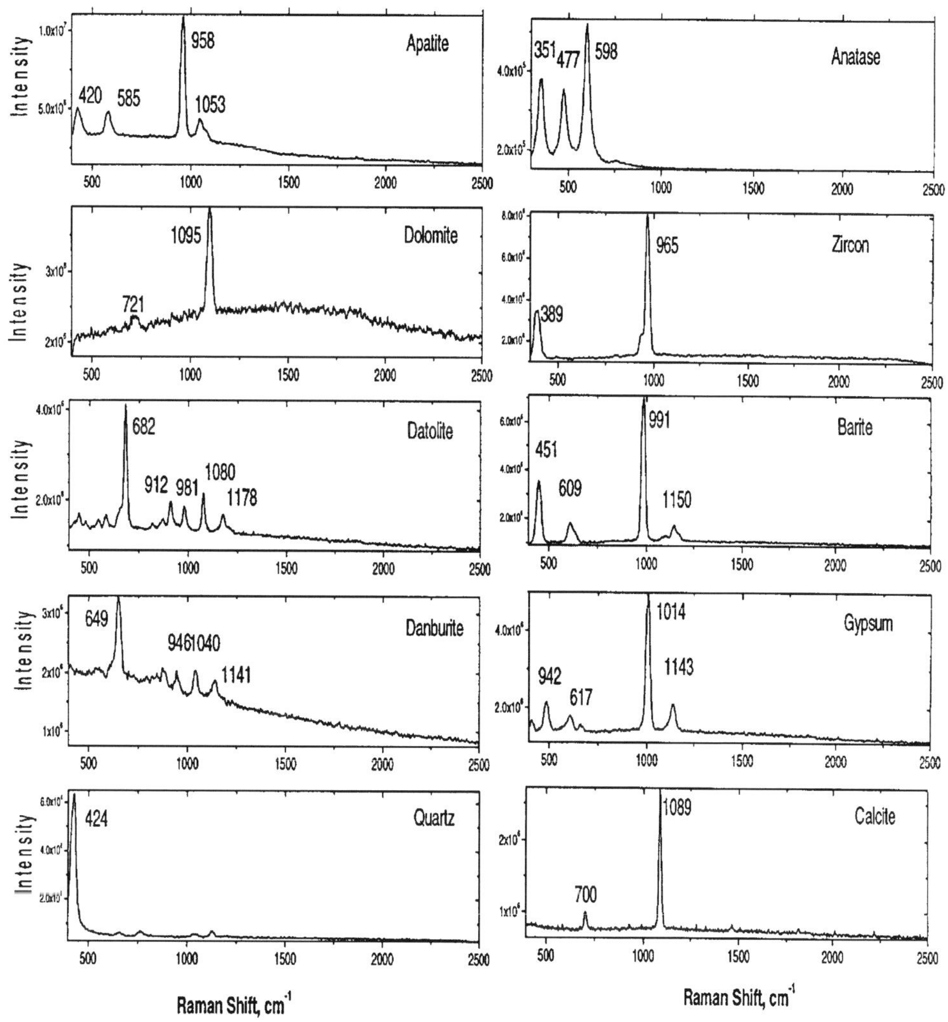

Figure 8 Raman spectra of several minerals that may be used for radiometric sorting

Gaft, M., Reisfeld, R., Panczer, G. 2004. *Modern Luminescence Spectroscopy of Minerals and Materials*, Springer, 2004.

Gao, Z., Zheng, S., Gu, Z. 2002. Review of Beneficiation Technology for Florida High Dolomite Pebble. In *Beneficiation of Phosphates, Fundamentals and Technology*, P. Zhang, H. El-Shall, P. Somasundran, R. Stana, eds. SME, Littleton, Colo. 247–259.

Gorobets, B., Litvintsev, E., Rogojine, A. 1997. Luminescence soting of non-metallic raw materials. *Proc. of the 5th Int. Congress on Applied Mineralogy and Mineral Industry*, Warsaw, Poland, 229-233.

Gudaev, I., Kanaev, V., Paterkin, I. et al. 1998. Optoelectronic unit for detection of diamonds by Raman light scattering. *J. of Mining Science*, V34, N6, 561–564.

Hugh, W., Hubble, H., Ghosh, M., Sharma, S., Horton, K., Lucey, P., Angel, M., Wiens, R. 2002. A combined remote LIBS and Raman spectroscopic study of minerals. Lunar and Planetary Science XXXIII.

Mokrousov, V., Lileev, V. 1979. Radiometric sorting of minerals. Moscow, Nedra (in Russian).

Ostapenko, M. 1990. *Technological Evaluation of Minerals*. Moscow, Nedra (in Russian).

Rusak, D., Castle, B., Smith, B., Winefordner, J. 1997. Fundamentals and applications of laser-induced breakdown spectroscopy. *Critical Reviews in Analytical Chemistry*, 27, N4, 257–290.

Salter, J., and Wyatt, N. 1991. Sorting machines in the mineral industry: problems or opportunities. In: *Proceedings of the 27th Int. Mineral Processing Symp.*, 5, 183–193.

Wills B. 1992. *Mineral Processing Technology*, Fifth Edition, Pergamon Press.

CHAPTER 28

Process Nuclear Magnetic Resonance (NMR) Analyzers—Benefits for Phosphate Beneficiation Plants

Vaughn E. Davis*

ABSTRACT

Nuclear magnetic resonance (NMR) provides accurate measurement of phosphorus (P_2O_5, BPL) and fluorine in solid and liquid states. On-line NMR phosphorus and fluorine measurements offer significant economic and environmental benefits to the phosphate industry and provide operational and economic enhancements to phosphate producers. Benefits of using process NMR include on-line sampling, no sample preparation, nondestructive quantitative analysis, and more precise control of the beneficiation process. Progression Inc.'s instrument provides an analysis every 5 to 10 minutes. This information is transmitted to the plant's distributed control system (DCS) for either closed loop or operator control.

NMR BACKGROUND

NMR has been utilized as a commercial analytical technique since the early 1960s when Varian produced the first available NMR spectrometer for sale. Since then, the technology has been an enabler for a multitude of industrial and medical technology developments in areas such as polymers, synthetic resins, petrochemical products, and drug discovery.

The first on-line process NMR analyzers were produced in the late 1980s and early 1990s by Auburn International and Foxboro. These analyzers were targeted to edible foods, polyolefins, and petrochemicals and were large, cumbersome, and expensive to install. Today, Progression, Inc. (Progression) is producing third generation process NMR instruments which are a fraction of the size of prior instruments, easy to install, and have been developed for a wide array of industries.

BENEFITS OF ON-LINE NMR IN PHOSPHATE BENEFICIATION

On-line instrumentation offers signification economic and operating benefits to mineral processor. Real-time or near real-time measurements can improve operating efficiencies, increase P_2O_5 recoveries, enhance process control, and allow for optimization of reagent usage. In many mining operations, current analytical techniques are run infrequently and provide little benefit to process control. Further, many sites rely almost solely on operator experience for estimation of ore quality. This reliance may lead to variations in plant operations and swings in P_2O_5 recoveries.

On-line NMR systems communicate directly with a facility's DCS system to allow for either

*Progression, Inc., York Harbor, Maine.

closed-loop or operator control. In either case, the near real-time ore quality feedback offers a significant opportunity for process improvement and cost savings.

BASIC NMR THEORY

NMR systems make use of a physical phenomenon common to many atoms such as phosphorus, fluorine, and hydrogen. These atoms will resonate, or absorb energy at a particular frequency, when placed in a strong magnetic field. This resonance occurs because the nuclei of these atoms have weak magnetic properties, typically referred to as the nuclei's "spin." In a strong magnetic field, the simplest case dictates that the "spins" align themselves with two energy states: with or against the magnetic field of the NMR instrument. The resulting bulk magnetization vector is described by the Boltzmann distribution (equation 1):

$$\frac{N_{upper}}{N_{lower}} = e\left[\frac{-\gamma_N(H)(h)}{k(T_K)}\right] \qquad (1)$$

Where:

N_{upper} = number of nuclei in the high energy state
N_{lower} = number of nuclei in the low energy state
γ_N = magnetogyroscopic ratio
H = external magnetic field (T)
h = Plank's constant (6.626176×10^{-34} J * s)
k = Boltzmann's constant (1.380662×10^{-23} J * K^{-1})
T_K = Sample temperature in kelvin (K)

In the case of phosphorus at 30° Celsius (303.15 K) and a magnetic field strength of 0.47 Tesla the distribution would be calculated as:

$$\frac{N_{upper}}{N_{lower}} = e\left[\frac{-10.8394 \wedge 10-7(rad * s-1 * T-1)(0.47T)(6.626176 \wedge 10-34(J * s))}{1.380662 \wedge 10-23(J * K-1)(303.15K)}\right] \qquad (2)$$

Where, the ratio of nuclei of the upper to lower energy states is 0.99999193457, or a slight excess in the lower energy state. In practical terms this means for every 5,000,000 spins there is a population difference of about 40 spins.

The population difference is inversely correlated with temperature and can be described by the Power law (see Figure 1).

This population distribution, like other spectroscopic techniques, has very important consequences. The signal intensity of any spectroscopic method depends largely on the population difference between the two energy levels involved. When a system is irradiated with a frequency, whose energy matches that separating the levels, transitions will be induced not only from the lower energy state to the higher state, but also in the reverse direction. Upward transitions absorb energy while downward transitions release energy. Because of this, the population difference gives rise to a net absorption or release of energy to observe.

When radio frequency (RF) energy is applied at exactly the right frequency to a sample in a magnetic field, resonance occurs, and the vector, or "spin," flips from one state to the other. As the spins are flipping, they absorb and then radiate energy. The exact frequency of the spin flips identifies the kind of atom involved (phosphorus, fluorine, etc.,) and the amplitude of the signal indicates quantity. NMR is one of the few nondestructive analytical techniques available for specific, elemental analysis and is able to measure bulk samples like phosphate pebble, slurries such as beneficiation plant feed and tails, and liquids.

Practically speaking, after the RF pulse is applied to the sample, a specific delay time is utilized before collecting the released energy from the sample. The resultant signal is referred to as a free induction decay (FID), and an example is illustrated in Figure 2.

A complete process NMR instrument is comprised of five basic components: magnet, probe, spectrometer, housing, and control software. Permanent magnet-based NMR instruments utilize a permanent magnet typically made from a rare-earth composition such as samarium cobalt (SmCo) or neodymium iron boron (NdFeB). Other choices are alnico, ceramic, or ferrite. For a typical phosphorus process installation, the magnet strength is 0.47 Tesla which corresponds

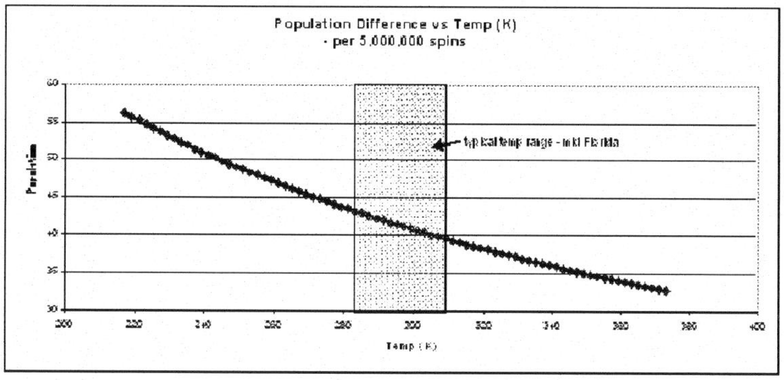

Figure 1 Population difference (per 5,000,000 spins)

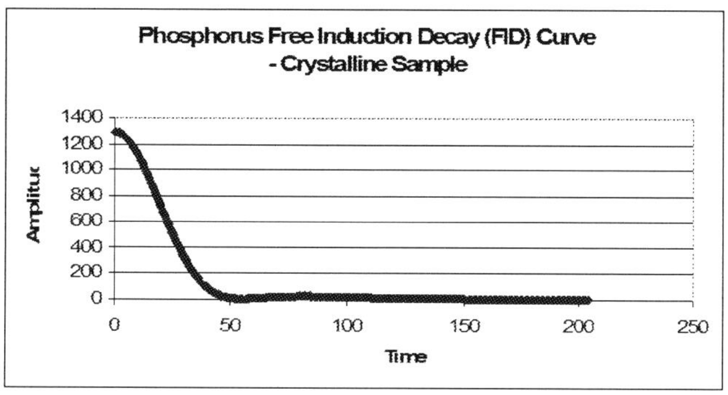

Figure 2 Phosphorus FID example

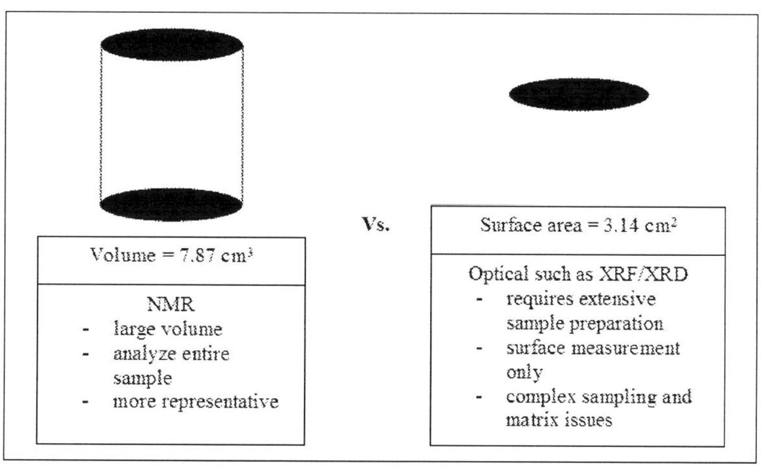

Figure 3 Comparison of NMR with optical techniques

Figure 4 Turbulent flow sampler *(photo courtesy of Mosaic Corporation)*

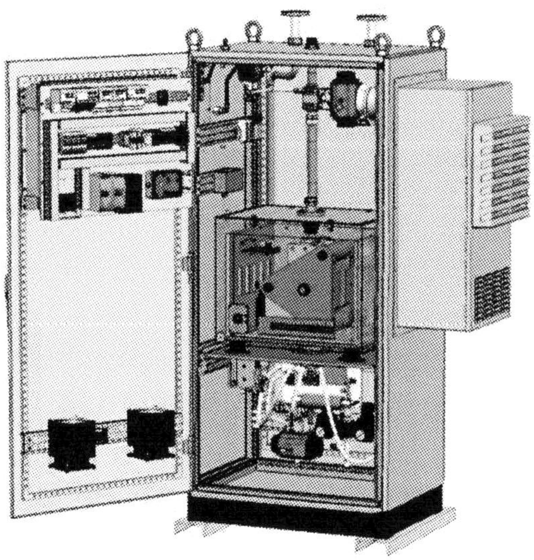

Figure 5 Cutaway of Progression, Inc. MagModule II

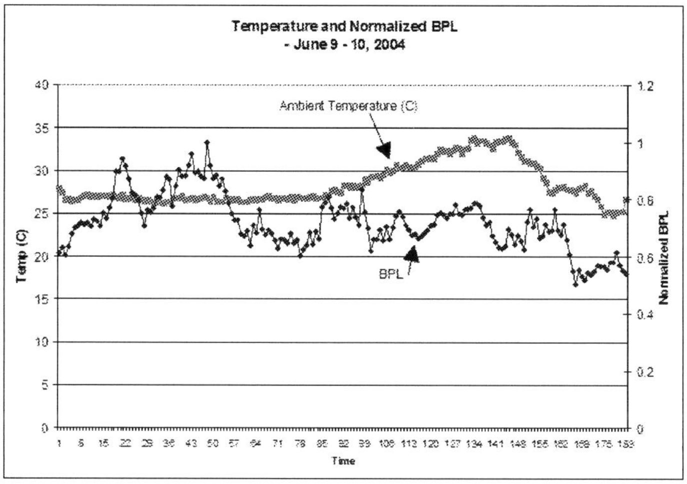

Figure 6 Trend of normalized BPL and ambient temperature

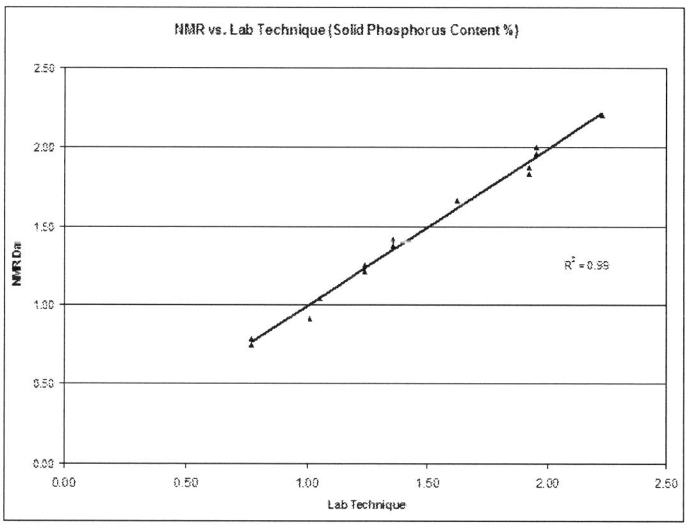

Figure 7 Calibration example of laboratory data vs. NMR

to nominal resonance frequency of 8.096 MHz for phosphorus.

NMR vs. OTHER ANALYTICAL TECHNIQUES

While NMR is not considered to be as sensitive an analytical technique as perhaps UV-VIS, NMR has the unique ability to look at a very large sample as illustrated in Figure 3 without any sample preparation. It is also a nondestructive technique as well.

PRACTICAL ANALYZER AND PLANT ISSUES

As is the case with most process analyzers, care must be taken with both sample presentation, or

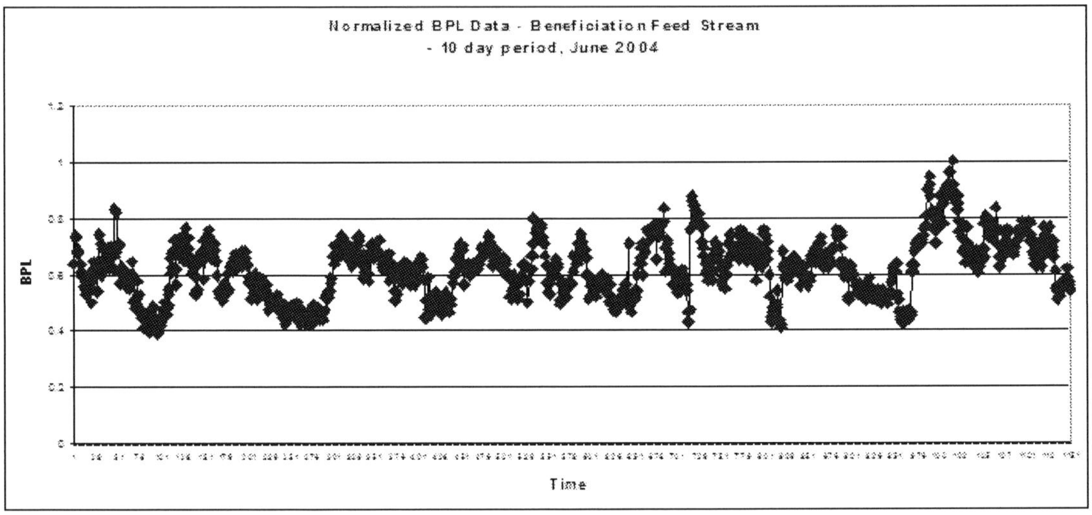

Figure 8 Normalized BPL data on a float plant feed stream

sampling, and the actual placement of the analyzer and attendant plant utility connections.

Sampling

Progression typically provides sampling systems for each of our process NMR installations—in the beneficiation process these can be from conveyor belts or turbulent or laminar flow streams. With an installed base of over 100 process NMR systems, designs are available for most applications. Figure 4 illustrates a sampling system from a turbulent flow line. It is designed to exclude "trash" from the sample line to minimize plant downtime due to plugging and to allow for easy, regular maintenance. The NMR analyzer itself is typically placed vertically below the sample point, but within 25 meters of the sampling point itself.

Analyzer

The analyzer itself is contained in a NEMA 4 rated air-conditioned cabinet. Within the cabinet are the magnet, NMR spectrometer, NMR probe, and all associated logic and control equipment. A cutaway of the NMR system is detailed in Figure 5.

Advances over the last 15 years have led to reduction in size of process NMR instruments of nearly 95%. Today, the MagModule II, offered by Progression, Inc., is only 1.5 m^3 and with proper maintenance is capable of uptimes greater than 97%. In addition due to the smaller size and rugged construction, these analyzers are available for placement in a wide variety of plant locations.

As we have previously reviewed, the NMR signal is affected by temperature. For that reason, compensating measures are built-in to the analyzer itself to neutralize the temperature effect. Figure 6 illustrates trends of BPL and ambient temperature during a 24-hr period in June 2004 for a location in central Florida.

Calibration

An NMR analyzer is typically calibrated against existing laboratory techniques over an analytical range of interest for a specified process stream. Figure 7 illustrates how this works in practice. Total error or standard deviation of the model can be explained by the following equation:

$$ModelSTD = \sqrt{NMRstd^2 + LABstd^2} \qquad (3)$$

Analytical Data

Once installed, process NMR data can be utilized to better control the beneficiation process for increased recovery. Figure 8 illustrates the data coming from an on-line NMR analyzer over a 10-day period in June 2004 on a feed stream to a float plant. All data has been normalized.

CONCLUSION

Process NMR should be strongly considered as an on-line analytical technique for beneficiation plants to measure phosphorus content in a variety of streams. It would also be a useful technique for phosphate chemical plants as well (acid streams, reactor feed, etc). Calibrations are simple to perform and need little adjustment, the uptime of the instruments as been shown to be excellent, and the data is extremely useful for increasing efficiencies.

NMR has several advantages to other analytical techniques when considering on-line phosphorus measurements: no sample preparation is required, a more representative sample can be looked at, it is nondestructive, the sample can be returned to the process if desired, there are no optical windows to foul, and uptime is excellent.

CHAPTER 29

New On-Stream Slurry Particle Sizer

Matti Kongas* and Kari Saloheimo*

ABSTRACT

A new type of on-line particle size measurement instrument has been installed at the apatite concentrator of Kemira Growhow at Siilinjärvi, Finland. It is based on laser scatter technology, commonly used in laboratories with manual sampling and sample dilution. Due to the fundamental nature of the analysis method, no calibration against a reference analysis is required. The analyzer automatically takes samples from one to three process streams and measures their particle size distribution in the range of 1–500 μm. An innovative automatic sample dilution system combined with multiple scatter analysis calculation techniques has contributed to reliable operation on a continuous 24-hour-a-day basis, with minimal manual maintenance needs. The PSI 500 particle size analyzer is used in controlling wet mineral processes, grinding, classification, and thickening. The paper describes the installation and results from the analyzer in operation at the phosphate concentrator. The first experiences on the performance of the analyzer are encouraging, as shown by comparisons to laboratory sieve analysis and responses to control actions.

*Outokumpu Technology Minerals, Espoo, Finland.

INTRODUCTION

On-line Particle Size Analysis—General Requirements

The particle size distribution of the feed to a flotation process has an influence on the flotation recovery through limited liberation of valuable minerals and reduced flotation rates of both too coarse and too fine particles. In addition, the particle size distribution has an effect to the flotation reagent consumption and to the dewatering of the slurry. That is why the control of the preceding grinding process usually includes constraints or target values set to the particle size distribution of the grinding product. In most cases the primary control variable for the particle size distribution is the fresh ore feed rate to the grinding circuit, and there is a tradeoff between the tonnage throughput and the fineness of the grinding product. With variable grindability of the ore, this gives a possibility for optimization—the softer the ore, the more tons can be pushed through the mill to achieve adequate liberation. The practical implementation of such control is often operator based, but there is a growing number of closed-loop expert system applications doing the job for the operators. In both cases an online particle size instrument is needed to give a feedback signal to the optimizing control.

As the purpose is process control, there is not an absolute specification on particle size distribution of the grinding product, and the needs for the measurement are quite different from those of a laboratory analysis. The general requirements for the online particle size measurement can be summarized as follows:

Reliability. This is the foremost requirement of online analysis from the operator's point of view. If false control decisions are done based on incorrect analysis data or if the analyzer needs continuous maintenance to keep working, the operators soon lose their confidence on the system altogether. Reliability is closely tied to equipment availability; the whole system needs to be designed for 24-hour, 7-day operation in the process environment.

Representativeness. The measurement results need to reflect the changes of the overall particle size distribution in the bulk process flow. Thus careful design is required in the sampling of the slurry and the presentation of the sample to the measurement technique to get a reliable measurement of especially the coarser (>100 m) fractions.

Precision. The analysis should be sensitive enough to indicate the process changes that are relevant for the purposes of control.

Speed.The analyzer needs to deliver new results frequently enough that the process control can react to the process changes. The typical average residence time of the material in the whole grinding circuit is about 10 minutes, and the analysis delay and cycle time should be less than or about half of that for effective control.

Ease of use. With the increasing demand of effectiveness in the industry, easy operation and maintenance are among major requirements. The personnel specialized in other tasks than the production is more and more rare at the plant, and there is not much room for extra maintenance items.

One fact of particle size analysis in general is that each technique describes a three-dimensional particle differently. The absolute result might be the same for perfect, homogeneous glass beads but the effect of particle shape, specific gravity variation, or other factors on the size measurement is dependent on the method used. In on-line analysis, absolute accuracy is not so critical, or the results can be correlated to a laboratory standard method. It is more the precision of the analysis that counts—the method must give a consistent response to process changes, and the repeatability should be in proportion to the needs of the control.

PSI 500 Particle Size Analyzer

To cover the process control needs that require the full particle size distribution information over the whole size range, Outokumpu recently developed a new analyzer, PSI 500, in collaboration with Malvern Instruments, England. The new analyzer is based on laser diffraction, as illustrated in Figure 1. Coherent light scattered from particles forms ring patterns, the width of which is dependent on the size of the particles. In a practical arrangement a representative sample of slurry is passed through a parallel laser beam, and the intensity distribution of the scattered light is measured using an optical detector. As the form of the scattering pattern is well known by theory presented by Mie (Jones, 2003), the size distribution can be calculated back from the measured intensity distribution.

Laser diffraction provides a consistent volumetric particle size analysis without any external calibration and with good precision over a wide size range. Typical relative precision of the distribution median over 1–500 μm size is 1–2% in the PSI 500 on-line analyzer. Also the speed, noncontact nature, and robustness of the method to ambient conditions make laser diffraction quite suitable for on-line mineral slurry applications. A large number of on-line laser diffraction systems have been installed in dry processes (Crawley, 2001).

Mineral process slurries are quite dense and need dilution to get any scattered light through the sample. Using multiple scattering calculation techniques (Crawley, 2001), only 10% of the laser light needs to pass the sample directly in order to get a reliable analysis. This condition is

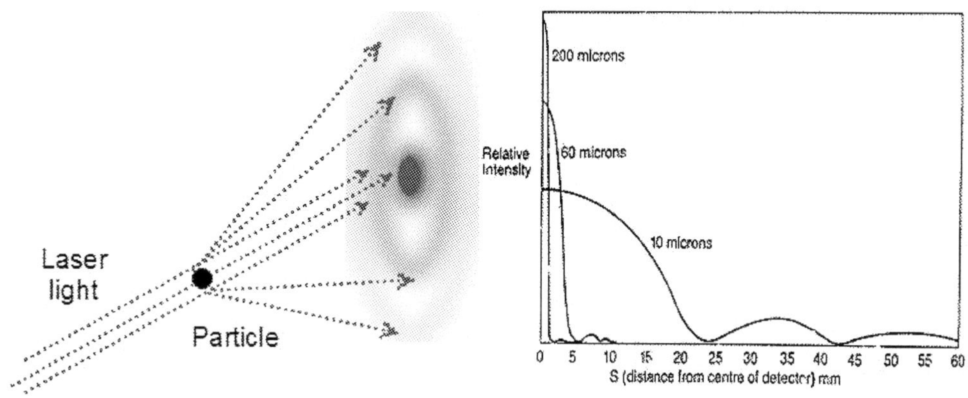

Figure 1 Principle of laser diffraction

met by diluting the solids content down to 0.1–1% by volume. The developed sample diluter does this by mixing sample cuts into a continuous flow of water. When the diluted slurry passes the flow cell, it is in well-developed turbulence to ensure good mixing and representation.

Figure 2 shows a schematic of the analysis system. The primary sampler in the process pipeline cuts a sample with a flow of 50–170 L/min in one or two steps. Static samplers are used, taking advantage of the mixing caused by process pumps and/or pipe geometry changes and isokinetic flow conditions (Braden et al., 2002). As most availability problems are related to primary sampling and sample transport, it is very important that the samplers and transport lines are engineered with reliable continuous operation in mind. Sample reduction by passing the flow across a cutter provides a representative sample to the diluter.

The system is controlled by a soft-plc application residing in an industrial PC. The touch screen display provides a local user interface, and the measurement and status data can be retrieved to the plant automation system through Ethernet or serial ports by various protocols.

PSI 500 can take up to three primary sample streams from various locations of the process. The measurement time for a fresh analysis is 2.5–3 minutes per sample.

INSTALLATION AT SIILINJÄRVI CONCENTRATOR

Siilinjärvi Phosphate Mine

Kemira GrowHow is the second largest producer of fertilizers, animal feed phosphates, and phosphoric acid in Western Europe. In 2003, the company had net sales of 1.13 billion Euros and 3,100 employees working at eleven sites in seven European countries. The phosphoric acid production capacity is 280 kt/a of P_2O_5. The phosphate mine and phosphoric acid plant are located in Siilinjärvi in eastern Finland (see Figure 3).

The capacity of the Siilinjärvi mining operation is around 10 Mt/a of ore. The main products are 0.8 Mt/a apatite concentrate, 0.1 Mt/a calcite concentrate, and 10 kt/a mica, as shown in Figure 4. The apatite concentrate production is used totally in Siilinjärvi as a raw material in the phosphoric acid and fertilizer production. The main minerals of the igneous carbonatite deposit are 10% apatite, 20% carbonate minerals (calcite + dolomite), and 65% phologopite mica. The carbonates/mica ratio varies throughout the deposit; the average is about 0.3 but it can range between 0.15 and 0.7. This ratio has a major influence to the grindability of the ore and the capacity of the concentrator.

The concentrator flowsheet, shown in Figure 5, consists of a three-stage crushing section, homogenization of the ore by a stacker–reclaimer system, two parallel rod mill–ball

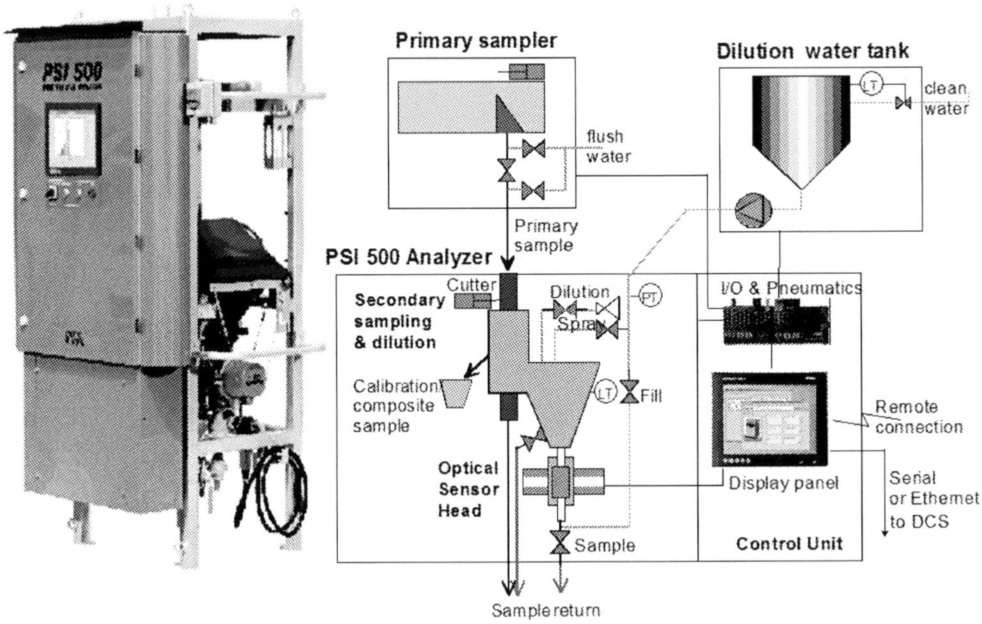

Figure 2 The PSI 500 analyzer system

Figure 3 Siilinjärvi Mine; open pit (middle) and concentrator plant (middle left)

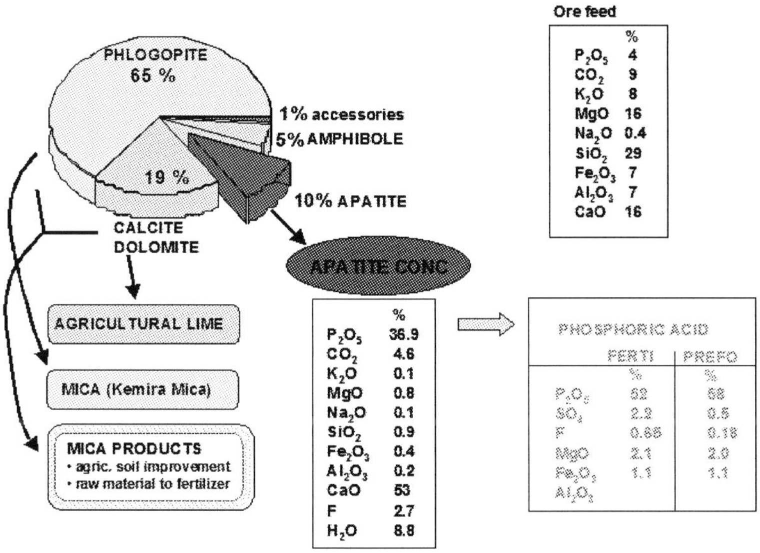

Figure 4 The minerals in Siilinjärvi Mine production

mill grinding sections, and an apatite flotation circuit. The original design capacity of the homogenization plant was 3 Mt/a of ore. As compared to the situation today, it is used to produce more stable feed to the grinding section, with a 0.18% carbonate content and a carbonate/mica ratio of 0.3. The most challenging task in the beneficiation is the separation of apatite and carbonate minerals in the flotation stage.

The apatite concentrate produced in Siilinjärvi is an excellent raw material for phosphoric acid production. The chemical composition is typically 36.5% P_2O_5, 0.8% MgO, 0.4% Fe_2O_3, 0.2% Al_2O_3, 0.9% SiO_2, and <0.3 ppm Cd. The gypsum from the phosphoric acid plant is very white, and it is used as a high quality raw material for paper coating pigment production at the site.

Motivation for On-line Particle Size Analysis and Control

According to current plans, the apatite production will be increased to at least 0.9 Mt/a in the next two years with minimum investments. In the first phase the focus is in improving the apatite recovery from the existing level of 83%.

The apatite recovery losses are today mainly in the coarse fraction—two thirds of the losses lay in the +210 μm fraction and one third in –40 μm fines. The apatite recovery will be increased especially by reducing the amount of oversize apatite in the flotation feed without increasing the amount of fines excessively. On the other hand there seems to be possibilities to improve the coarse apatite recovery to a certain extent by modifying the flotation units and configuration.

The particle size distribution of the flotation feed is 25% –74 μm as an average, and at the same time roughly 30% of apatite is in oversize fractions. The particle size distribution of apatite and carbonates is much finer than that of phologopite. The energy consumption in phologopite grinding is high and on the other hand flaky phologopite particles will deteriorate the sharpness of the separation in hydrocyclones.

The recovery of apatite has been above 83% even if the fineness of the feed has varied between 22 to 30% –74 μm, depending on the ore type and conditions in the flotation stage. With the history described above, it has been difficult to specify for each ore batch the optimal

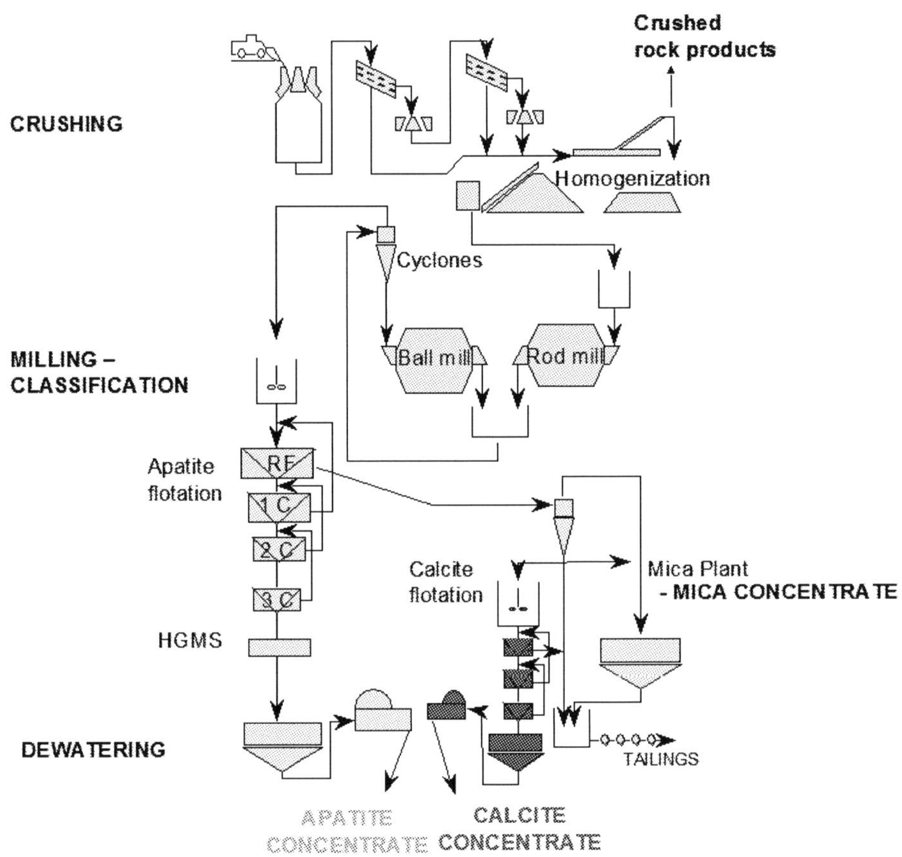

Figure 5 Flowsheet of the concentrator plant

particle size distribution and feed level that would provide the best economical and metallurgical results. This is especially challenging in conditions where the quality of ore feed may change significantly every second day without possibilities to characterize the new ore adequately before it enters the grinding section.

The targets for the test work with the PSI 500 particle size analyzer were set as follows:

- Get reliable and real-time information about the changes in particle size distribution especially when new ore batch enters the grinding process; there are no means to characterize the ore and its grindability in advance only by chemical analysis, for example:
  - Analyze and control the particle size distribution of both grinding circuits separately especially because of the different size of the grinding lines and differences in the particle size distributions of the feed to the rod mills.
  - Analyze the whole distribution, not only the percent passing 74 μm screen but also for example the amount of solids between 20 and 250 μm, or d95 and d50 sizes. With this information it should be easier than today, together with the response from the flotation circuit, to adjust the feed level fast so that maximum amount of apatite would be in the size range from 20 μm to 250 μm.

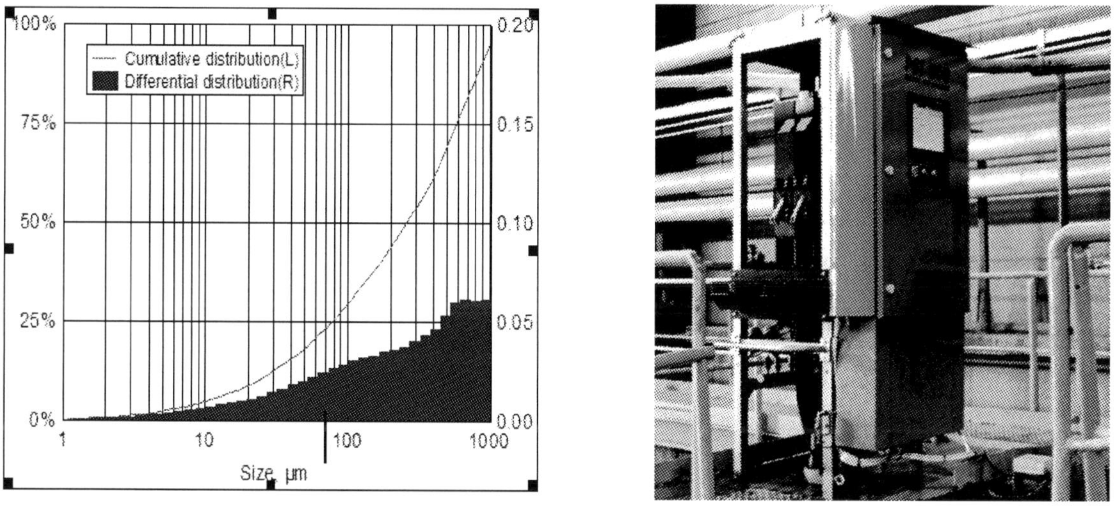

Figure 6 PSI 500 installation and a typical particle size distribution

- Fine-tune the grinding process by optimizing for example the pulp densities in the mills and the circulating loads by changing the pulp density in cyclone feed and number of cyclones in use.

On-Stream Analyzer Installation

The PSI 500 analyzer was installed at Siilinjärvi in the grinding circuit in March 2004. Two launder box samplers were installed in the cyclone overflows of each grinding circuit. Figure 6 shows the analyzer located on a service platform about 5 meters below the cyclone battery. The primary sample lines flow by gravity, and the sample return flows to a conditioner tank directly below the analyzer. The samplers are automatically backflushed with water every 4 hours to make sure that the cutters keep open. At the same time, the analyzer makes a background check in the measurement system by running only dilution water through the flow cell. The analysis cycle with the two sample lines is 5.5 minutes.

The plant automation system (Metso Damatic XD/XIS) reads two fractions of each sample line: % –74 µm and % +210 µm. The Modbus serial link also carries status information and remote commands for the control room operators to switch the samplers and the analyzer to standby at shutdowns.

Verification Tests Results

The large and flaky phlogopite mica particles make a special challenge to the measurement. The largest flakes are several millimeters in size, beyond the range of the analysis. On the other hand, the behavior of the flakes on the laboratory sieves is probably quite different from how laser diffraction sees the randomly oriented particles in the sample flow. Figure 6 shows a typical particle size distribution measured from the samples.

To compare the measurement techniques, a sampling campaign was carried out. Forty-two samples measured by PSI 500 were brought to the laboratory to be analyzed by sieves. Another comparison was done between the average reading of PSI 500 on both lines and daily samples taken from the combined feed. The averaging was weighed by the process flows and solids contents. In the first phase especially % passing –74 µm and 210 µm screens were studied. The results of the comparisons are shown in Figures 7 and 8. There was a significant improvement in the accuracy in the beginning of June, after

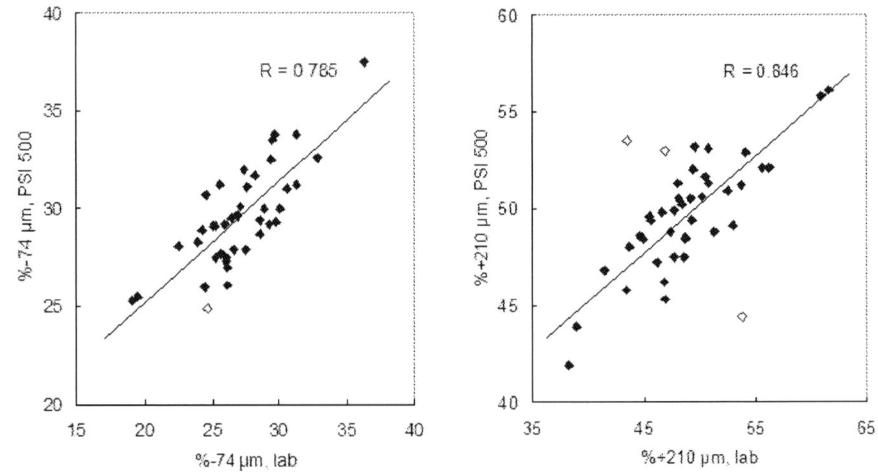

Figure 7 Comparison of sieve and PSI 500 results of spot samples, March–August, 2004

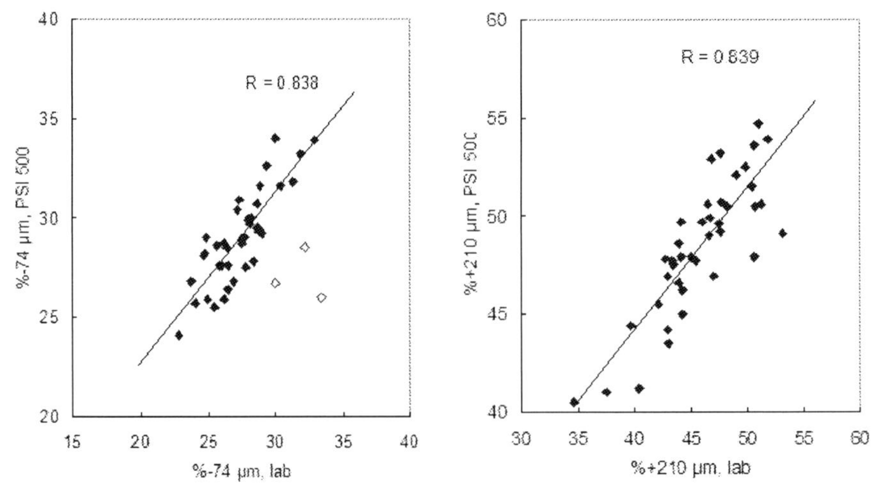

Figure 8 Comparison of sieve and PSI 500 results of daily samples, June–August 2004

updating the software by Outokumpu, to respond correctly to sample behavior specific to this application. Apart from a few outliers, the correlation is amazingly good when one takes the challenges of the analysis into account.

The precision of the analysis was estimated by recording the analysis results over a 30 min period of stable process conditions. Standard deviations of successive analyses are given in Table 1. Since all analyses are independent and no filtering is performed on the signals, the numbers present the overall short-term random error component from sampling to the actual analysis.

To see how the analyzer tracks process changes, a 7% step increase was made to the first grinding section, while keeping the second one constant. The on-line measurements were recorded, and spot samples were taken from each of the lines for laboratory analysis both before the feed increase and after the response had settled. The results are shown in Figure 9. In grinding section 2 there was only a slight change, whereas in grinding section 1 there was a noticeable shift towards coarse sizes, indicated by both techniques.

The most striking changes in the particle size distribution occur when the ore in the homogenization stack changes. Figure 10 shows one of these instances. By the change, the grinding product gets finer, and the operator has increased the ore feed to the mills. The new ore allowed 15% more tons into the grinding circuit and it still produced a finer distribution. The particle size distributions before and after the ore change are also of different shapes.

Table 1 Statistics of successive sample measurements (over 30 minutes)

Sample line	Analysis	Average, %	Std. Deviation, %(abs)
J1	%+210 µm	57.1	0.86
	%-74 µm	33.0	0.63
J2	%+210 µm	53.5	0.81
	%-74 µm	30.3	0.55

CONCLUSIONS

Without any calibration, PSI 500 measurements were in good correlation with the routine screen analysis made by standard methods in the laboratory. It was concluded that there is no need for external calibration or modify the results of the analyzer by any means.

The mechanical reliability of the sampling system and the analyzer has been very good. There has been no insuperable problems with the coarse and fast settling mica particles as was the experience with earlier on-stream particle size analyzers at the site.

The real exploitation of the new particle size information is still in progress but already now it is obvious that the information from PSI 500 is useful together with the other measured information of the grinding circuit in monitoring the performance of the grinding process and for example in observing the status of hydrocyclones.

As for future development, the particle size information including % –74 µm, % –210 µm, d50, d95, and the fraction between 20 µm to 250 µm will be transferred to the automation system.

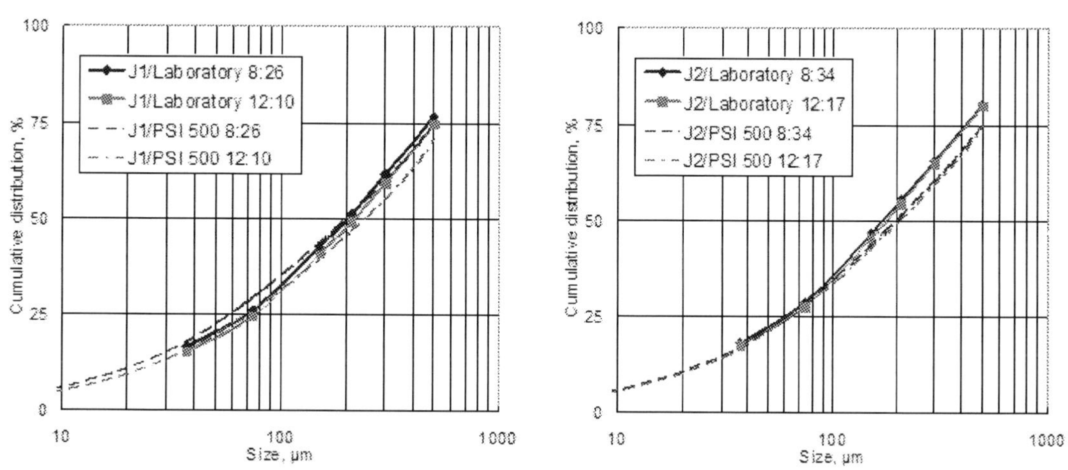

Figure 9 The particle size distributions of grinding section J1 and J2 before and after the feed increase in J1, measured both in the laboratory (solid lines) and by PSI 500 (dashed lines)

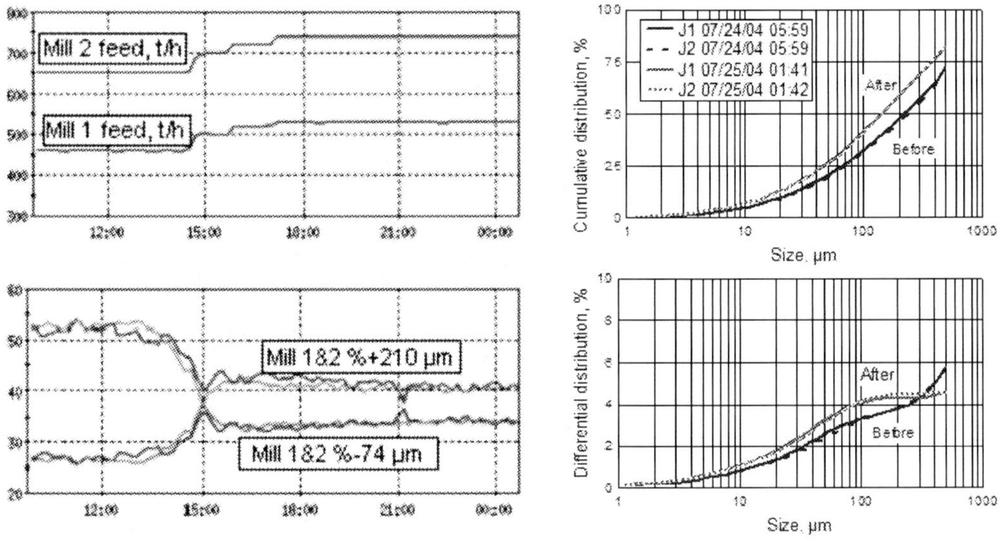

Figure 10 Ore change in July 24; process trends on the plant automation system and particle size distributions

The maintenance procedures, including cleaning of the sampling system, will be automated as much as possible so that the manual operation and the possibilities for human error could be minimized.

The fine-tuning of the grinding process will be started in the beginning of 2005 by checking the optimum pulp densities and circulating loads resulting to the best capacity with an acceptable particle size distribution.

Reliable and fast information about the apatite losses together with the information of the particle size distribution would be very useful in controlling the feed rate to grinding section. Together with the introduction of PSI 500 particle size analyzer, there are ideas to develop and test a new on-line P_2O_5-analyzer which could analyze the P_2O_5 content of the final tailings and the feed in the 0.3–5% P_2O_5 range. The P_2O_5 content of the final concentrate can be measured satisfactorily today by Outokumpu's Courier 40 diffraction analyzer.

ACKNOWLEDGMENTS

The authors gratefully acknowledge the valuable contributions made to this paper by Jarmo Aaltonen and Jari Turunen of Kemira GrowHow, Helsinki, Finland.

REFERENCES

Braden, T., Kongas, M. and Saloheimo, K., 2002. On-line Composition Analysis for Mineral Slurries. In: *Mineral Processing Plant Design, Practice and Control*. Mular, A.I., Halbe, D.N., Barratt, D.J. (editors). SME, Littleton, Colo., pp. 2020–2047.

Crawley, G.M., 2001. Particle Sizing Online, *Powder Metallurgy* 44, No. 4, pp. 304–306.

Jones, R.M., 2003. Particle Size Analysis By Laser Diffraction: ISO 13320, Standard Operating Procedures, and Mie Theory, *American Laboratory*, January 2003, pp. 44–47.

CHAPTER 30

Beneficiation Automatic Control System: The Next Generation Status Report

R. Stana*

ABSTRACT

Up until a few years ago, control systems in a beneficiation plant consisted of a few control loops. These control systems added the reagents based on simple feedback like pH or were added in proportion to a feed rate or flow. Recently, some companies have tried to make further improvements in the controls by adding on-line instruments that looked at the phosphate in various streams. While these have had some limited success, their main benefit has been the collection of data that could lead to a significantly better control system. The development and status of such a control system is described.

INTRODUCTION

The Florida phosphate industry is the largest mining industry in Florida, typically mining over 30 million tons of phosphate rock. About half of this rock is recovered from the phosphate matrix simply by washing the rock to remove the clay fraction and then screening the rock to remove all particles greater than 16 mesh. The finer phosphate is separated from the fine sand particles by flotation processes. Typically soda ash is added to the phosphate slurry to adjust the pH and then a fatty acid and fuel oil are added to selectively float the phosphate. This typically gives a phosphate concentrate that is 40–60 BPL. This is further concentrated to typically 65 plus BPL by first acid stripping the fatty acid and fuel oil from the phosphate and then adding an amine to float most of the remaining sand from the mix.

The operation of the floatation processes requires continual adjustment of the various reagent streams as the rock flow rate, quantity of sand in the mixture, and the type of rock changes. The steps requiring pH adjustment typically are controlled by feedback from a pH probe installed in the system. The fatty acid, fuel oil, and amine are controlled by the operator, making adjustments as he makes observations on the quality of the various concentrate and tail streams. In some cases he is aided by off-line analyses of the streams. There are also some systems that vary the reagents in some fashion with the feed rate of rock. However, since the operator cannot continuously look at all streams in the process and make adjustments to the reagents, most of the time either a surplus of reagent is added or not enough reagent is added, giving a less than optimum recovery of phosphate rock.

*R. Squared S Inc., Lakeland, Florida

Up until about 10 years ago, most plant data was collected by the use of paper log sheets. The operator would note the various flows, pHs, densities, amps, pressure, etc., of the system every hour. However this data does not really characterize the operation in that the feed rate (flow rate, % solids in the feed) is constantly changing and the operator may make several changes in reagent addition between log sheet entries. Fortunately, in the last 10 years, a number of companies have added computerized data collection systems to the process. These systems continuously collect the data and store it in a manner that makes it easy to analyze.

Using this data, it should be possible to determine how best to add the reagents to economically optimize the recovery of the phosphate rock. Presumably, a model could be developed that will control the various reagents dependent on the changes in key input variables. The model might also be somewhat intelligent in that if it receives feedback that the expected phosphate recovery is not achieved (because rock floatability characteristics have changed or reagent strength has changed, etc.), it would change the operating conditions in order to improve performance.

Clearly, some sort of online phosphate analysis would make the control system somewhat easier to develop. However, the instruments developed to date either take too long to give a good analysis, or are not accurate enough, or require extensive maintenance. Thus, it would be good if a control system could be developed that did not require any on-stream analysis.

The erosive and rapid settling nature of the rock slurry tends to cause errors or drifts in most flow measuring or flow density devices with time. Any use of this information by a control system can't rely on the absolute values, but must adapt to the changing values.

While these are all significant obstacles to the development of a control system, the benefits can also be very significant. One measure of potential improvement is to compare the average performance with the average of the top 25% performing shifts. To do this, data from 4,000 shifts of operation at a small beneficiation plant were obtained. The data were first sorted on feed BPL and then broken into 10 groups of 400 shifts each. Then each group was sorted on the BPL recovery. The average recovery for the highest 100 shifts was then compared with the average recovery for all 400 shifts for each BPL range. Figure 1 shows a plot of the recovery vs feed BPL for the fine rougher circuit.

The average difference in recovery between the top 25% and the average performance is 8.0%. More importantly, the recovery for the top 25% is much flatter (not as large a drop with lower feed BPL). Could the improved performance be due to increased or significantly different reagent consumption? Figure 2 shows the fatty acid consumption vs the BPL for the same average and top 25% shift performances. As can be seen, there is no significant difference in fatty acid use between top 25% performances and the average performance.

Figure 3 is a similar plot for the fuel oil consumption. Again, there is no significant difference in fuel oil use between top 25% performances and the average performance.

Figures 4, 5, and 6 are similar plots for the coarse rougher section of the plant. Figure 4 is very similar to Figure 1 except that the average difference is 6.0%. Figures 5 and 6 show slightly more reagent use for the top 25% compared to the average performance.

Figures 7 and 8 are similar for the amine (cleaner) section of the plant. The average difference in Figure 7 is 3.4%. If the plant could just operate at the same performance as the top 25% of the shifts, concentrate production would increase by 12.6%. At a rock production cost of $30/ton, this represents a cost savings of almost $4.00/ton.

While the objectives of this program may seem aggressive, they actually were achieved to a degree in the phosphate industry in a control system developed by IMC for the wet process phosphoric acid reactors. It is described in US patents 5,262,963, 5,395,603, and 5,188,812. The systems have been installed in several reactors and have been in operation since 1991. The same program has been operating for 14 years with no changes in spite of numerous process

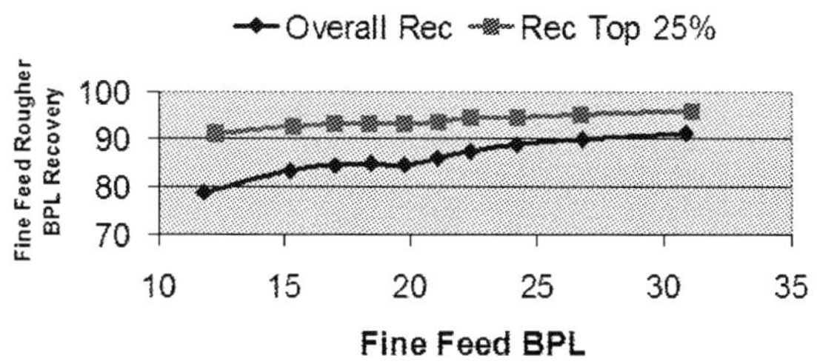

Figure 1 Fine rougher recovery vs. fine feed BPL

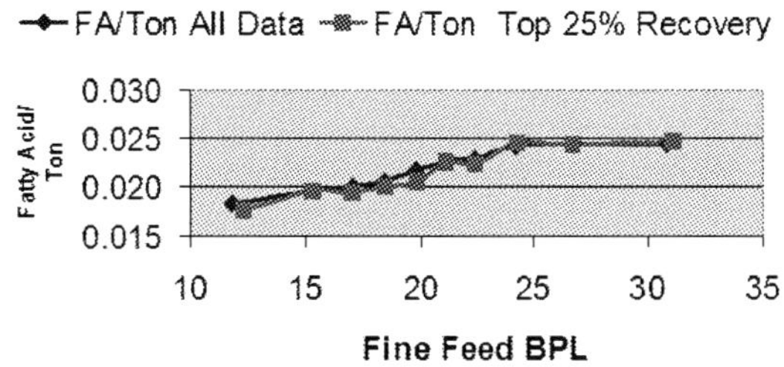

Figure 2 Fatty acid/ton vs. fine feed BPL

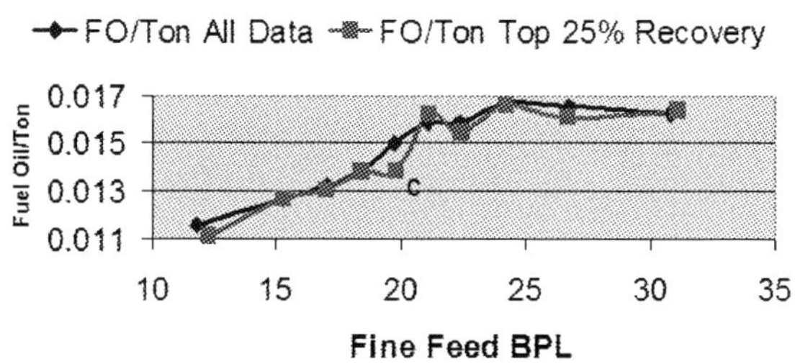

Figure 3 Fuel oil/ton vs. fine feed BPL

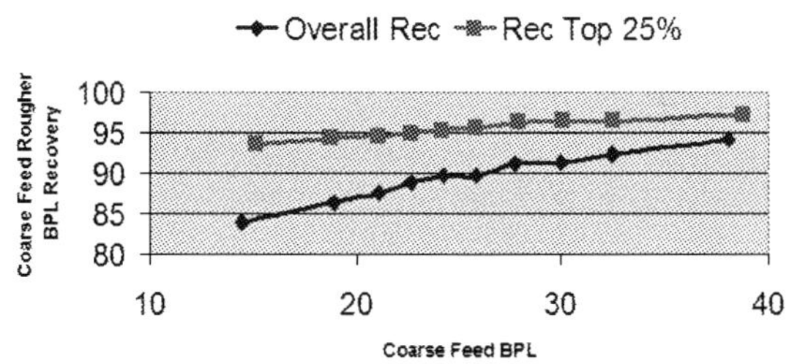

Figure 4 Coarse feed rougher recovery vs. feed BPL

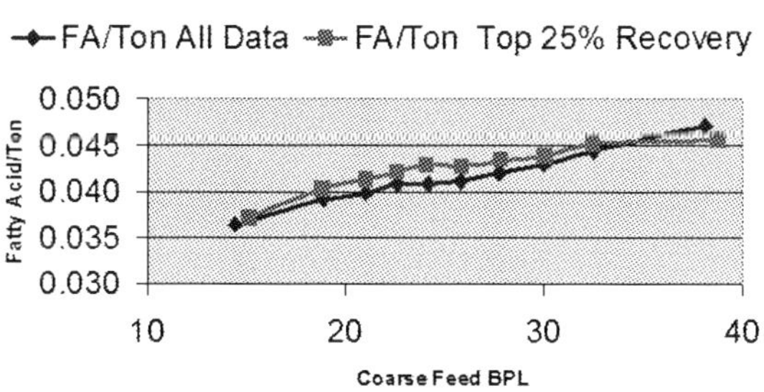

Figure 5 FA/ton vs. coarse feed BPL

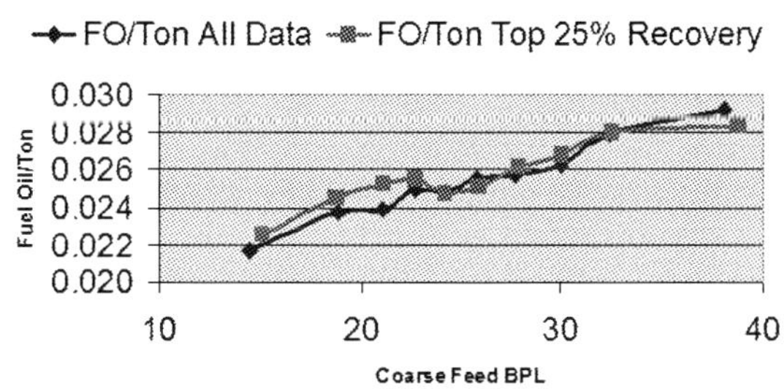

Figure 6 FO/ton vs. coarse feed BPL

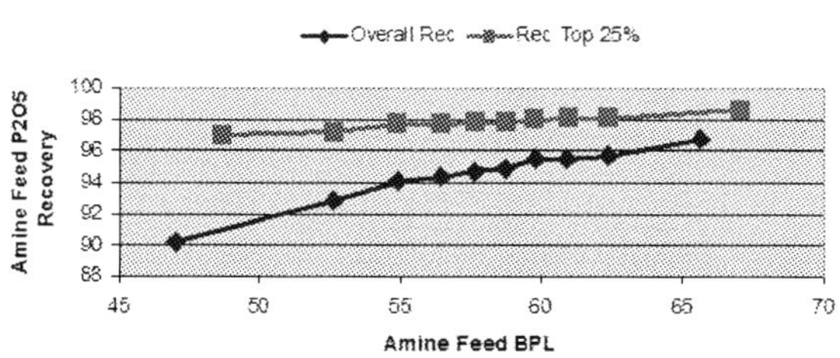

Figure 7 Amine P_2O_5 recovery vs. amine feed BPL

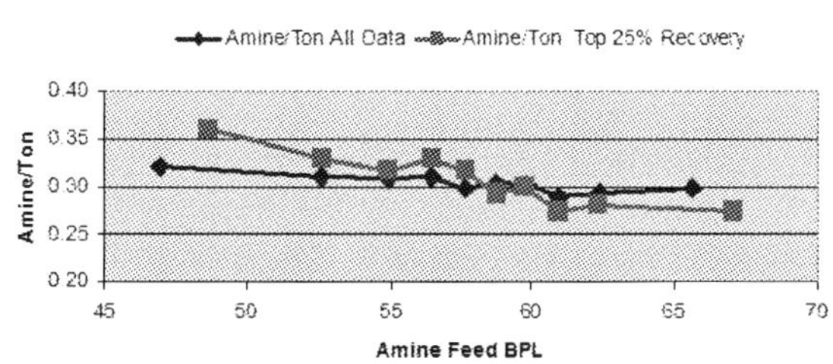

Figure 8 Amine/ton vs. amine feed BPL

changes, instrument changes, and recalibrations. This was achieved because the program is adaptive. It is constantly evaluating how it is performing and will change itself when it detects any change in performance. It is this same type of program that is now being developed for the beneficiation process.

To develop the control system, it is first necessary to validate, and in many cases correct, all the data used by the program. An example is given in Figure 9. A flow meter and density meter calculates the final concentrate production for this plant continuously. The final concentrate is also measured each shift by the operators by measuring the volume of rock in the product bins. Since the operators do not make their measurements at exactly the same time each shift, the ratio of the two would be expected to vary somewhat from shift to shift. However, these variations should be minimal when the ratio is compared for a week's worth of production. Figure 9 shows both the ratio of the daily and weekly tons measured by the process to the tons measured by the operators over a period of two years.

As can be seen, the ratio varies from a low of 0.6 to a high of 1.7 and an average deviation from one of 0.23. This means that at the same measured flow of concentrate and the same reagent rate, the actual dose as pounds per ton could vary as much as a factor of three. Clearly, any control program must know the correct flow rate if it is to work. It is not important to understand why the changes occur. It is only important that the data be corrected continuously in a feed forward manner so that the rates measured by the process match the actual rates. Fortunately, this can be achieved by some logic applied to the data giving the corrected data shown in Figure 10.

Here, the maximum 7-day ratio varies from a low of 0.87 to a high of 1.11 with an average deviation from one of 0.029. Similar corrections can be applied to all the flow rate data so that it is possible to compare performance of the system over along periods of time, knowing that the reported rates are reasonably accurate.

The data must also be filtered to eliminate "absurd data." Typically when a computer collection system or the instrumentation is started or stopped, the data is corrupted and it's not unlikely that absurd data such as a pH of 26 or a specific gravity of 926.124 is seen. All these values must be carefully filtered from the inputs to the control system.

The control system can be best developed from the actual plant data. The data analysis technique that generally works best where there is a lot of plant data is "clumping." This technique uses clumps of data that are relatively similar in operating conditions but is collected over a long period of time. An example follows.

Several thousand shifts of corrected operating data is first sorted by feed rate. It is then broken into 10 equal subsets. Each of these subsets is then sorted by BPL recovery. The top 25% of the recoveries are then sorted on feed BPL. Clumps of this data (generally 25–50 sets of data) are picked from each subset that have the same feed BPL. The data is then plotted with the reagent dose vs the feed rate. Plots made this way for the fatty acid and fuel oil consumption per ton for the top 25% highest performing shifts are shown in Figures 11 and 12. For both Figures 11 and 12, each point represents 50 shifts of data with similar operating conditions. Figures 13 and 14 are similar except the feed BPL is higher.

When clumping is used as shown, it is not as sensitive to outlying data as regression analysis, which uses the square of the difference to develop the correlation. The plots of the clumps of data make it easy to see any nonlinear effects. The same techniques can be used to look at the effect of other variables, such as sorting on feed BPL first and then matching on feed rate to determine if there is an effect of feed BPL.

After the effects of all the significant variables have been determined, the clumps of data can then be used to determine the correlation for the fatty acid and fuel oil dosage that will give the maximum recovery under any set of conditions. Figure 15 shows the correlation for the fatty acid dose rate versus the actual fatty acid dose to give the maximum recovery.

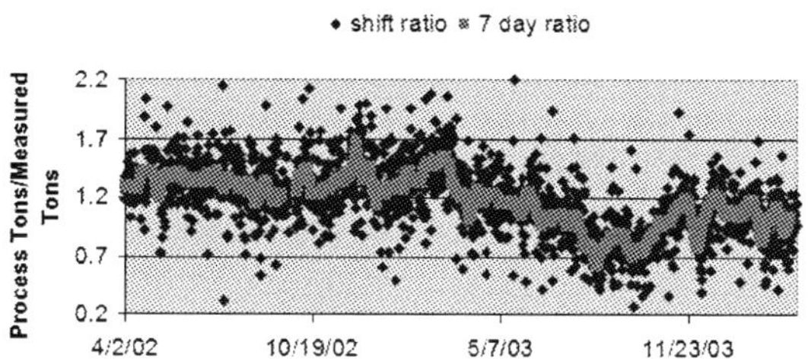

Figure 9 Process tons/measured tons vs. time

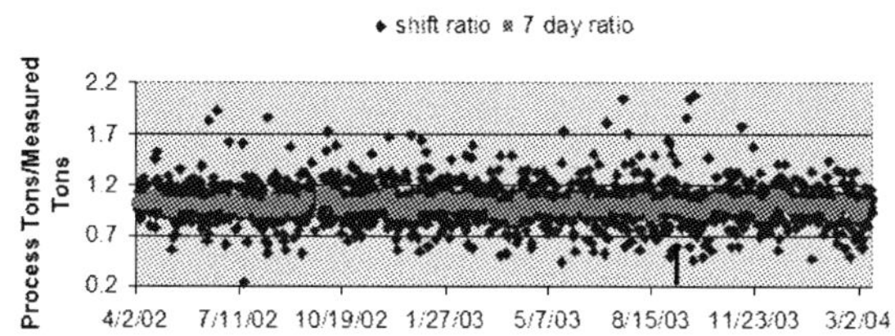

Figure 10 Corrected process tons/measured tons vs. time

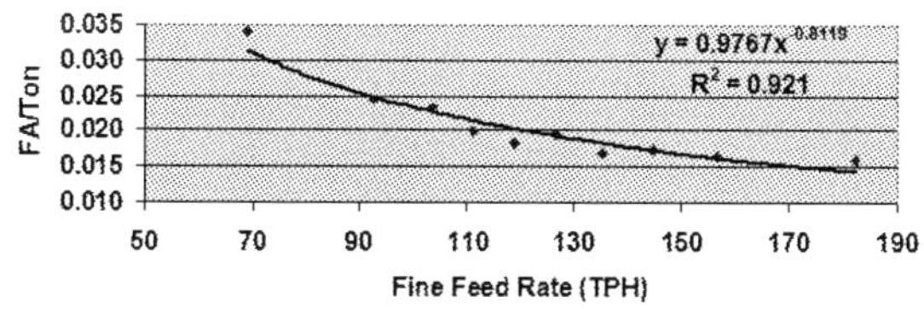

Figure 11 Fatty acid/corrected TPH vs. fine feed rate for maximum recovery at 20.5 BPL

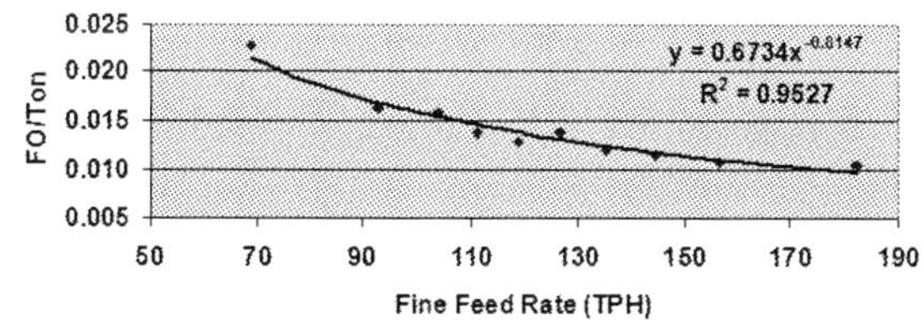

Figure 12 Fuel oil/corrected TPH vs. fine feed rate for maximum recovery at 20.5 BPL

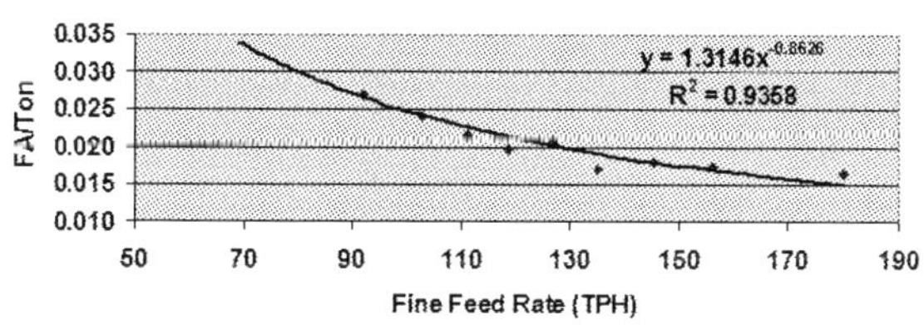

Figure 13 Fatty acid/corrected TPH vs. fine feed rate for maximum recovery at 23.4 BPL

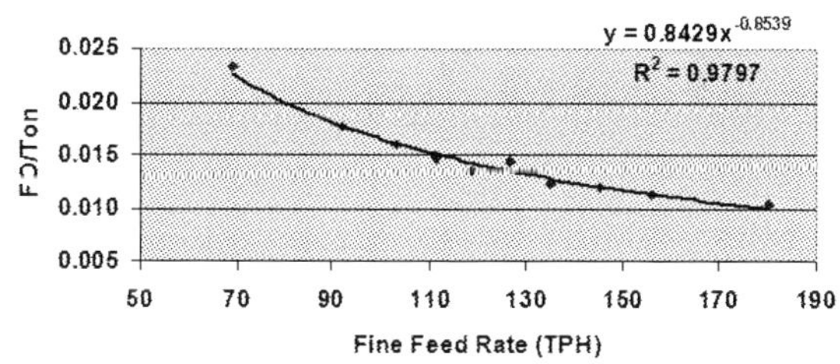

Figure 14 Fuel oil/corrected TPH vs. fine feed rate for maximum recovery at 23.4 BPL

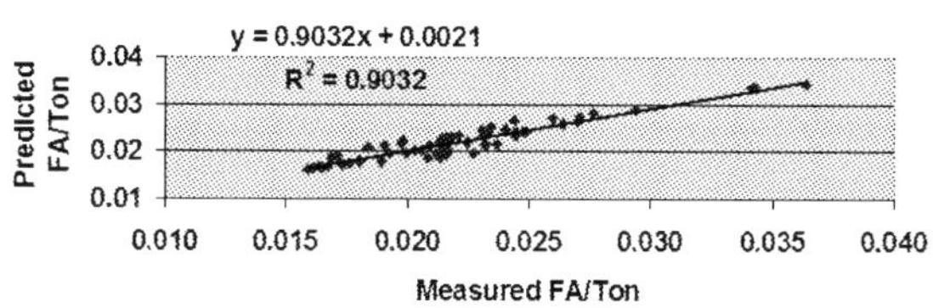

Figure 15 Predicted FA/ton vs. actual FA/ton at maximum recovery for fine feed roughers

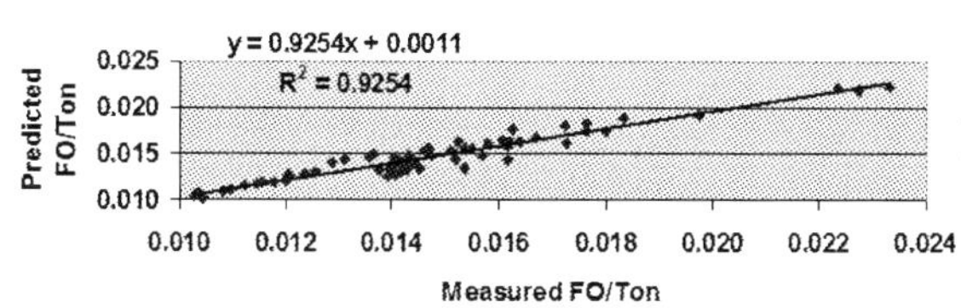

Figure 16 Predicted FO/ton vs. actual FO/ton at maximum recovery for fine feed roughers

Figure 16 is a similar relationship for the fuel oil dose. Again, each point represents a clump of similar data. While these regressions give R squared values above 0.9, regressions using the individual points give similar (but not the same due to the stronger effects of some outlier data) results with R squared values in the 0.3 to 0.4 range.

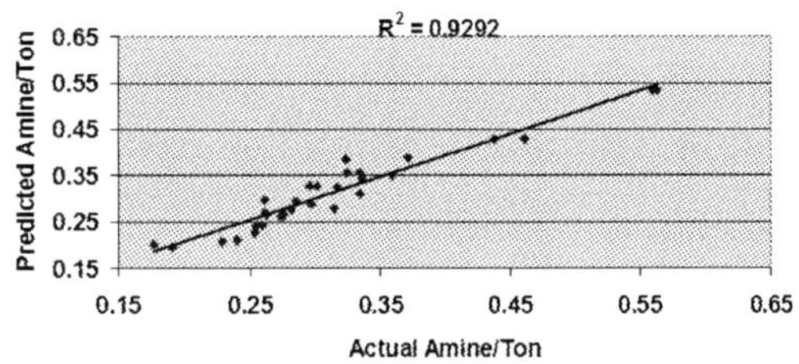

Figure 17 Predicted amine/ton vs. actual amine/ton for maximum recovery

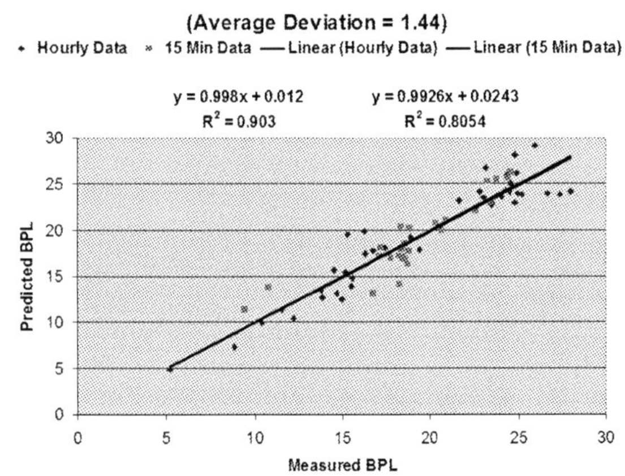

Figure 18 Predicted feed BPL vs. actual, 11/26/03–04/28/04

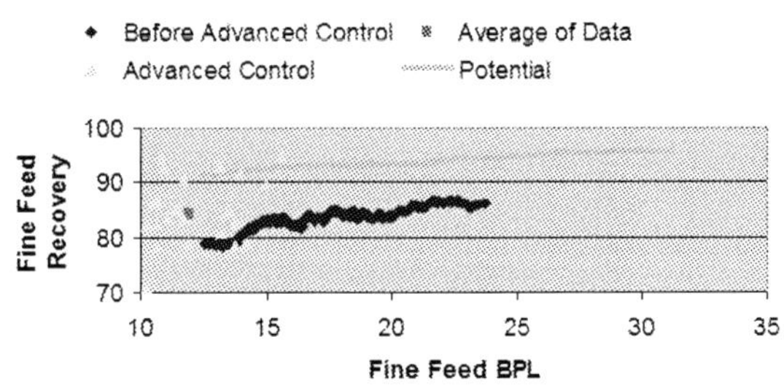

Figure 19 Fine feed recovery vs. fine feed BPL

Figure 17 is a similar correlation for the amine (cleaner) section of the plant. Again, the correlation has a R squared value above 0.9.

The relationships for the coarse roughers are still being developed. While all the variables that were found to be significant will not be revealed, one that is significant is the feed BPL. While this would suggest that an on-line analyzer would be necessary for a good control system, this may not be the case. It is possible to develop a "virtual analysis" by using process data and some fancy logic. Figure 18 shows some test data for such a virtual analysis for fine feed BPL.

As can be seen, the virtual analyses are fairly close to the actual analysis. The virtual analyses are calculated continuously and are available instantly.

The fine feed control system was installed in mid-2004. It replaced a system that was based on a neural net algorithm. No changes were made in the new control system after installation. Figure 19 shows the results to date.

As can be seen, while the results are encouraging, the overall performance is still not up to the potential. However, most of the operating data is in a feed BPL range that was never seen in the 11 years of available operating data. More encouraging is the reaction of the operators to the new control system. When the previous control system was in place, the operators could add a bias to the set point given by the program. In the three months prior to the installation of the new program, the operators made an average of 1.14 changes in the fatty acid bias per shift (8 hrs) and 0.81 changes in the fuel oil bias per shift. After the installation, the number of bias changes per shift dropped to 0.54 for the fatty acid and 0.29 for the fuel oil. More importantly, the number of changes has been steadily dropping and the recovery has been improving.

It is planned to install similar control systems on the coarse rougher and amine circuit. This should be completed in 2005.

Index

S

T

U

W